Physics for Philosophers
An Introduction

Tomasz Bigaj

Studies in Philosophy

Warsaw 2025

The publication was created under the patronage of
Faculty of Philosophy, University of Warsaw

Typesetting and layout: *Tomasz Bigaj*

Cover: Artika studio, *Dariusz Pieńkos*

Cover art: *Ewa Bigaj*

ISBN-13: 978-83-971439-3-7

Table of contents

PREFACE

The title of this book may be met with disbelief. Why do philosophers need to know anything about physics? Aren't these two branches of knowledge separated from each other, one belonging to the humanities, the other being an empirical science? Perhaps it is true that philosophers should possess, above all, a general understanding of the role and place of science, with physics at its forefront, among many creations of humankind – literature, arts, politics. However, even a cursory glance at the subsequent chapters of the book shows that they penetrate deeper into the conceptual and even mathematical structure of basic physical theories. The dominant question posed here is what this fundamental science teaches us about the world, as well as what methods it uses and what can guarantee that they will not lead us astray. These are philosophical questions *par excellence*; those concerning reality belong to ontology, and those about the cognitive method and its legitimacy belong to epistemology. The ultimate goal is to better understand, within reasonable limits of course, the complexities of the physical world and our efforts to get to know it. The lovers of wisdom (*phileo sophia*) should gladly welcome such an understanding.

Our journey through physics will be primarily chronological: we will start with ancient cosmological concepts, focusing on Ptolemy's geocentric model and its ingenious, if immensely complicated elements. A turning point in the history of physical sciences was the development of the heliocentric model in astronomy, associated with a no less radical breakthrough in mechanics, thanks to Galileo. In this way, we will come into contact with the first important theory of physics, namely classical mechanics, also known as Newtonian mechanics. The conceptual development of this theory, from the version proposed by Newton himself, to the mathematically and conceptually advanced formalisms of Lagrange and Hamilton, will be the subject of the second chapter. At the same time, we will not forget about typical philosophical problems, such as the question of the empirical nature of the basic laws of dynamics, the problem of determinism and causality, and the question of the status of time and space.

The subsequent chapters will be devoted to other great classical theories of physics: thermodynamics with statistical mechanics, and the theory of electricity and magnetism. Philosophical topics will again recur here, concerning both the status of the basic laws of these theories, such as the second law of thermodynamics or Maxwell's equations, as well as the ontological claims about reality. We will thus ask about the objective direction of time expressed in the law of entropy increase, as well as about the status of physical fields and the question of the unification of electricity and magnetism. This brings us to the three great physical theories of the early twentieth century: special and general relativity, and quantum mechanics. First of all, we will show what the sources of these radically non-classical approaches were. We will identify some interpretative difficulties arising within the theory of electromagnetism, which gave rise to a new concept of time and space in the special theory of relativity. The desire to equate the status of all observers regardless of their state of motion led to a radical reconstruction of mechanics and the theory of gravity in the form of the general theory of relativity. Finally, the wealth of empirical material on subatomic phenomena convinced researchers that an entirely new quantum theory based on the irreducible use of probability and indeterminacy of measurable parameters was needed.

My intention is to present these and other topics in the most accurate, precise and comprehensive, yet accessible way possible. This is a difficult task, because physical problems – especially those discussed in modern theories – often require an advanced formal apparatus. Many of the truths discovered about the physical world are hidden behind a veil of specialized terminology and puzzling mathematical concepts and symbols. We will try to slightly lift this veil, but this requires cooperation and some effort on the part of the reader. Hence, in the text intended primarily for philosophers, mathematical formulas and equations will unfortunately appear. However, they do not aim to intimidate the readers, but are intended to convince them that mathematics can express facts about the physical world in a precise and elegant way, which would be very difficult to adequately convey in everyday language. Mathematics is not a torture device for hapless students, especially those with humanistic inclinations, but it is an extremely useful tool with which we can discover and formulate regularities that are difficult to grasp purely qualitatively.

One of my goals is to teach the reader how to "decipher" the mathematical formulas that are used to express many important laws and principles in physics. For example, one of the basic interpretive rules is the principle that the derivative of a given quantity with respect to a certain parameter (for example, time) determines the rate of change of this quantity with the variation of this parameter. (Of course, I will also remind you what a derivative is – no worries.) Surely, there will be more such rules, often more complicated. In order not to burden the readers excessively, in most chapters I relegate more difficult fragments to additional sections marked with an asterisk, which can be omitted without much detriment to the understanding of the basic ideas of the chapter. They will contain more detailed and advanced information, mainly about the mathematical apparatus of a given theory. However, I encourage you to try your hand at reading these parts, even if at some point you decide that the difficulties become too great.

Is it reasonable to expect philosophers to be able to understand the basic methods and concepts of mathematical physics? Shouldn't I give up this perhaps too ambitious enterprise and present the main ideas of the book in a purely descriptive language, as is the case in many popular science publications? I think this is too pessimistic. After all, the basic curriculum of university philosophy includes subjects closely related to mathematics, such as logic, in which students learn the basic methods of mathematical logic or set theory. In later years, students can choose a course in advanced logic, where they will learn how to formulate and prove very technical theorems, such as the completeness theorem for first-order logic, Cantor's theorem in set theory, or Gödel's incompleteness theorem. I do not think that understanding the laws of electromagnetism in their differential version (with the help of such concepts of vector calculus as rotation and divergence) is more difficult than understanding the basic idea of the proof of Gödel's theorem based on the numbering of formulas and the appropriate use of the liar paradox. I also hope that logically proficient philosophers will not be deterred by the basic concepts of differential geometry used in general relativity, or the operators in Hilbert spaces used in quantum mechanics.

I would also like to emphasize something which is probably obvious, that, contrary to what the title may suggest, this is not a book for philosophers only (so if you are not one, don't turn away). It can be used by any reader interested in the foundations of modern physics. I think that even students of physics will find some material here to reflect on issues they know well. Perhaps under pressure to quickly improve their computational skills, they have missed some hidden assumptions or controversial problems that appear at the foundations of physical theories. I must admit that the approach to physics proposed in this study is not dogmatic. Many assumptions made in textbooks without second thoughts will be the subject of critical analysis, which may slightly surprise the reader unaccustomed to the attitude of philosophical skepticism. In fact, the work of the philosopher of physics consists to a large extent in finding and possibly removing gaps and ambiguities or inconsistencies in the conceptual framework of physical theories.

Much of the material in this book forms the basis of a semester-long lecture that I have been giving to first-year philosophy students for years. To make it easier for my students to prepare for the exam, I have attached a set of questions and problems to each chapter, checking how well they assimilated the presented material. These questions are closely correlated with individual sections, so they can also serve as a guide during reading. In addition, following each chapter I have placed a list of recommended further readings from the literature on the subject, expanding on or supplementing the subject matter. In the main text of the book, I have also inserted here and there comments and remarks on related philosophical topics or on certain physical details, highlighted by a greyed box and a different typeface. In many cases, the text is illustrated with diagrams, which, although perhaps not always fully satisfactory in terms of elegance and aesthetics, will, I hope, be helpful in understanding the main presentation.

CHAPTER 1. FROM ANCIENT ASTRONOMY TO GALILEO

1.1 Ancient astronomy and Ptolemy's model of the universe

Astronomical phenomena have attracted attention of humankind since time immemorial. Sunrises and sunsets, the annual variability of the Sun's position in the sky, the phases of the Moon, constellations and their revolution in the night sky – all this stimulated people's imagination and made them ponder. Difficult to predict eclipses of the Moon and the Sun, or the sporadic appearances of comets, caused terror, but also piqued curiosity. The religious significance of celestial phenomena is well known – from the ritual celebration of the solstices in megalithic buildings of the Stone Age to the Egyptian pyramids and their alignments with the constellations. Gradually, however, the mythological and anthropomorphic approach to astronomical phenomena began to be replaced by naturalistic considerations. Bold hypotheses have been developed about the nature of the bodies observed in the sky, their relationships with each other, and the hidden mechanisms responsible for their behavior. In this section we will learn the basics of the most advanced system of ancient astronomy. This title undoubtedly belongs to the geocentric system of Claudius Ptolemy (a Greek scholar from Alexandria, who lived and worked in the second century AD). We will trace the path that ancient astronomers had to traverse, from singular observations and facts to the theoretical hypotheses and generalizations of the Ptolemaic model.

We will begin our journey in time to the inception of astronomy by presenting the basic empirical data that ancient stargazers had at their disposal. A cursory observation of the night sky teaches us that the celestial sphere is constantly rotating around a certain point in the sky occupied by the North Star (also known as Polaris). All objects in the sky visible to the naked eye: the stars, the Sun and the Moon, as well as ephemera such as occasional comets, participate in this movement. Moreover, ancient travelers observed that the elevation of Polaris changes depending on the position of the observer. The further north we go, the higher Polaris appears in the sky. On the other hand, traveling south (towards the equator) results in lowering the apparent altitude of the North Star, until it meets the horizon. Another important fact, well known to the observers, was the existence of a different type of motion for some celestial bodies beyond the diurnal (i.e. daily) rotation of the entire sphere. For now, we will skip the movements of the Moon, which, along with the change in its observed shape (the phases of the Moon), posed a great challenge. In the case of the Sun, on the other hand, it can be seen that it makes a slow motion against the background of stars in the opposite direction to the direction of rotation of the entire celestial sphere (Fig. 1.1). In other words, the Sun "retreats" relative to fixed stars, completing a full cycle during one calendar year (so the solar day is about 1/365 day longer than the sidereal, i.e. based on the stars, day). By tracing the path of the Sun's year-round motion in the sky, we get a circle today called the ecliptic. This circle is located in a belt dubbed the zodiac, known for the fact that it is divided into twelve constellations, from the constellations of Aries, Taurus, Gemini to Pisces. The Sun resides in each constellation for one month.

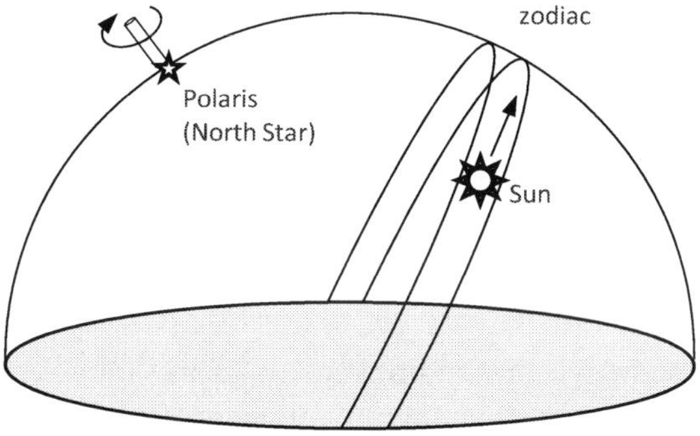

Fig. 1.1 The celestial sphere with the diurnal rotation and annual motion of the Sun.

The plane of the ecliptic is inclined with respect to the "celestial equator", i.e. the large circle perpendicular to the axis of diurnal rotation. This fact is of paramount importance for life on Earth, since its consequence is the annual variability of the point of the highest daily position of the Sun – from the highest in the summer months to the lowest during the winter. Determining the key points of the equinoxes and solstices for the annual motion of the Sun was of great significance from a practical and ritual point of view. However, far more crucial for the development of astronomy was the behavior of a special category of objects in the night sky – the so-called wandering stars, that is planets. The planets are characterized by a significant brightness compared to the rest of the stars. The ancients distinguished five planets named after Roman gods: Mercury, Venus, Mars, Jupiter and Saturn. Their characteristic feature is that, like the Sun, they change their position in relation to the fixed stars. Each planet makes a slow motion in the zodiac belt in the opposite direction to the daily rotation. However, the speeds of this movement vary from planet to planet. Mercury and Venus are the fastest, Mars is slower, and Jupiter and Saturn are the slowest.

However, the complexity of planetary motion does not end there. First, the migration of individual planets against the rest of the stars is not uniform, as in the case of the Sun.[1] At certain times, the planets move faster, and at other times they slow down. Moreover, these slow changes in pace are compounded by rapid changes in both speed and direction of movement. Careful sky observers have noticed that each planet goes through a period of slowing

[1] In fact, the Sun does not move at a constant speed during the year, due to the elliptical shape of the Earth's orbit. However, the ancients were unable to observe this small effect.

down, stopping and going backwards, after which it returns to normal motion. In other words, the planets draw small "loops" in the sky. Such a movement is called today "retrograde". It is characteristic that Mars, Jupiter and Saturn (called the outer planets for reasons we will soon see) perform retrograde motions no more than once a year, while the inner planets – Mercury and Venus – are in retrograde several times a year. To these observational facts, let us add that the inner planets always stay close to the Sun – they never stray too far from it, as if they were tied to it (they appear from one side of the Sun and then from the other). The outer planets, on the other hand, move freely, from time to time being in the so-called opposition to the Sun (exactly on the other side of the celestial sphere). Finally, let us mention the well-known and easily noticeable fact that the brightness of the planets, especially the outer ones, varies greatly in time. In some periods, for example, Jupiter is extremely bright, and in others its light seems to dim. This, of course, stands in stark contrast to other stars whose brightness remains the same.[2]

As you can see, the ancients collected a lot of empirical material regarding the behavior of individual celestial bodies. The next step should be to put all these various facts into a coherent whole. Such a whole, called a theory or model, should provide us with a uniform explanation of the individual observable regularities. The question "Why?" is one of the fundamental questions that people have asked since the dawn of time. It is not easy to determine what kind of requirements we should impose on scientific explanations. Instead of formulating a ready-made philosophical concept of explanation in science, we will simply look at how explanations functioned in astronomy, and later also in physics. In any case, we can assume that the explanations of individual unrelated facts should, in a sense, go beyond a mere collection of these facts – they should contain some "added value". This added value may be an appropriate generalization or synthesis, but perhaps more important than the generality itself is the use of certain new concepts and hidden mechanisms. Such new concepts are often described as "theoretical", as opposed to observational concepts and terms, which refer to objects and properties directly accessible to the human senses.

The path leading from the multitude of astronomical facts to the full Ptolemaic model was not easy. It is not the case that Ptolemy, in a moment of revelation, formulated his theory from scratch, as it were. The development of his conception was influenced by the work of many earlier generations of astronomers and researchers. Their first important hypothesis was that the Earth was spherical. It was supported by a variety of empirical evidence, such as the aforementioned dependence of the elevation of Polaris above the horizon on the position of the observer on Earth. An alternative model of the Earth as a flat disk "covered" by a celestial sphere also implies certain dependence of that kind, but not to the extent observed (e.g. there would be no position in which Polaris is seemingly on the horizon). In addition, we have the argument from lunar eclipses, which is mentioned by Aristotle. Assuming that the darkening

[2] One more type of motion that the ancient astronomers were able to observe should be mentioned here. It is a "global" movement, i.e. it involves all celestial bodies to the same extent, just like the daily rotation of the celestial sphere. Careful observations of the equinox points (the position of the Sun when days and nights are equal in length) showed that these places move very slowly, corresponding to an extremely slow movement of the apparent axis of daily rotation in a circle. This axis completes a full rotation in about 26 thousand years. This movement is called precession today. Due to the precessional motion, the North Star will not always be stationary on the sky – in a couple of thousand years it will move away from the center of rotation of the celestial sphere.

of the Moon's face is caused by the Earth's casting a shadow, the circular shape of this darkening unambiguously indicates the spherical shape of the Earth.

It is worth noting that the sphericity hypothesis was not only qualitative, but also quantitative. Eratosthenes' famous measurements (comparison of the angle of incidence of the Sun's rays at the same time in two different locations) allowed him to estimate the length of the Earth's radius, surprisingly close to the currently known value. The first astronomical model reflecting the thesis of the sphericity of the Earth was the so-called model of two spheres: the Earth's sphere and the sphere of celestial bodies surrounding it at a distance. However, this model was unable to explain the relative motions of the planets, the Sun and the stars. It was therefore quickly replaced by Eudoxus' model assuming many misaligned spheres, performing independent rotations. Each planet (and also the Sun) was to be attached to its own sphere, which in turn was affixed to the sphere of fixed stars, which made a full rotation in 24 hours. Thus, the movements of the individual planets and the Sun were a combination of the rotation of the fixed stars and the rotation of the individual spheres. Such an otherwise very ingenious solution did not reflect the complexity of planetary motions (retrograde motion, global changes in speed and brightness, etc.).

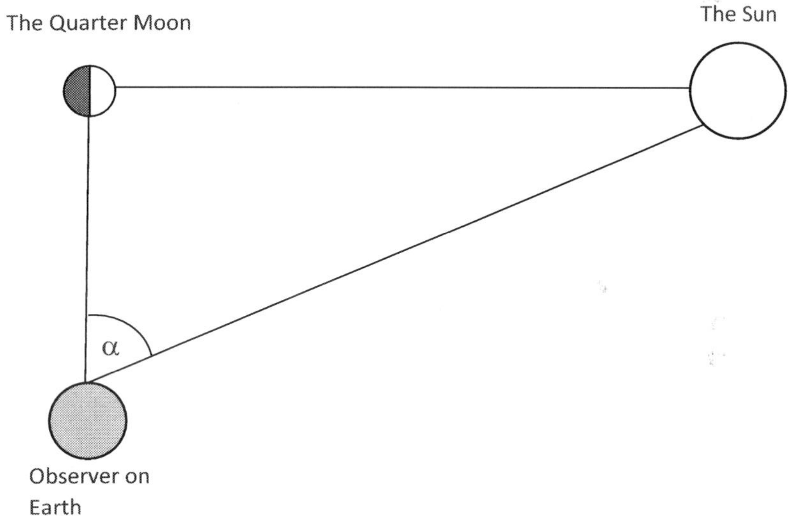

Fig. 1.2 Estimation of the distance to the Sun by Aristarchus. Determining the angular separation α between the Sun and the Moon during the quarter moon allows one to calculate the ratio of the distance between the Earth and the Sun to the distance between the Earth and the Moon.

The formation of Ptolemy's full model was also undoubtedly influenced by extremely subtle methods of estimating the distances from the main celestial bodies: the Moon and the Sun. The naïve belief that all celestial bodies are at the same distance from Earth on a single celestial sphere was quickly challenged by geometric arguments. The comparison of the distances between the Earth and the Sun and the Earth and the Moon was carried out by Aristarchus of Samos. He took into account the situation in which the Moon's phase is the quarter ("half Moon"), i.e. exactly half of its disk is illuminated by the Sun's rays (Fig. 1.2). This means that the angle between the direction from the observer to the Moon and the line connecting the Moon to the Sun is a right angle. By then determining the angular separation between the Moon and the Sun, we can calculate the ratio of the sides of the Earth-Moon-

Sun right triangle from simple trigonometric relations. Although Aristarchus made a mistake in his measurements, the result he obtained was astonishing to the contemporaries: the distance to the Sun turned out to be almost twenty times greater than the distance to the Moon (the real ratio is 389.2). Aristarchus also estimated the size of the lunar orbit based on observations of the Earth's shadow during a lunar eclipse. Knowing the size of the Earth from Eratosthenes' measurements, we can assume that the Earth's shadow is the same size. By calculating the time of the Moon's passage through the total shadow and comparing it with the duration of the lunar month, we can estimate the length of our satellite's orbit, which gives us the size of its radius. On this basis, we can then determine the absolute distance to the Sun using the proportion calculated earlier.

Now let us move on to the complete model of Ptolemy, starting with its basic version and then making the necessary corrections and additions. Ptolemy's basic idea was to use two concepts to describe the motion of the planets and the Sun: the deferent and epicycle. The deferent of a given planet and the Sun is a circle in the center of which is the Earth. The deferents of individual planets have different sizes. The planet Mercury had the smallest deferent (not counting the Moon), followed by Venus, the Sun and the three outer planets: Mars, Jupiter and Saturn. It was not possible to determine the exact radii of planetary orbits, so the principle of proportionality to the period of the planet's orbit was applied.

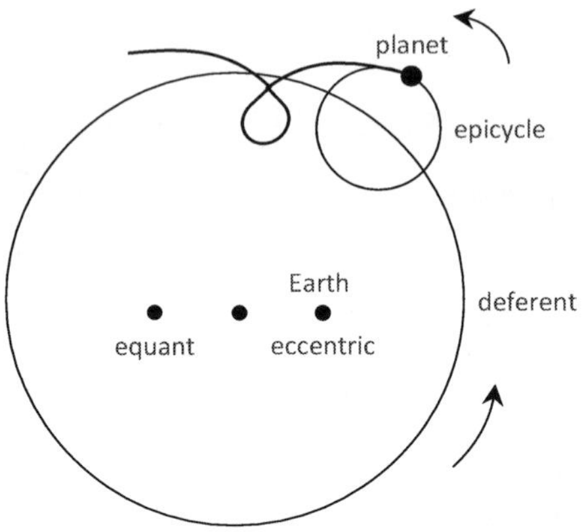

Fig. 1.3 Formation of the retrograde motion by the composition of motions on the epicycle and deferent. The Earth is outside the center of the deferent in the eccentric.

The epicycle of a given planet is a small circle whose center is on the deferent. A planet that revolves around the Earth participates in two motions: a rotational motion on an epicycle and along a great deferent wheel. The combination of these two motions results in the observed retrograde motions, i.e. periodic retrogressions of the planets (Fig. 1.3). Only the Sun in the Ptolemaic model did not have its epicycle for a simple reason – it does not show the retrograde effect. Of course, it was a huge challenge to match the size of the epicycles, the deferents and the speed of movement along these circles to the observed astronomical data. However, it turned out to be possible, although extremely difficult. In this way, Ptolemy's model made it possible to replicate the basic characteristics of the observable behavior of the planets.

The use of epicycles and deferents had not only a practical, but also a theoretical and even ideological justification. In Ptolemy's time, it was commonly believed that the sphere of celestial bodies (the so-called supralunar sphere) was an area of perfection, unparalleled on Earth. Perhaps it was a remnant of the religious approach to astronomical phenomena, in which the celestial sphere was identified with the realm of the gods. Since circles are the most perfect geometric figures, celestial bodies should move only in circles. However, experience does not confirm this belief – the planets perform complex, chaotic movements, which do not resemble a perfect circular orbit. Ptolemy's genius was to notice that the observed movements of the planets could be reproduced by an appropriate combination of two ideal circular motions, which saved the assumption of perfection. Unfortunately, Ptolemy's idea required additional refinements and corrections, which were departing further and further away from the ideal of uniform movements in circles.

The basic model relying on deferents and epicycles did not reproduce many of the known empirical facts we mentioned earlier. For example, one challenge was the observed variability in the brightness of the planets, which could not be explained by the movements on the epicycles and deferents. To solve this problem, Ptolemy assumed that the Earth is not exactly in the center of a given deferent, but slightly to the side, at a point called the eccentric.[3] He explained the change in the brightness of the planets by changes in their distance from the Earth. The price to pay for this explanation was the added complication of the model. However, this is not the end of the list of fixes and additions. Another problem to be solved was the "global" variability of the planet's speed, independent of its retrograde motion. Observations show that, for example, Jupiter moves across the sky much slower during periods of lower brightness than when its brightness is high. This effect could be partially explained by the eccentric, since the angular velocity of a planet depends on its distance from the observer, which, as we already know, fluctuates due to the assumed eccentricity of the orbit. However, it turned out that the eccentric effect is too weak to explain the observed speed fluctuations. So we are dealing here with a very important aspect of explanation or prediction in exact sciences, such as astronomy or physics. It is not enough for a theory to have consequences that are qualitatively consistent with empirical data—we also need quantitative agreement.

In order to ensure the quantitative accuracy of the predictions of the Ptolemaic model, an additional assumption of the so-called equant was introduced. The equant is a point located exactly on the opposite side of the deferent's center in relation to the Earth. It is an abstract point for which it is assumed that the planet's velocity is constant relative to it. In other words, the planet sweeps equal angles with respect to the equant in equal times. This assumption implies that from Earth's perspective, the observed speed of the planet will fluctuate much more strongly than if the orbital velocity were constant. During the period when the planet is closer to the equant, its actual orbital velocity must be reduced to maintain the constancy of its angular velocity relative to the equant (Figure 1.4). By combining this effect with the effect of variable distance from Earth, we can fit the model to the actually observed behavior. It may be argued, moreover, that the assumption of the equant does not entirely destroy the

[3]To be precise, Ptolemy assumed that the Earth was still in the center of the universe, while the planets' deferents were slightly off center, hence the term "eccentric" for him denoted the centers of the deferents moved away from the Earth. Either way, we may simply state that the eccentricity resulted in the shift between the position of the Earth and the center of the deferent, as depicted on Fig. 1.3.

conviction of the perfection of the uniform motion of the planets. The planets still move uniformly, but not relative to the Earth or the center of the deferent, but relative to the equant.

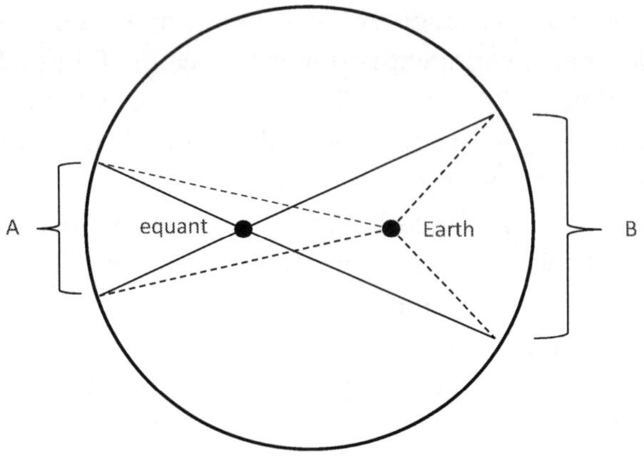

Fig. 1.4 Explanation of the function of the equant. The marked sections A and B of the orbit are covered by the planet in equal periods of time. From the Earth's perspective, the angular velocity of the planet is greater in segment B than in A.

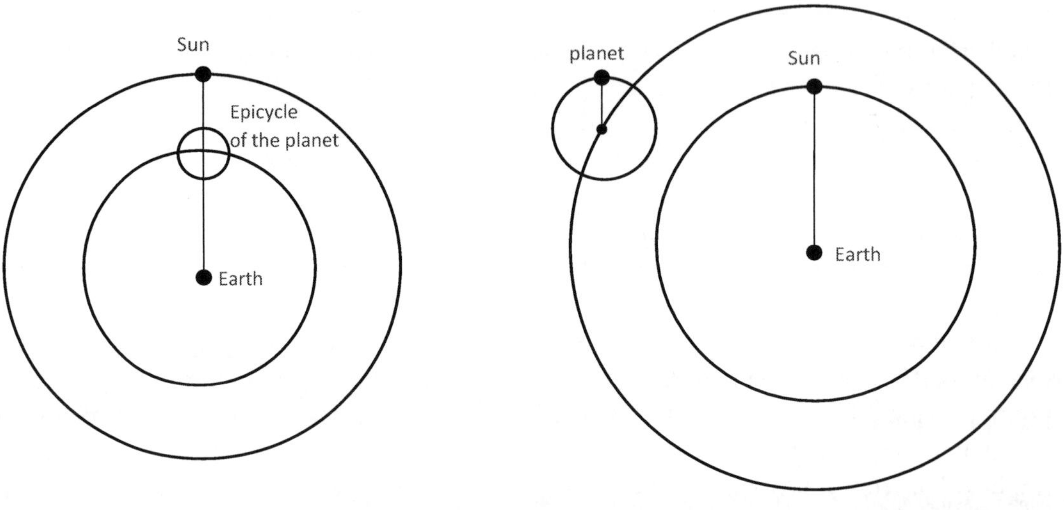

Fig. 1.5 Explanation of the observed behavior of the inner planets (left diagram) and outer planets (right diagram).

We have two additional facts to explain: the proximity of the inner planets to the Sun and the frequency of retrograde periods for the outer planets (no more than once a year). Note that the Ptolemaic model does not in any way force these effects. The inner planets, by the mere fact of being closer to the Earth than the Sun, do not have to "stick" to the Sun – they could move on their deferents completely independently. The same applies to the regularities in the retrograde motion of the outer planets. The epicycles of these planets could make any number of revolutions in a given period. It is not known why this number would be limited by the solar year. So we need some additional explanatory mechanism. Ptolemy made two assumptions, each to explain the corresponding effect. For the inner planets, he introduced the hypothesis that the center of their epicycles should always stay on the line connecting the Earth with the Sun. With this assumption, Mercury and Venus will be able to move away from the Sun at most by the distance of the radius of their epicycle. (Of course, this imposes an additional limit on the postulated size of epicycles, but we trust that Ptolemy succeeded in staying in agreement with all the observational facts.) The problem of the outer planets was solved by accepting the postulate that the radius of the epicycle of each planet must be parallel to the radius connecting the Earth with the Sun (Fig. 1.5). Then for one full rotation of the Sun around the Earth, there is one revolution of the epicycle, and therefore also one retrograde cycle.[4]

Summing up, we can observe that the Ptolemaic model consists of several independent postulates. The fundamental assumption regarding the circular motions of the celestial bodies around the Earth is supplemented by various auxiliary hypotheses. Each of these hypotheses is formulated in order to explain a separate phenomenon. The table below lists main Ptolemaic postulates together with their corresponding astronomical phenomena. A conjecture posited exclusively for the purpose of explaining one phenomenon is called *ad hoc* (literally: "to that"). It is commonly believed that the presence of *ad hoc* postulates and explanations in science is a sign of weakness of a given theory. Good theories should avoid "manual" introductions of assumptions just to make them "right". The ideal of science is the search for universal laws and propositions that should be able to explain a series of independent facts.

Hypothesis (element of the Ptolemaic model)	**Explained phenomenon**
Epicycle	Retrograde motion
Eccentric	Variation of the planet's brightness
Equant	Variation of the planet's velocity
The location of the planet on the Earth-Sun line.	The proximity of the inner planets to the Sun
The parallelism of the radius of the epicycle and the Earth-Sun line.	The frequency of the retrograde for the outer planets (once a year)

Tab. 1.1 Elements of the Ptolemaic model and their explanatory role.

One may ask why we should prefer theories based on a small number of independent hypotheses with a wider range of applicability. Certainly, the elusive issues of elegance, simplicity or aesthetics in general play a role here. A multitude of hypotheses, each of which has little explanatory power limited to a narrow type of phenomena, gives the impression of chaos and disorder, as opposed to, for example, a single hypothesis explaining a whole range of different phenomena. However, there is also an argument referring to the main goal of science, which is the pursuit of truth. Hypotheses in science are evaluated in terms of their probability, i.e. how high the chance is that they accurately reflect

[4] In fact, the observational data show that the period between the retrogrades of the outer planets is slightly longer than the solar year. In the Ptolemaic model, this is due to the fact that during the solar year the outer planet keeps moving on its deferent, and therefore the retrograde period will recur after a time slightly longer than one rotation of its epicycle.

reality. From elementary probability calculus we know that the probability of a joint occurrence of independent events equals the product of their individual probabilities. Thus if a theory consists of several independent hypotheses, then the likelihood that they will be jointly true is equal to the product of their probabilities, and hence will be rather low. A smaller number of independent postulates means that they have a greater chance of being true together.

1.2 The heliocentric theory of Copernicus

Throughout the centuries separating Ptolemy from Copernicus, the Greek scholar's model was constantly supplemented and improved. With the influx of new, more accurate empirical data on the positions of celestial bodies (obtained through improvements in observational techniques and instruments), Ptolemy's original model became more and more out of step with reality. The solution to the emerging inconsistencies between theory and experience was the repeated use of the method that worked so well in the original model: the motions along independent circles (epicycles). By introducing more and more wheels, scientists were able to maintain some kind of consistency with the data, albeit with considerable difficulty. However, the complexity of the model, significant at its very beginning, grew to gargantuan proportions. It became clear that new, radical solutions were needed.

The solution proposed by Copernicus was not completely new. Already in antiquity it was hypothesized that the Sun occupies a central place, and the Earth, along with other planets, revolves around this center. Ptolemy even devoted an entire chapter of his work *Almagest* to discussing and criticizing this concept. The primary argument against heliocentrism was the lack of observed variability in the relative distances between stars that should be visible due to the motion of the Earth (the so-called stellar parallax problem). We will return to this issue after discussing Copernicus's conception. For now, let us take a look at the unquestionable and immediate benefits of the assumption of the Earth's revolution around the Sun.

The main tenets of the Copernican model are widely known. In its basic, simplified version, this model assumes that each planet moves in a circle with the Sun in the center, in the following order counting from the Sun: Mercury, Venus, Earth, Mars, Jupiter, Saturn. Only the Moon remains in motion around the Earth, at the same time participating in the revolution around the Sun together with the Earth. In addition, the Earth rotates around its axis, which explains the observed daily rotation of the entire celestial firmament. However, this school version of the Copernicus model does not take into account the fact that he continued to use epicycles and eccentrics in order to better fit the model to the data. (The number of wheels used in Ptolemy's original model and Copernicus's model was approximately the same.) Unlike Ptolemy's model, the Copernican model used epicycles not to recreate the retrograde motions of planets, but to ensure better fit with the data. This was due to the fundamentally incorrect assumption of the circularity of the orbits. As we know today, planetary orbits are not circles, but ellipses. It turns out that the ellipsoidal shape can be recreated by superimposing appropriately selected movements on deferents and epicycles. Imagine a body orbiting on a large wheel, the center of which in turn makes one small revolution in the opposite direction. It is easy to see that this will result in an elongated orbit resembling an ellipse. This is the method of approximating ellipses by circles unwittingly used by Copernicus, thus preserving the idea of perfection of the movements of celestial bodies.

However, let us skip this detail of Copernicus's original construction and briefly discuss how his model reproduces known empirical data. The explanation for most of the observational facts, which was quite difficult to achieve for ancient astronomers, is the combination of two motions in the heliocentric system: the "true" motion of a given planet and the motion of the Earth and the observers associated with it. Retrograde movement of a given planet against the background of fixed stars occurs when the Earth begins to "overtake" this planet on its orbit (Fig. 1.6). This is an effect analogous to the effect observed when overtaking a slower moving vehicle, when it appears to be reversing relative to its surroundings. It is not difficult to see that overtaking of the outer planets can take place only once during the entire revolution of the Earth around the Sun, which automatically explains the known regularity without the need to introduce additional assumptions such as the parallelism of the epicycle's radius and the radius of the Sun's deferent. In contrast, the inner planets display a retrograde effect once during their solar year, which is shorter than the Earth's, so the observed retrograde frequency is higher than that for the outer planets.

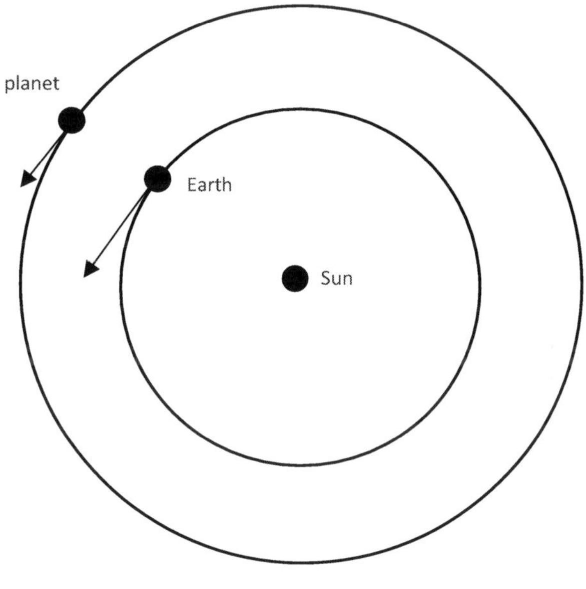

Fig. 1.6 Retrograde in the Copernican model. The Earth overtakes the planet during the period of closest approach, which produces the effect of backward motion.

It is also easy to explain the variability of the planets' brightness and their global observed velocity. It results from the difference in distance between a given planet and Earth. For the outer planets, the maximum distance from the Earth is $R_E + R_p$, where R_E is the radius of the Earth's orbit, and R_p is the radius of the orbit of the planet. The minimum distance is $R_E - R_p$ (Fig. 1.7). In addition, we immediately have an explanation for the fact that the planet's minimum brightness and minimum angular velocity coincides with the period when it is seemingly close to the Sun (in "conjunction"), while the maximum values are when it is in opposition. In Ptolemy's case, this had to be entered by hand, properly aligning the eccentrics and equants of a given planet in relation to the Sun. Finally, let us mention that the fact that the inner planets remain near the Sun is easily explained by the fact that their orbits are entirely contained within the Earth's orbit (Fig. 1.8).

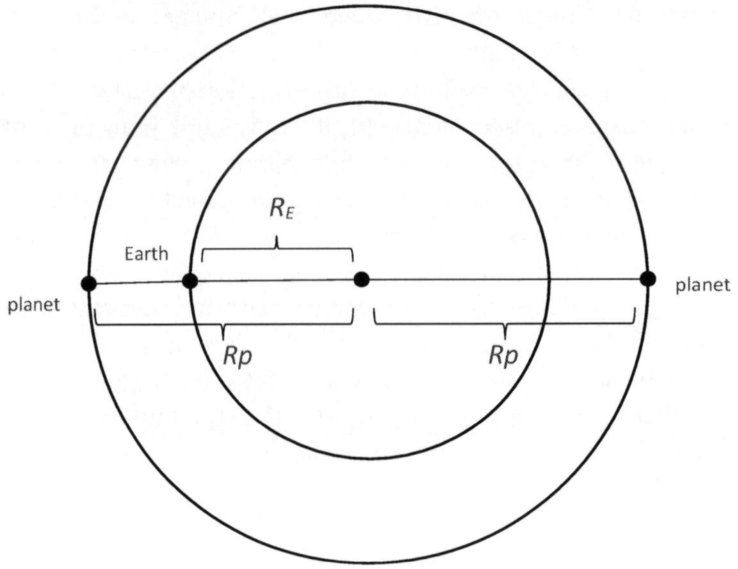

Fig. 1.7 Explanation of the variability of the brightness of the outer planets in the Copernican system.

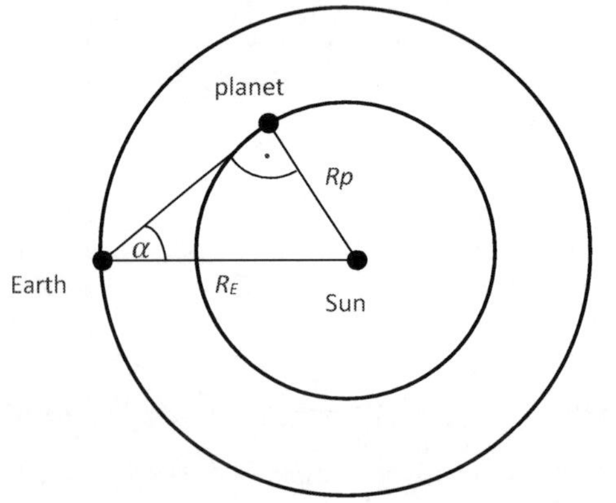

Fig. 1.8 Explanation of the proximity of the inner planet to the Sun and calculation of its orbital radius. The angle α is determined during the maximum separation of the planet and the Sun. The radius R_p will be equal to R_E times the sine of the angle α.

So we have the indisputable theoretical advantages of the Copernican model: it is definitely simpler than the Ptolemaic model, even if we take into account the presence of additional epicycles. This simplicity is revealed not only in a smaller number of additional assumptions, but above all in the mechanism of explanation. A single assumption (the motion of the Earth around the Sun) is able to explain many seemingly unrelated facts, without the need for *ad hoc* hypotheses. In such a situation, we talk about the unificatory role of a given hypothesis, because it gives one basic mechanism of the formation of various effects (their common cause). Copernicus's conception had other theoretical advantages in relation to the Ptolemaic model. For example, in the heliocentric model, it is possible to determine the radii of the orbits of individual planets using simple astronomical observations. In the case of the inner planets, this determination is based on measuring the maximum angular separation of a given planet from the Sun and solving the corresponding right triangle. For the outer planets, the procedure is a bit more complicated, because it requires two observations: one when the planet is opposing the Sun, and the other when the angular separation of the planet and the Sun is equal to 90°. Details of the geometric reasoning in both cases are given in the diagrams (Figs. 1.8 and 1.9). Ptolemy's model did not provide such a theoretical possibility except for estimating the size of orbits using orbital periods (this was, of course, incorrect for fundamental reasons, since the orbital velocities of individual planets are not the same).

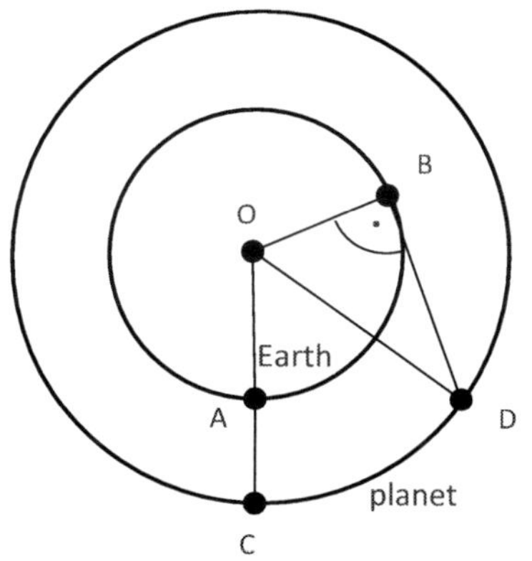

Fig. 1.9 Determination of the radius of the orbit of the outer planet. We measure the transition time t of the outer planet from the opposition to the Sun (point C) to the moment when the angular separation between the planet and the Sun is 90° (point D). From the ratio t to the respective periods of the planet's and Earth's revolutions, we determine the angles $\angle COD$ and $\angle AOB$ and then $\angle DOB$, which allows us to solve the right triangle DOB.

However, it should be emphasized that the new theory was not without its flaws. Considering it from the point of view of the state of knowledge in Copernicus's time, we can understand why many scientists of the time were skeptical about the proposed solution. First of all, it was believed that the movement of the Earth, both rotating around its axis and revolving around the Sun, should be noticeable to observers on its surface. Especially in the case of rotational motion, significant effects were expected. It was imagined that, like on a spinning merry-go-round, it would be extremely difficult for us to stand on our feet if the Earth rotated at an incredible speed (which at the equator equals about 1600 km/h). On top of that, we would feel a huge rush of air, knocking over trees and buildings. Falling objects should deviate from the vertical towards the west, due to the "escape" of the Earth's surface during the fall. These problems were completely ignored by Copernicus. It was the Italian scientist Galileo Galilei who found a correct answer to the problem of the lack of observable effects of the Earth's rotation, but it required the rejection of Aristotelian physics presupposed in Copernicus's time. We will talk about this in more detail in the following sections.

On the other hand, the Earth's orbiting around the Sun should have an observable consequence in the form of the stellar parallax effect. In general, parallax is a phenomenon that involves a change in the apparent distance between objects from the perspective of an observer in motion. For example, if you observe two trees from a moving car, you will notice that from one point on the road they appear very close to each other, and from another point their angular separation is greater. This effect is stronger, the closer the observed objects are to us. It seemed obvious to the astronomers that the change in the position of the Earth due to the annual motion should allow us to observe the parallax of stars on the celestial sphere (Fig. 1.10). However, no such effect was found. Copernicus's response to this objection was very simple: apparently the stars are too far away for stellar parallax to be noticeable. However, let us note that Copernicus did not have any data independently confirming his hypothesis. In Ptolemy's model, the fixed stars were the farthest from Earth, but certainly not at a distance that would make it impossible to observe the parallax effect if the Earth really moved in an orbit around the Sun. Copernicus's solution was therefore clearly *ad hoc*. It doesn't matter that his intuition turned out to be right – in the light of the data available to him, it was pure speculation, formulated only to get rid of the problem. By the way, the effect of stellar parallax involving the stars closest to the Solar System was finally confirmed, but only in the nineteenth century. For the majority of stars, even the most accurate instruments are not able to detect their apparent changes in position, due to the enormous distance separating us from them. Thus, Copernicus's theory gained widespread acceptance long before the parallax problem found a satisfactory solution.

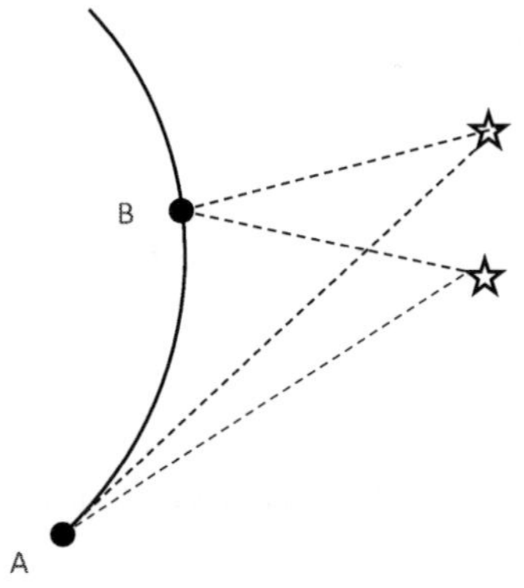

Fig. 1.10 Stellar parallax. At point A, stars are seen at a lower angle than at point B, so their relative position in the sky changes.

The situation of the new heliocentric theory would be much clearer if there was an empirical confirmation of it, which at the same time would undermine Ptolemy's competing geocentric theory. Copernicus pointed to the possibility of such an empirical fact. Copernicus's theory predicts that the inner planets should go through four phases, similar to those of the Moon: the new moon, first quarter (half moon), full moon, and second quarter (half moon again). These phases are the result of different orientations of the planet relative to the Sun: a new moon occurs when the planet is roughly on the line connecting the Sun with the Earth, quarters (half moon) occur when the planet is at the greatest angular separation from the Sun, and a full moon happens when the planet is about to set behind the Sun. In the Ptolemaic model, on the other hand, there are only three phases: the new moon, the first quarter, the new moon, the second quarter, because the inner planets are never farther from the Earth than the Sun (Figure 1.11). Empirical verification of both predictions, e.g. in the case of the planet Venus, would allow us to decide which of the two theories should be rejected. Unfortunately, in Copernicus's time, it was impossible to accurately observe the shape of the planets. The situation changed with the introduction of telescopes. The pioneer of using the telescope for astronomical observations was Galileo. He also observed the phases of Venus and unequivocally confirmed the existence of the full moon phase, thus tipping the scales in favor of the Copernican theory.

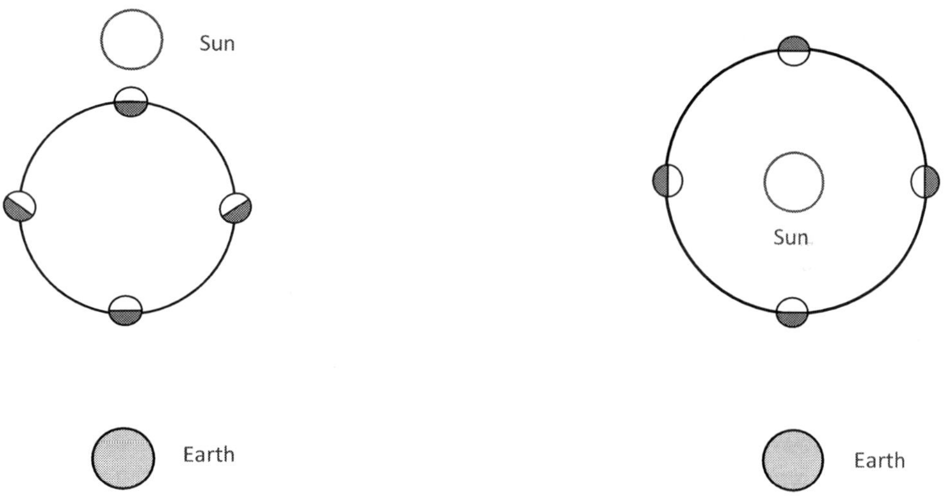

Fig. 1.11 Comparison of the phases of the inner planet in the geocentric (left) and heliocentric (right) models.

1.3 Philosophical aspects of the Copernican revolution

The breakthrough that took place in astronomy in the sixteenth century provides rich material for philosophical reflection. In this section, we will list three philosophical issues inspired by the Copernican revolution. These are: the dispute between the instrumentalist and realist interpretations of theories, the issue of comparing competing theories in science, and the concept of scientific revolutions. Starting with the first of them, let us quote fragments from the introduction to Copernicus's work *De Revolutionibus Orbium Coelestium* by the Protestant theologian Andreas Osiander:

> Astronomical hypotheses "need not be true nor even probable. On the contrary, if they provide a calculus consistent with the observations, that alone is enough."
>
> "For this art [i.e. astronomy] is completely and absolutely ignorant of the causes of the apparent nonuniform motions."

"Since different hypotheses are sometimes offered for one and the same motion [...], the astronomer will take as his first choice that hypothesis which is the easiest to grasp."

Osiander's position is today commonly known as instrumentalism. It states that the primary goal of science is not to discover truths about reality, but to ensure agreement with observational data (the so-called "saving of phenomena"). There is no point in asking whether the planets "really" orbit around the Sun, or whether they orbit the Earth following the deferents and epicycles. The only important thing is which of the competing theories is better in line with empirical evidence, or has other qualities, such as simplicity. A contemporary instrumentalist, Bas van Fraassen, introduced the notion of "empirical adequacy". Instead of asking whether a scientific theory is true, we should focus on assessing whether it is empirically accurate, i.e., whether it reproduces observable facts. The empirical adequacy of a theory means that the part of it containing only observational terms is true. The remaining part, including the so-called theoretical terms, does not have to reflect the reality.

Osiander's motivation was, of course, theological. As a staunch supporter of Protestantism, he viewed the Bible as the only source of revealed true knowledge. Since there is no mention of a specific astronomical model in the Bible (the concepts of epicycles, deferents, or Copernican orbits around the Sun do not appear in it), we cannot settle the question of the truth of these models with our finite minds. Naturalistic versions of instrumentalism, on the other hand, emphasize above all the fundamental undecidability of the theoretical assumptions of models in science. In the case of astronomical models, we are not able to rise above the Solar System and from this perspective assess whether the planets really orbit the Earth, or whether, as Copernicus wants, they revolve around the Sun along with the Earth. All that remains is to analyze the available data on the observed position of celestial bodies. In the case of other scientific theories, such an empirically inaccessible sphere may be, according to instrumentalists, the domain of unobservable objects, for instance molecules, atoms, genes or elementary particles and their properties. These objects are postulated to explain the complexity of the observed physicochemical or biological phenomena, but they themselves are not available to our direct observation due to their small size.

The opposite position to instrumentalism is taken by scientific realism. Realists emphasize that the main goal of science is and should be to discover how things really are. They refer to the practice of scientists, who very rarely adopt a skeptical or agnostic attitude towards the theoretical assumptions of a given theory. It sometimes happens that scientists openly introduce a certain auxiliary concept as an instrument that facilitates or simplifies the description of phenomena. As an example we can quote the familiar concept of electron holes, used in semiconductor theory to denote the absence of an electron in a given energy band of a semiconductor. It turns out that it is convenient to describe the flow of the current in a semiconductor as if there were positively charged particles (holes) in it. On the fundamental level, however, only electrons exist. The status of electrons or other particles is completely different from that of instrumentalistically treated holes – electrons are attributed objective, independent existence.

Realists believe that instrumentalists adopt too restrictive epistemic standards of truthfulness in science. For the instrumentalist, only direct sensory perception is a guarantee of the truth of the relevant statements. However, the notion of direct observation is vague in itself. Is the image of a mirage, disturbed by the hot air in the desert, a direct perception that ensures the truth of an appropriate perceptual sentence (such as the sentence "There is an oasis in front of me")? To rule out such cases, we must make the criteria of truth so strict that the result is a skeptical consequence—nothing can be known with certainty. Realists believe that this is turning the matter on its head. The truth of a given hypothesis may be supported by indirect evidence. If a theory is a good representation of empirical facts, this counts as an argument for its truth. Realists refer to the so-called argument to the best explanation. The best explanation for the empirical success of a theory is simply its truth. The empirical successes of Copernicus's theory should convince us that Copernicus was closer to the truth than Ptolemy.

We can now move on to the second philosophical issue, which is the question of comparing competing theories. If we have two incompatible theories that explain a similar set of phenomena, on what basis should we choose one of them? In general, we can distinguish two aspects of the evaluation of scientific theories: the theoretical and empirical. Theoretical evaluation includes such features of a theory as elegance, simplicity, range of applicability, the way of explaining known phenomena and predicting new ones. As we have already seen, a comparison of the theories of Ptolemy and Copernicus in this respect clearly indicates the advantage of the latter. The simplicity and elegance of the explanations offered in the heliocentric theory is indisputable. A wider range of predictions, including parameters such as the radii of planetary orbits, is also worth mentioning. However, it is generally accepted that the most important aspect of comparing theories is their empirical testing. The ideal of such testing is the so-called crucial experiment (*experimentum crucis*), which in principle should allow for the definitive choice of one of many competing theories.

The structure of a crucial experiment is very simple. For two theories T_1 and T_2 we need to find an empirical sentence e such that e follows from T_1, while its negation results from T_2. Then we should empirically verify the occurrence or non-occurrence of the fact described by statement e. In this way, the falsity of the sentence e allows us to choose the theory T_2, and the truth of e – the theory T_1. An example of a possible *experimentum crucis* in the case of astronomical theories is the question of the existence of the four phases of the inner planets. The statement about the existence of four phases results from the Copernican theory, while its negation – from the theory of Ptolemy (because it predicts the existence of three phases). By making an appropriate observation of, for example, the phases of the planet Venus, we choose the right theory, which turns out to be the Copernican theory.

A potential crucial experiment for both astronomical theories could also be the phenomenon of stellar parallax. Here, however, the problem of additional tacit assumptions is revealed. On the surface, it seems that the *experimentum crucis* based on the phenomenon of parallax suggests the choice of the Ptolemaic theory. Copernicus's heliocentric theory implies the existence of parallax effect, while the geocentric theory implies its absence. Since we do not observe the apparent movement of stars relative to each other during the solar year, this speaks against the Copernican theory and for Ptolemy's theory. However, the problem is that the heliocentric theory implies the existence of an observable parallax effect only with the additional assumption that the fixed stars are close enough to us. Since we do not know whether this is the case, the crucial experiment in this case does not give an unambiguous solution.

Many philosophers believe that the problem of additional auxiliary assumptions in fact affects every case of empirical choice between theories. Even the seemingly unmistakable case of the phases of the inner planets is based on additional assumptions, such as the laws of geometric optics (propagation of light rays along straight lines, the law of reflection, etc.).

With a little invention, you can always come up with an argument that will undermine the result of a given crucial experiment. French physicist and philosopher Pierre Duhem formulated the main slogan of his view, called conventionalism, in the form of the thesis "There is no such thing as *experimentum crucis*". The conventionalists believe that in the end, the choice of a given theory is determined not so much by empirical considerations as by the previously mentioned theoretical advantages: simplicity, elegance, etc.[5]

The replacement of Ptolemy's theory by the Copernican theory sheds new light on the problem of the development of science. There is a popular view that science develops gradually and linearly by accumulating more and more facts, data, and finally generalizations. Such a model of scientific development is often called cumulative, because it assumes that scientific progress is made by expanding the existing body of knowledge. On the other hand, the changes that took place in astronomy in the sixteenth century suggest a different model. Copernicus did not complement Ptolemy's theory, but completely rejected it and replaced it with an entirely different approach, based on new concepts and assumptions. Thomas Kuhn, a twentieth-century philosopher and historian of science, describes such changes in science as "scientific revolutions". According to him, the history of the development of any science can be divided into two types of periods: the so-called periods of normal science and crises, resulting in violent, revolutionary changes. Normal science is characterized by a gradual development, consisting of collecting data and improving existing models and theories and solving emerging difficulties and problems. In astronomy, the stage of normal development falls on the centuries separating Ptolemy from modern astronomy. During this period, Ptolemy's model was subject to changes and corrections, but its basic core remained the same – it was based on the assumption of geocentrism and the widespread use of deferens and epicycles to reconstruct the observed planetary movements.

The crisis in Ptolemaic astronomy consisted in the increasing complexity of the model while at the same time the mounting difficulties in fitting it to the more accurate empirical data. Kuhn points out that when a crisis in a given field reaches a tipping point, scientists rapidly change the methods used and create a new theory. He describes such a change using the concept of paradigm. A paradigm is a set of assumptions and methods of solving problems that are not questioned by the scientific community. The paradigm of the Ptolemaic astronomy included an assumption of the immobility of the Earth, as well as a method of reproducing planetary motions by means of composing circular motions. The Copernican revolution shifted this paradigm to heliocentric, replacing the composition of planetary motions with a combination of true planetary motion and the observer's motion on the Earth's surface. A controversial element of Kuhn's doctrine was the assumption of the so-called incommensurability of paradigms before and after the revolution in science. Incommensurability means that different paradigms cannot even be compared with each other in a single conceptual framework. Adopting a new paradigm makes the old one not so much false as meaningless.

---

[5] Another problem with *experimentum crucis* is that it tacitly assumes that there are no other options besides the theories under consideration. However, it may happen that there are many theories that have the same empirical consequences, confirmed by experience. In such a situation, the crucial experiment does not give us the opportunity to choose between these theories. Such situations are referred to as "underdetermination of theory by data".

1.4 Kepler's laws of planetary motion

The heliocentric model of Copernicus was not immediately accepted by the scientific community. Resistance to the doctrine of heliocentrism was partly due to political and ideological issues (the influence of the Church's teaching), but as we pointed out in the previous section, there were also serious objections of a scientific nature. Danish astronomer Tycho de Brahe proposed a compromise solution, combining elements of Copernicanism with Ptolemy's model. He assumed, in accordance with the Ptolemaic model, that the Earth rests motionless at the center of the universe. However, he proposed that the other planets revolve not around the Earth, but around the Sun, and only together with the Sun do they orbit the Earth. Thanks to this, de Brahe avoided problems with the lack of perceptible effects of the Earth's motion (including the lack of stellar parallax). It is perhaps worth noting that with properly selected orbits of the planets around the Sun, his model is basically "kinematically" equivalent to the Copernican model, if we ignore the fixed stars.[6] It is obtained by shifting from the frame of reference in which the Sun rests to the frame associated with the Earth. Of course, de Brahe himself did not yet have the concept of different frames of reference or the idea of the relativity of motion, which only started to appear in Galileo.

The real progress in the heliocentric model was made by Johannes Kepler. This astronomer was truly obsessed with the mathematical description of the regularities in the movement of the planets. Adopting the Copernican position, he tried to discover mathematical regularities in the movements of individual planets around the Sun. One of his ideas, which turned out to be completely misguided, was an attempt to use Plato's perfect solids to express the proportions between the orbits of successive planets. While considering various alternative options, he came up with the idea of using the so-called conic sections to describe planetary orbits. Conic sections are curves created by cutting a cone with planes at different angles. One such curve is, of course, the circle, but by changing the angle of the plane in relation to the altitude of the cone, we will get ellipses. (Other conic sections are the parabola and hyperbola.) Kepler, in a stroke of genius, tried to assume that the orbits of the planets are not circles, but ellipses. It turned out to be the proverbial bull's eye. Thanks to this assumption, he was able to precisely reproduce the observed astronomical data without the need for Copernican epicycles.

However, Kepler's contribution to astronomy does not end there. Describing the shape of planetary orbits is just the beginning. Kepler also wanted to discover regularities characterizing the velocity of moving planets, as well as regularities describing the size of individual orbits. He formulated the results of his research in the form of the three well-known laws of planetary motion, which we will quote in their modern form.

1. All planets move in ellipses with the Sun in one of the foci.
2. Every planet's radius (attached to the Sun) sweeps equal surface areas in equal periods of time.
3. The ratio of the square of the orbital period to the cube of the average radius is the same for all planets.

---

[6] However, these models are not dynamically equivalent due to the occurrence of apparent forces (centrifugal, Coriolis) in rotating systems. We will talk about this in the next sections.

The foci of a given ellipse are two points such that the sum of the distances of any point on the ellipse from the foci is always the same. Kepler's second law bears some resemblance to the principle of the equant in the Ptolemaic model. In Ptolemy, the planets sweep equal angles with respect to the equant in equal periods of time. Kepler achieved a better match with experience by replacing angles with surface areas and the equant point with the Sun. It is worth noting that Kepler's second law follows from the principle of conservation of angular momentum in classical mechanics. Kepler's third law, on the other hand, is the result of his attempts to find regularities in the sizes of the orbits of individual planets. His efforts to express the regularity in terms of radii alone failed. Instead, he discovered that although in principle any planet can take an orbit of any size, fixing the radius of the orbit R determines its period of revolution T, in accordance with the proportion $\frac{T^2}{R^3} = \text{const}$.

Kepler's laws organized astronomical knowledge, ensuring perfect match with experience, unattainable in previous models. However, their contributions to science go beyond astronomy alone. Kepler's laws enabled Newton to discover a mathematical formula that describes universal gravitational interactions, both between the Sun and the planets, and between any objects endowed with mass. We will talk about this in more detail in the next chapter.

The next great figure in the development of astronomy was Galileo, already mentioned several times. The achievements of the scientist from Pisa go far beyond the systematization of astronomical phenomena. Galileo paved the way for a huge breakthrough in science, which was the replacement of Aristotelian mechanics by Newtonian mechanics. His contribution to astronomy consisted primarily in the use of an optical instrument, the telescope, in detailed observations of celestial bodies. Thanks to this technological advancement, he made several spectacular discoveries, expanding our knowledge and definitively breaking with the image of the universe shaped by Ptolemy, and even partially adopted by Copernicus. Galileo challenged the belief in the perfection of celestial bodies by observing irregular craters and darkened "seas" on the surface of the Moon. He observed and correctly identified Jupiter's moons, the existence of which was not even suspected by his predecessors. As we have already mentioned, with his telescope he confirmed the existence of the four phases of Venus, as predicted by the Copernican theory.[7]

Galileo was, of course, a staunch supporter of the heliocentric Copernican system, perfected by Kepler. Although he did not introduce any additional theoretical improvements to this system, he undertook the extremely important task of refuting the objections against heliocentrism, based on the lack of perceptible effects of the Earth's motion. This was a very fortunate coincidence, because thanks to his determination, Galileo noticed the fundamental errors in the perception of kinematic phenomena that had been passed down to us from antiquity. His defense of Copernicus took the form of a frontal attack on the principles of Aristotelian physics. This attack set the right direction for further developments of modern phys-

[7] It is perhaps worth mentioning that Galileo has laid the foundations of scientific methodology, for instance taking up the issue of the reliability of the use of optical instruments, such as the telescope or the spyglass. In response to the accusations that astronomical observations with telescopes are unreliable due to the possible creation of artifacts and optical illusions, he proposed a simple test. He pointed his optical instrument at a distant terrestrial object, and then compared the image of the object observed by the telescope with its actual appearance, convincing himself that no significant distortions or disturbances were created in the telescopic image.

ics. However, in order to present these matters a little more precisely, we must go back in time to the physical theorizing of the famous philosopher from Stagira.

1.5 Aristotle's physics vs. Galileo's physics

In Aristotle's time, physics was almost exclusively limited to mechanics. In mechanics we distinguish three branches, the names of which indicate their Greek origin. The branch of mechanics dealing with the general description of motion is known as kinematics (from the Greek *kinesis* – motion). Kinematics classifies and describes motions using concepts such as position, velocity, and acceleration. Within this classification, we distinguish uniform, rectilinear, curvilinear (including circular) and accelerated motions. On the other hand, dynamics (Greek *dinamis* – force, power) looks for the causes of movement in the form of acting forces. The third branch of mechanics is statics (Greek *statos* – permanent, unchangeable), describing objects in the equilibrium position, with an emphasis on equilibrium conditions given in terms of acting forces (Archimedes' law of the lever).

Aristotelian dynamics uses the distinction between natural and forced movements. Natural movements result from the essence of the objects involved and do not require an external (efficient) cause. They can at most be described in terms of final causes (referring to the end of the motion). The natural motion of bodies near the Earth is vertical upward or downward motion. Heavy bodies fall downwards because their nature is earthly, and so they strive to reach their natural place in the center of the Earth (final cause). Light bodies, such as smoke or steam, on the other hand, rise upwards because they are dominated by the celestial element. Aristotle also distinguished two spheres of the world: sublunar (terrestrial) and supralunar (heavenly). In the supralunar sphere, natural movements are circular – as we remember, this belief was the basis for the Ptolemaic and even Copernican theory.

Forced motions are those that require effort (cause) to support them. We know from experience that in order to move an object – e.g. a cart – we have to push or pull it. When we stop exerting force on it, the cart returns to its resting state. Aristotle generalized this observation to all possible situations of unnatural motion. Thus, the basis of Aristotle's dynamics was the belief that the amount of motion is proportional to the force applied – when the force is zero, the motion stops. Of course, Aristotle did not use the modern concept of force, which was introduced by Newton, but it would probably not be a great anachronism to interpret his position with today's concepts. Aristotle's proposed law of motion for forced (unnatural) movements would therefore have the following form today:

$$v \sim F,$$

where F is the force applied to the body, v its velocity, and the symbol $\sim$ stands for proportionality. Aristotle was also aware that the movement of objects is influenced by the resistance of the environment, including air. The greater the resistance, the lower the velocity of the object under a given force. Denoting the measure of resistance by R, we have the following version of Aristotle's law of motion:

$$v \sim \frac{F}{R}.$$

Aristotle noticed a disturbing consequence of his law – according to the above proportion, when the resistance of the environment R decreases to zero, the velocity should increase to infinity. This is clearly inconsistent with experience, and therefore some correction must be made. The philosopher drew the conclusion that eliminating resistance is impossible for fundamental reasons. In particular, it is impossible to completely remove air from a given spatial region. Aristotle believed that a vacuum could not exist (hence the medieval term *horror vacui* – fear of a vacuum). The celestial bodies move in a perfect etheric substance (the so-called fifth substance – *quinta essentia*).

It is perhaps worth noting that if Aristotle had used modern algebra instead of ordinary proportions, he could have easily dealt with the problem of "an escape to infinity". It would be enough to make the following modification of the above formula:

$$v \sim \frac{F}{R+c},$$

where c is a constant. In such a situation, the velocity of the body in a medium without resistance (e.g. vacuum) would approach the maximal but finite value $\frac{F}{c}$. Unfortunately, centuries passed before the development of mathematics made it possible to use similar tricks.

As a critical observer, Aristotle was aware of another serious difficulty of his theory. It was a problem of projectiles: thrown stones, arrows fired from a bow or missiles hurled from catapults. In all these cases, the force acting on the projectile ceases when it is launched, and yet it continues its flight until it hits its target or falls to the ground. Thus, it seems that motion without a sustaining force is possible. In order to deal with this difficulty, Aristotle came up with the following solution. A body in flight interacts with air, so perhaps it is the air that keeps it moving. Considering this problem more closely, Aristotle suggested that the projectile displaces air in flight and creates a momentary vacuum behind it. However, as we already know, nature does not tolerate vacuum, so the air immediately flows into this place, which exerts pressure on the body, keeping it in flight.

Aristotle's ingenuity cannot be denied, although there are also visible holes in his argument. His explanation is inconsistent – on the one hand, the presence of air creates resistance to motion, reducing the velocity of the projectile, but on the other hand, it exerts a force, supporting the projectile in motion. But let us wait for Galileo's penetrating argument to criticize Aristotle's theory. For now, we will mention his theory of free fall. As we have already seen, free fall is a natural motion, depending on the amount of the terrestrial element in a given body. This earthly element has a simple quantitative interpretation in the form of weight. No wonder that Aristotle accepted without hesitation the law of free fall, according to which heavier bodies fall faster than light ones.[8] In other words, there is proportionality

[8] In any case, such a law of free fall was attributed to Aristotle by his successors, including Galileo himself. It is interesting, however, that the fragment of Aristotle's writings (in Book IV *of Physics*), in which he formulates the principle of proportionality of the velocity of a body to its weight, does not mention fall at all, but only an unspecific motion in the air (probably caused by a given initial force). Aristotle explains his proportionality relation quite convincingly by the effect of air resistance (the heavier the body, the less resistance the air will exert on it – today we know that it is rather the size or shape of the body that plays a role here). Moreover, the Stagyrite unequivocally states that in a space devoid of any resistance, bodies would move at the same speed! Thus, two thousand years before

between the time of the fall from a given height and the weight of the body. It seems that experience confirms such a principle – it is enough to compare the fall of a stone and a bird's feather.

Galileo critically analyzed the basic theses of Aristotle's physics. Characteristically, although Galileo himself proclaimed the necessity of empirically verifying all statements about reality, even the most obvious ones, many of his arguments are *a priori* in nature, revealing the internal inconsistency of Aristotle's assumptions. Let us start with the problem of free fall. A cursory observation makes us realize that it is not possible for the rate of free fall to be proportional to the weight of the body. A body weighing ten kilograms certainly does not fall ten times faster than a one-kilogram weight. It is possible, however, that the dependence of the rate of fall on the weight is of a different kind than simple proportionality, described by a linear function. Galileo used an ingenious argument to show that the velocity of descent cannot be any increasing function of weight.

Consider two bodies with weights G_1 and G_2 where $G_1 > G_2$. By design, body number 1 falls faster than body 2. Let us now connect these bodies, e.g. with a rigid rod. On the one hand, we will then get an object with a total weight $G_1 + G_2$, which should therefore fall even faster than each component individually. On the other hand, the body with a weight of G_2, as slower, will slightly delay the descent of the body G_1, while the heavier body will accelerate the slower, so the "resultant" velocity should be between the two velocities of each of the objects considered separately. So we have a contradiction. Galileo drew the surprising conclusion that the time of free fall cannot depend on weight at all. All bodies fall at the same speed.[9]

So how to explain the fact that there are observable differences in the time of fall between some objects (the example of a stone and a feather)? Galileo gave the answer we know from school textbooks: the cause of this difference is not the intrinsic characteristics of falling objects, such as weight, but their interaction with the environment, i.e. air resistance. Experiments with dropping objects in sealed containers emptied of air unequivocally confirm this assumption. In the absence of air resistance, the time of the fall of objects is indeed the same. Let us note, however, that Galileo did not have the possibility of empirically verifying his law of free fall. Apparently, he decided that the theoretical merits of his hypothesis were more important than the inconsistency with observable facts. The explanation of this incompatibility by the effect of air resistance had all the features of an *ad hoc* solution. As you can see, *ad hoc* arguments can play a positive role in science, and clinging to empirical evidence without theoretical assumptions can lead us astray.

Galileo's argument can be generalized as follows. Suppose the rate of free fall is a function of some parameter *m*, which is assumed to be an additive quantity. An additive quantity is one whose value for the physical sum of two objects is the sum of the values for each of these objects separately. The weight of an object falls into this category (the weight of two bodies is the sum of their weights). Other examples of additive quantities

---

Galileo, he formulated an essentially correct law of free fall. However, Aristotle found such a consequence absurd, from which he drew the already known conclusion that the vacuum does not exist.

[9] Of course, today we know that it is not the speed, but the acceleration that is constant in free fall. Galileo was aware of the difference between the two concepts, but in his argument he used the imprecise term of speed (probably understood as the average speed over the entire stretch of road). You can also use another parameter in the form of the time of the body's fall from a given height.

are mass, volume, or electric charge. If we now assume, with Galileo, that the function f representing the rate of descent, satisfies the following condition:

$$f(m_1) \leq f(m_1 + m_2) \leq f(m_2) \text{ for } m_1 \leq m_2,$$

then from the assumption of continuity it can be proven that f must be a constant function.

However, let us pay attention to the extremely important assumption that the rate of descend depends on exactly one parameter. It could happen that f would be a function of more independent parameters. In particular, consider the case when f is a non-decreasing function of the quotient of two independent additive quantities: $f\left(\frac{q}{m}\right)$. It turns out that in this case, the Galilean condition

$$f\left(\frac{q_1}{m_1}\right) \leq f\left(\frac{q_1+q_2}{m_1+m_2}\right) \leq f\left(\frac{q_2}{m_2}\right) \text{ for } \frac{q_1}{m_1} \leq \frac{q_2}{m_2}$$

will be automatically satisfied, regardless of the specific form of the function f. Therefore, the constancy of the free fall time is not true by necessity – it results from the fact that there is exactly one additive parameter on which this time depends. There is a deeper physical truth behind it. As it turns out, the acceleration of a body in free fall can be represented as a function of the quotient of two quantities: gravitational mass, characterizing gravitational interactions, and inertial mass, occurring in Newton's law of motion. These two quantities turn out to be equal, which implies the constancy of acceleration in the gravitational field. However, the equality of gravitational and inertial mass is an empirical fact that does not follow from *a priori* assumptions.

Galileo's equally ingenious argument shows that the Aristotelian explanation of projectile motion has consequences that are manifestly inconsistent with experience. If it were true that a body's flight was sustained by the pressure of the air pushed out by the flying object, blunt-tipped objects should move faster than pointed objects, because the former push out more air. Ironically, Galileo notes that in this case, archers should place the arrow perpendicular to the direction of flight, not parallel. It is therefore necessary to look for another explanation for the existence of motions without a visible sustaining force. Already in the Middle Ages, it was noticed that objects have the ability to maintain the amount of motion imparted to them, as long as they do not encounter any obstacles. This property is called inertia. Galileo accepted the principle of inertia in the form of the thesis that bodies tend to remain in motion at a constant speed until some external factor acts. Such a factor can be the resistance of the environment – air or a rough surface on which we slide an object – which we already remember from the analysis of free fall.

Again, then, we are dealing with a situation where observations alone are not enough to unambiguously determine the laws governing certain phenomena. In the case of the relationship between force and motion, we have two opposing observations: on the one hand, the experience of bodies sliding on the surface of the earth, which require a constant force to keep moving, and on the other hand, the observation of projectiles in flight. Aristotle took the first situation as the basis for his law of motion, and treated the second case as an anomaly that required additional explanation. Galileo, on the other hand, did the opposite – he decided that the behavior of missiles in flight reflects the true nature of the phenomena, and the first, "anomalous" situation should be explained by additional disturbing factors, i.e. friction. The success of Galileo's conception is due, among other things, to the fact that Aristotle's explanation of the anomaly was unsatisfactory. This shows that theorizing and hypothesizing is a cognitive activity that goes beyond a simple analysis of empirical data.

1.6 Galileo's principle of inertia and principle of relativity

Galileo's principle of inertia was of paramount importance for his polemics with the opponents of the heliocentric theory of Copernicus. As we remember, one of the main problems for Copernicus was the lack of noticeable effects of the Earth's rotation. For example, it was claimed that if the Earth rotated on its axis, an object dropped from the tower would deflect from the base of the tower by the distance that the tower and its entire surface would travel at that time. Galileo first noted that the suggested effect does not occur, for example, when an object is dropped from the mast of a ship sailing at a certain constant speed. Objects dropped from the mast fall at the base of the mast, regardless of the uniform movement of the ship (provided that we ignore the rocking on the ship). It is not clear whether Galileo made appropriate observations to confirm this empirical fact, which of course holds true. On the other hand, his theoretical explanation was based on the principle of maintaining a constant velocity of the body in motion. Both the body at the top of the tower and the body attached to the mast of a moving ship participate in the motion (of the Earth or the ship) and possess a certain horizontal velocity. Dropping these objects does not deprive them of this velocity, so even if they lose contact with the tower or mast, these objects still move in the same direction as the other objects.

Galileo did not stop at a simple generalization of empirical observations to justify the principle of inertia. In accordance with his tendency to theorize, he proposed an interesting *a priori* argument for the truth of this principle. Consider a plane inclined at a certain angle and a ball at the foot of this plane (Fig. 1.12). By imparting a certain initial velocity to this ball, we will cause it to roll up onto the plane, but at the same time it will decrease its speed until it comes to a complete stop (and rolls down). When we decrease the angle of inclination of the plane, the ball will lose its speed more and more slowly (i.e. it will cover greater and greater distances). By rotating the plane even further, we will reach a "negative" angle (downward inclination), which will cause the ball to accelerate. Therefore, for a certain angle, we must have a situation in which the ball will neither accelerate nor slow down, i.e. it will move at a constant speed.[10] It is not difficult to see that this should happen for an angle of inclination equal to zero, and the movement of the ball will be inertial in this case. Extending the horizontal path of its inertial motion, we can assume that this motion is rectilinear. Unfortunately, Galileo, taking into account the spherical shape of the Earth, erroneously concluded from his argument that inertial motion would in fact be circular, at a constant distance from the center of the Earth. This was partly due to the fact that he did not yet have the concept of the force

[10] Galileo's argument can be reconstructed in the modern language of mathematical physics. We assume the existence of a functional dependence of the acceleration of the ball on the angle of inclination of the plane: $a = f(\alpha)$; we also assume that the function *f* takes negative values for positive angles α and positive values for negative angles. From these assumptions it follows, if we additionally assume the continuity of the function *f*, that for angle $\alpha = 0$ the acceleration is 0.

of gravity, which appeared only in Newton. Galileo's concept of gravity was still closer to Aristotle's idea of natural motions without an external cause.

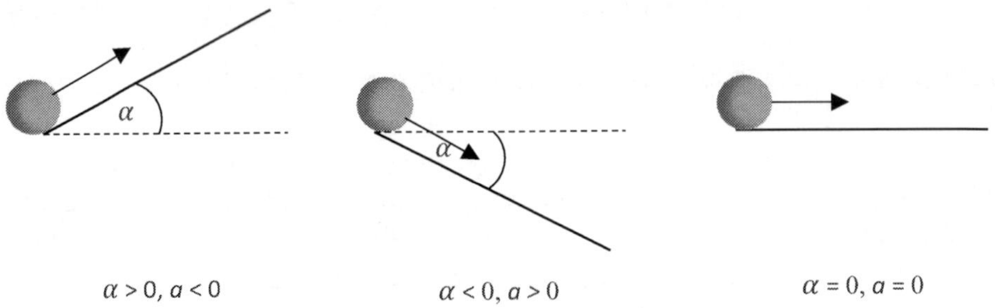

Fig. 1.12 Galileo's argument from inclines for the principle of inertia

The lack of the concept of gravity also influenced the way Galileo tried to refute the objection of the lack of perceptible centrifugal force from the assumed rotation of the Earth. In the times of Copernicus and Galileo, people were well aware of the existence of a force that "pushes" objects in rotation in the opposite direction to the axis of rotation. Therefore, one would expect that an analogous force should act on all objects on the Earth's surface, which under this influence would therefore fly away into space at high speed. Galileo put a lot of effort into refuting this argument. Unfortunately, his solution was completely wrong – he tried to show that because the motion along a circle can be divided into very small segments close to tangent lines, the body participating in the rotation of the Earth will not be subject to a momentary repulsive centrifugal force. It is difficult to understand why Galileo's keen mind failed to notice that if his argument were correct, it would deny the existence of all centrifugal forces, e.g., in the ejection of projectiles from a slingshot. The correct solution to the centrifugal force problem, as we well know, is that it is compensated by the force of gravitational attraction. More precisely, the net force acting on a given object on the Earth's surface arises from the vector subtraction of the centrifugal force (in the direction away from the axis of rotation) from the force of gravity directed towards the center of the Earth. This means that when we are at the Equator, we weigh slightly less than at the Pole, but this difference is practically imperceptible.

Let us mention another type of force that arises in rotating systems, the existence of which can be easily established empirically. This is the so-called Coriolis force, which acts only on bodies moving at a certain speed in a rotating system. A body that moves towards or away from the axis of rotation is subjected to a force perpendicular to the direction of its motion whose strength depends on both the speed of rotation and the speed of the body's movement. The existence of Coriolis forces is responsible for the occurrence of hurricanes or cyclones, which are formed when air masses move from the hot equatorial region to the north or south. This creates a "whirlpool" that always rotates clockwise in the southern hemisphere and counterclockwise in the northern hemisphere. Another evidence of the Coriolis force is the behavior of the so-called Foucault pendulum, which changes the plane of its swinging motion in a rotational fashion. The effects of the Coriolis force are strong evidence in favor of the rotation of the Earth – unfortunately, Galileo did not know this effect, so he could not use this argument in his polemic with the Aristotelians.

It is interesting to observe that the existence of both centrifugal and Coriolis force can be qualitatively justified by Galileo's principle of inertia. Consider an object in rotation around a certain point. At each moment of its motion, this object has a certain linear velocity, directed tangentially to the circle along which it moves. Due to the principle of inertia, this object has a tendency to continue its flight without changing its speed (its value and direction), so if it is not held, it will "shoot" along the tangent line to the circle. To keep the object moving in a circle, we must act on it with a centripetal force that is exactly equal in value, but opposite to the centrifugal force (Fig. 1.13).

In order to derive the Coriolis force, consider a situation in which an object on a rotating disk begins to move in the direction of the axis of rotation. As it gets closer and closer to the axis of rotation, this object encounters points that have a lower linear velocity (the linear velocity of the points on the rotating disk depends on their distance from the center). According to the principle of inertia, the linear velocity of the object tangent to a circle remains unchanged, and therefore it will begin to "overtake" the points encountered, which an observer on a rotating disk will interpret as a manifestation of a force acting perpendicular to the direction of its motion.

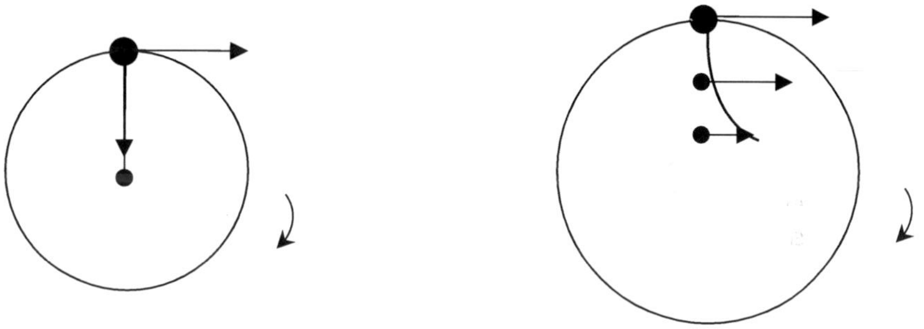

Fig. 1.13 Explanation of the formation of centrifugal and Coriolis forces in rotational motion using the principle of inertia.

Returning to Galileo's favorite example with a ship, we can see that he also used it to discuss two related principles: the principle of composition for velocities (independence of motions) and the principle of relativity. Consider again the example of a ship moving steadily and a heavy object dropped from the top of the mast. As Galileo argued, from the perspective of the ship, the fall of a body will be no different from the fall of a stationary object. On the other hand, from the point of view of an observer on the shore, the dropped body will draw a curve (today we know that it is a parabola). Galileo noted that this motion can be thought of as consisting of two independent motions occurring simultaneously: the vertical free fall and the horizontal movement of the ship and all the objects associated with it (Fig. 1.14). This was a completely innovative discovery – never before had movements been "decomposed" in this way.

Galileo also drew attention to the fact that one of the components of complex motion can in a sense be "turned off" without disturbing the other, by changing the perspective (today we would say: by changing the frame of reference). He used this observation, e.g. to describe the motion of a cannonball fired horizontally in relation to the ground. He argued that by taking the perspective of an observer moving horizontally with the projectile, we will get nothing more than a free fall of a cannonball. Considerations of this kind probably gave him an idea of paramount importance for the later development of physics. He came to the conclusion that mechanical phenomena look essentially the same regardless of the state of motion of the observer, as long as the observer himself moves in a uniform and rectilinear way. This thesis, which we now call the principle of relativity, was illustrated by Galileo again with the example of the ship. This time, however, imagine that the observer was locked below deck in a windowless cabin. As before, suppose that a ship glides on a calm sea at a constant speed, without rocking or accelerating. Will the observer in the cabin be able to determine whether the ship is moving or standing in port? Galileo stated that any mechanical processes taking place inside the cabin would not depend on the state of the ship's motion. All movements of objects enclosed in the cabin, whether free or forced, will look the same on both the resting and moving ships. To determine that a ship is moving, you need to go on deck and compare its position with the external environment. In other words, uniform and rectilinear motion is a relative concept, not an absolute one.

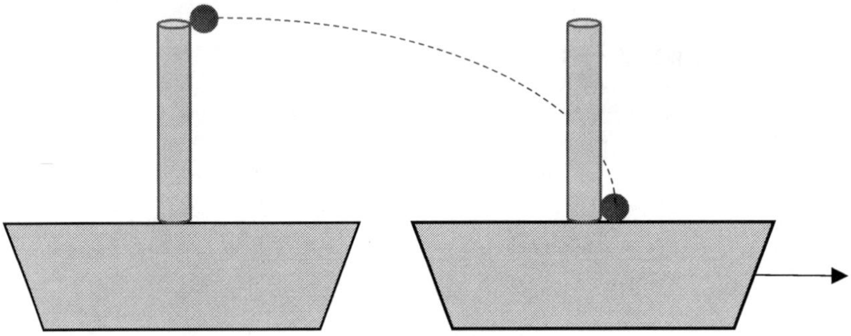

Fig. 1.14 Galileo's principle of composition of motions

In the next section, we will see how Galileo's principle of relativity can be presented in the framework of Newtonian mechanics in the form of a thesis about the invariance of the laws of mechanics with respect to a certain type of transformation, called Galilean transformations. We will also need to introduce the important concept of an inertial frame of reference. For now, however, let us conclude the review of Galileo's remarkable achievements by mentioning his research on uniformly accelerated motion. He noticed that the motion of a free-falling body cannot be uniform, because it begins with zero velocity and ends with a certain non-zero value. Therefore, the falling body will not travel equal distances at equal intervals. To investigate this matter further, the scholar from Pisa observed balls rolling down inclined planes (contrary to a popular myth, he did not study the times of fall for objects dropped from the leaning tower of Pisa). Measuring the distance that these balls traveled in equal time intervals (e.g. in each successive second), he noticed an interesting regularity:

these distances remained in ratios corresponding to successive odd numbers: 1 : 3 : 5 : 7, etc. This proportion can be expressed slightly differently by comparing the cumulative distances traveled in one, two, three, etc. seconds. In this approach, the distances will relate to each other like squares of consecutive numbers: $1^2 : 2^2 : 3^2 : 4^2$ ... (This is due to the algebraic fact that the sum of n consecutive odd numbers is the square of the number n.) So Galileo *de facto* discovered that distance in uniformly accelerated motion increases in proportion to the square of time. In this way, he extended the mathematical methods used to describe kinematic phenomena, going beyond simple Aristotelian proportions. Further improvements of the mathematical apparatus of physics will be accomplished by the great Isaac Newton.

Questions and problems

1. List the basic observational facts about the motions of celestial bodies (stars, including Polaris, the Sun, and the planets) known to the ancients.
2. How were the ancient astronomers able to estimate the absolute distance between the Earth and the Sun? Note the auxiliary role of Eratosthenes' calculation of the radius of the Earth and the estimation of the distance from the Earth to the Moon.
3. List the elements of Ptolemy's geocentric model and their role in explaining the observable regularities of planetary motions. Why are these elements considered *ad hoc*?
4. What relationship must hold between the velocities of the planets on the epicycle and the deferent for Ptolemy's model to correctly reproduce the observed phenomenon of the retrograde motion? What kind of assumption about the orbital velocity of the Earth and the orbital velocities of the outer and inner planets must be made in Copernican theory to achieve the same effect? Which of the assumptions – Ptolemaic or Copernican – is more *ad hoc*?
5. Discuss the main advantages and difficulties of the Copernican model compared to the geocentric Ptolemaic model. What scientific reasons were there for treating Copernicus's theory with a certain degree of skepticism?
6. What is an *experimentum crucis* (crucial experiment)? Can we find crucial experiments in their pure form in science?
7. What is the instrumentalist interpretation of a given scientific theory? What kind of division within the language of a given theory does instrumentalism postulate? Illustrate this issue with the example of the Copernican theory.
8. How can we know that science has undergone a revolution? What are the characteristics of scientific revolutions?
9. Discuss Kepler's three laws of planetary motion. Do these laws apply to all objects in the universe?
10. Formulate Aristotle's basic law of motion. How can we express it in the language of modern algebra? In Aristotle's approach, what is the necessary condition for the body to continue its movement (not to stop)?
11. Discuss Galileo's argument against Aristotle's law of free fall, as well as his argument against Aristotle's explanation of projectile motion.
12. Present Galileo's argument from the inclines for the principle of inertia. What erroneous conclusion about the path of a body moving inertially near the Earth's surface did Galileo draw?
13. How did Galileo use his principle of inertia to refute objections to the heliocentric theory from the unobservability of the effects of the Earth's rotation?

14. Discuss the principle of velocity composition using the example of a body dropped from the top of a moving ship.

15. Present Galileo's ship argument for the principle of relativity. What exactly does this principle state? What is the limitation of its applicability?

Further reading

A very accessible and illuminating comparison of the Ptolemaic and Copernican models can be found in the popular textbook: E.M. Rogers, *Physics for the Inquiring Mind.* Part Two. *Astronomy: A History of Theory*, Princeton University Press, Princeton 1960.

The development of astronomy from Ptolemy to Copernicus and Galileo is the main theme of Part II of the book: J.T. Cushing, *Philosophical Concepts in Physics*, Cambridge University Press, Cambridge 1998.

The philosophical aspects of the Copernican revolution, with particular emphasis on the problem of revolution in science, are discussed in the classic book: T. Kuhn, *The Copernican Revolution: Planetary Astronomy in the Development of Western Thought*, Harvard University Press, Cambridge Mass. 1957.

A widely commented defense of a variant of instrumentalism (in the author's terminology "constructive empiricism") can be found in the book: Bas van Fraassen, *The Scientific Image,* Clarendon Press, Oxford 1980.

I recommend reading at least some parts of the beautiful piece of scientific literature that is Galileo Galilei's *Dialogues Concerning the Two Chief World Systems: Ptolemaic and Copernican*, (various editions).

CHAPTER 2. CLASSICAL MECHANICS

Classical mechanics, also known as Newtonian mechanics, provides the foundation for all physics, including modern physics. Without the concepts analyzed in mechanics, such as force, momentum, energy, mass, etc., it is impossible to imagine the development of even the most advanced physical theories. Although officially Newtonian mechanics has been abandoned and replaced by more empirically and conceptually adequate theories (both theories of relativity and quantum mechanics), its presence is clearly visible in almost every branch of physics. Mathematical methods originating in mechanics are still successfully used to describe a variety of phenomena, including processes at the quantum level which involve both particles and quantum fields. From a philosophical point of view, Newtonian mechanics offers a wealth of methodological, ontological, and epistemological problems. We will begin our philosophical analysis of classical mechanics by presenting the fundamental laws of motion in a form similar to Newton's original formulation. We will reflect on the methodological status and empirical content of these laws, addressing the controversial question of whether they are merely hidden definitions (terminological conventions). Next, we will show how the laws of motion work in practice, enabling us to determine the future mechanical behavior (motion) of bodies subjected to various forces. This leads directly to an extremely important ontological and epistemological issue, which is the problem of determinism. It is widely believed that classical mechanics is a model example of a deterministic theory, as opposed to, for example, indeterministic quantum mechanics. We will examine this view in more detail, emphasizing the precise characterization of different variants of the deterministic claim. It will turn out that the thesis of the deterministic nature of mechanical phenomena is not as indisputable as it might seem.

There is a close connection between classical mechanics and the question of the nature of time and space. The description of the mechanical behavior of bodies requires an introduction of parameters characterizing their position in space and time (spatial and temporal coordinates). But what do these parameters refer to? Are time and space substances that exist independently of the physical objects that occupy them? The dispute on this subject took place between two outstanding minds of the era, Newton and Leibniz, the first of whom represented the position of absolutism and the second relationalism. We will analyze two typical arguments used in this dispute, as well as the formal aspect of this issue in the form of the so-called Galilean transformation and Galilean invariants. We will also discuss the derivation of the universal law of gravity from astronomical observations and Kepler's laws of planetary motion based on them, as well as the philosophical problem of the so-called action at a distance. Finally, we will mention newer and more mathematically advanced approaches to classical mechanics: Hamiltonian mechanics, the principle of least action, and Lagrangian mechanics. The principle of least action particularly deserves the philosopher's attention, as it seems to rehabilitate the notion of a final cause, which has been all but discredited in the natural sciences.

2.1 Newton's laws of motion

The foundations of modern classical mechanics were laid by Isaac Newton in his groundbreaking work *Principia Mathematica* (its full title is *Philosophiae Naturalis Principia Mathematica* – Mathematical Principles of Natural Philosophy), published in 1687. Newton's biography deserves a separate chapter, but we do not have space for it here. Born into a poor family, raised by his mother and stepfather, for whom he did not have particularly warm feelings, he quickly showed extraordinary academic skills and ambition, as well as a difficult character. His scientific activity included work on the basics of mechanics, mathematics (differential and integral calculus), the theory of gravity, and optics. He also showed interest in problems of a more esoteric or theological nature. To top it up, Newton had considerable organizational talents, serving as the overseer (the master) of the royal mint and the president of the Royal Scientific Society. He was involved in bitter personal conflicts, with Robert Hook and, of course, with Leibniz over the priority of the discovery of calculus. However, Newton's enormous achievements in the field of science, which set the direction for the development of physics for the following centuries, allow for a more lenient treatment of the shortcomings of his character.

Newton's *Principia* are written in a way that reflects his preferred scientific method, based on strict definitions, axioms and theorems. At the beginning of the work, he gives eight definitions of fundamental concepts, including the concepts of mass, the amount of motion (today called momentum) and force. Then, in the fragment under the name *Scholium* he presents his views on time and space. Finally, concluding the first part, *Principia* formulates the famous three laws (axioms) of dynamics, from which we will start our encounter with classic mechanics. This is the form in which we find them in Newton:

1. Every body continues in its state of rest, or of uniform motion in a right line, unless it is compelled to change that state by forces impressed upon it.
2. The change of motion is proportional to the motive force impressed; and is made in the direction of the right line in which that force acts.
3. To every action there is always opposed an equal reaction; or, the mutual actions of two bodies upon each other are always equal, and directed to contrary parts.

The first law of motion is recognized as the principle of inertia, already anticipated by Galileo and adopted by the Cartesians. Breaking definitively with the Aristotelian paradigm, Newton emphasizes in this law that motion is a natural state that does not require any cause, as long as it takes place in a uniform manner and in a straight line. Newton's formulation also contains an additional hint about the role of forces in mechanics – the result of forces is not to "sustain" motion, but to change it. This idea is developed in a quantitative way in the second law, which was Newton's original and innovative contribution. According to this law, the change in the amount of motion, defined as the product of the mass of a body and its velocity, is proportional to the magnitude of the force applied to the body. It should be added that proportionality also covers the duration of force – the longer this time, the greater the change. As a result, we get the following simple mathematical relationship:

$$\Delta(mv) = F\Delta t\,,$$

from which we derive, assuming the invariance of mass, the formulation of the second law known to all of us:

$$F = m\frac{\Delta v}{\Delta t} = ma\,,$$

where a is the acceleration of the body.[1]

The third law of motion is the well-known principle of action and reaction. It is perhaps worth stressing that the forces of action and reaction do not "cancel" each other, although they are equal and oppositely directed, since they are applied to different objects. If, on the other hand, we applied such forces to the same body, their result would be zero. This brings us to another principle adopted by Newton, which some call the fourth law. This is a well-known parallelogram rule, allowing us to add forces acting in different directions. Newton realized that quantities such as force, velocity or acceleration are characterized not only by numerical values, but also by their direction. Today, such quantities are called vectorial. To add two vectors together, we cannot simply add their lengths – we must plot a parallelogram with sides equal to the values of these vectors and directions consistent with the directions of the vectors, and their sum will be the vector connecting the opposite vertices of the parallelogram (Fig. 2.1).

We will now deal in greater detail with the problem of the methodological status of the laws of dynamics. It is sometimes suggested that Newton's laws of motion are in fact implicit definitions of terms such as force or mass, and are therefore true by virtue of meaning rather than of non-linguistic facts. Even if the axioms of dynamics are not merely definitions, i.e. they have some empirical content, the question remains whether it is possible to definitively empirically verify their truth. Newton himself certainly did not conduct experiments to verify

[1] It is worth noting that Newton never used the equation $F = ma$. The first scholar to introduce this formulation of the second law was Leonhard Euler. In addition, the question can be asked whether the first law is not already included in the second. Substituting $F = 0$, we have the conclusion that the acceleration of the body is zero, which can be interpreted as implying that the body moves in uniform and rectilinear motion. However, the matter is not obvious. To begin with, Newton's second law of dynamics in its original form says nothing about how a body moves in the absence of forces. Substituting a value of zero force is, strictly speaking, going beyond the scope of applicability of the second law (although this step can be justified by the assumption of continuity when considering incrementally smaller values of force). Moreover, the assumption that the acceleration of a body is zero does not yet imply that a body can have a non-zero velocity. A staunch Aristotelian might argue that the absence of force results in an immediate halt of the body, with the outcome that both velocity and acceleration turn to zero, in accordance with the second law. Of course, further arguments are possible to show the fallacy of such reasoning – if at the moment when the force ceases to act on the body, it had a non-zero velocity, then according to the second law this velocity cannot change, so the body must move in a uniform motion. In any case, for at least historical reasons, the first law is formulated as a separate principle.

the laws he formulated. One may doubt whether such verification (or falsification, i.e. refutation) is feasible at all. If it turned out that no experimental procedure could demonstrate the falsity of the laws of motion, their acceptance would be a matter of convention rather than empirical research.

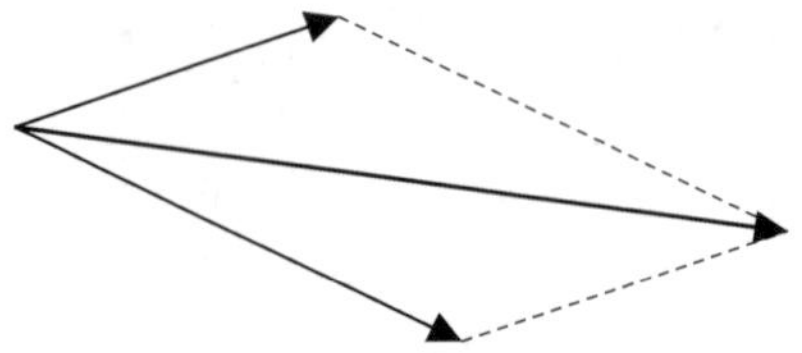

Fig. 2.1 Parallelogram rule used to calculate the resultant force

It may be surprising why we think that the impossibility of overturning a law is a problem. Isn't it the case that in science we want to accept only true statements? Philosopher Karl Popper pointed out that the possibility of refuting (falsifying) a given statement is as important, or perhaps even more important, as its truth. Sentences that cannot be refuted in any way are uninteresting from the point of view of empiricism, because they say nothing about reality (they do not limit the range of possibilities in any way). Popper's falsificationism is the position that the primary goal of a scientist should be to overturn recognized laws by constantly testing them. Only if a given hypothesis passes such tests, it is accepted.

But let us start by introducing some basic epistemological concepts that will help us investigate in depth the above problems. Philosophers distinguish between empirical and *a priori* statements (propositions), as well as between analytical and synthetic ones. Empirical statements (also known as *a posteriori*) are ones whose truth (or falsity) depends on certain facts given to us in experience. For example, the proposition "This rose is red" is evidently empirical, since we ascertain its truth on the basis of direct observation. There can, of course, be empirical statements whose truth is not based on direct observations, such as the proposition "Electrons possess a negative electric charge." However, even in this case, the verification of such a sentence takes place through often complex observations with the use of appropriate instruments. Propositions *a priori*, on the other hand, do not have "verifiers" in the form of observation. No sensory experience is needed to ascertain their truth. The most typical examples of statements *a priori* are theorems of mathematics such as "The sum of all angles in a triangle is equal to a straight angle". We verify this claim by giving its proof, not by appropriate observations.

The distinction between analytic and synthetic sentences is a bit more subtle. It was introduced by the famous German philosopher of the Enlightenment, Immanuel Kant. An analytic sentence, in Kant's view, "decomposes" (i.e. analyzes) the subject in the predicate. In other words, in an analytic sentence we state about the subject what is already contained in it. Nowadays, it is more common to interpret analytic statements as true by virtue of the meaning of words, which in turn are usually given by definition. The sentence "Triangle is a figure with three angles" is analytical, because the expression "triangle" is defined as a figure with three angles. Synthetic propositions, on the other hand, are simply those that are not analytical. Their truth must be based on something other than the meanings we have given to the words.

From the above characteristics it follows that an analytically true sentence must be *a priori*, because assigning meanings to words is not a procedure based on observations or, more broadly, on sensory experience. However, the question of the existence of synthetic *a priori* propositions is open to debate. Kant believed that such statements can be found for instance among the theorems of mathematics, which are not based on definitions, but this is a controversial position (cf. the table below). Without resolving this philosophical problem, let us now turn to the question of what category of propositions the laws of motion should be classified in. There are two possibilities: either they are synthetic *a posteriori* propositions, i.e. not resulting from meanings, but based on experience, or analytic *a priori*. To rule out the latter possibility, we need to look at how the basic terms included in them are interpreted.

	analytic	**synthetic**
a priori	"This triangle has three angles"	?
a posteriori	X	"Grass is green"

Table 2.1 Categories of propositions.

We will begin with the first law, i.e. the principle of inertia. It mentions the notion of force, which we should set aside for the moment until the second law is examined. In addition, it speaks of uniform motion. Uniform motion is one in which a body covers equal distances in equal periods of time. The concept of equality of spatial segments does not pose any special difficulties – measurements of spatial distances are widely known and well understood. Time measurements, on the other hand, are afflicted with a certain problem. The point is that the principle of a good clock is its regularity – each subsequent clock cycle should have exactly the same duration as the previous one. But how can we know that our clock, for example, does not slow down imperceptibly? To determine this, we need a second clock to check the first one. But the same question can be asked with regard to the control clock, and thus the problem is repeated indefinitely.

Most philosophers of science agree that the solution to this problem must be based on the adoption of a certain convention. One possibility that arises is directly related to the first law of motion. Why couldn't we simply take as a standard time-measuring device any body that has not been subjected to any force? Equal periods of time would then be defined as those in which the body travels equal distances. Of course, this is not an idea to be implemented in practice for reasons that we will talk about in a moment. For now, however, let us consider the question whether the adoption of such a convention deprives the first law of its empirical character, reducing it to an analytic *a priori* proposition, true by definition. On the surface, it seems that this is the case – from the accepted definition of the equality of time intervals, it automatically follows that a body not subjected to a force will move uniformly, as required. However, we have missed an important detail: the first law of motion applies to all "free" objects in the universe, not just one model body. There is still a possibility that a clock based on the motion of the selected body will show that other objects in similar situations accelerate.

The empirical content of the first law is revealed in the statement that bodies on which no forces act do not accelerate relative to each other. By choosing one particular free body, we can arbitrarily determine whether we call its motion uniform or not. However, we cannot ensure that all free bodies in the universe will behave in the same way as the model one.

> In our discussions on the status of the laws of motion, it is worth paying more attention to the issue of defining terms. Philosophers distinguish many types of definitions, the most typical of which is an identity definition: we define a term by giving its equivalent (e.g. water is a substance with a chemical composition of H_2O). However, in general, the definition of a term can be any sentence containing that term which we assume to be true by virtue of the meaning of the term being defined (sometimes such definitions are called semantic postulates). Semantic postulates must meet certain conditions of correctness, the most important of which is the so-called condition of conservativeness. It states that a postulate containing a defined term *t* should not have any consequences that do not contain the term *t*, except for tautologies, i.e. statements that are true by virtue of logic. The point is that the truth of our postulate should not depend on accidental facts unrelated to the meaning given to a given term – facts that could turn out to be false. If, for example, we decided to adopt the sentence "Every body on which no force acts travels equal distances at equal intervals of time" as a semantic postulate determining the meaning of the term "equal time intervals", then this would be an example of a non-conservative, i.e. incorrect, postulate. It implies that for every two bodies not subjected to the action of forces, at the times determined by the equal paths traveled by the first of them, the second body will also travel equal paths (the bodies do not accelerate with respect to each other). However, as we have mentioned, this is an empirical fact that might not have occurred. On the other hand, the postulate "Body A, on which no force acts, travels equal distances in equal times" is conservative. It does not imply anything about the behavior of other bodies in an analogous situation.

It seems, then, that we have defended the thesis of the empirical nature of the first law. However, there is another challenge. The principle of inertia says something about the behavior of bodies in the complete absence of acting forces. However, this is a difficult or even impossible situation to create, if only because of the omnipresent force of gravity, from which there is no escape. In principle, therefore, we cannot empirically check whether the first law applies, because we will not eliminate all the forces acting on a given body. In other words, it is impossible to empirically refute the first principle. Some conclude from this that the first law is a kind of convention, not so much semantic as methodological, dictated by considerations of elegance or simplicity. It is simply convenient for us to describe the observed mechanical phenomena, taking for granted that if we could eliminate all the forces acting upon an object, it would move in a perfectly uniform, rectilinear manner.

However, the conclusion that the first law is un-empirical due to the above-mentioned problem seems too hasty. First, while it is impossible to eliminate all acting forces, they can be approximately balanced by creating a situation in which the resultant force is equal to or close to zero. In the case of the force of gravity, its perfect balance with the apparent force of inertia occurs in systems in free fall (e.g. on a space station orbiting the Earth). It is true that we will not be able to eliminate other forces (e.g., the influence of unevenly distributed pressure of the surrounding air, etc.), but within the limits of error, with a good approximation, we can put the first law to the test. Second, even under ordinary terrestrial conditions, we can ignore the force of gravity if we consider motion perpendicular to its direction.[2] Finally, we can say that realistic testing of the first law occurs in any situation in which we use Newton's equations to determine the motion of a given object (e.g., a planet or a space rocket). The first law is, in a sense, part of every model in classical mechanics that describes a particular situation. This shows that testing a given scientific hypothesis in practice almost always takes place "collectively", in conjunction with other assumptions.

Now let us move on to the second law of motion. There are two key terms present in it: force and mass. We will talk more about mass on the occasion of the third law, but for now let us deal with force. Is there a way to characterize this concept independently of the equation $F = ma$? Newton, in his *Principia,* defined force in general as an action exerted on a body in order to change its state of rest or uniform rectilinear motion. Such a characterization is dangerously close to the second law itself, which may lead to the conclusion that this law is in fact a quantitative definition of force: the resultant force acting on a body is by definition the product of its mass (inertia) and the rate of change of velocity. However, in the further part of his definitional explanations, Newton mentions special types of force: centripetal force, gravitational and magnetic forces. It is therefore likely that his intention was to indicate the possibility of independent characteristics of forces. Indeed, we can characterize many mechanical forces by giving the appropriate formulas. For example, using Hooke's law, which links the extension of metals to an applied force, we can calculate the force coming from a stretched-out spring. Newton himself, on the other hand, derived the famous formula expressing the law of universal gravitation (we will talk about it in one of the next sections), determining the gravitational force acting on a given body and coming from another body. Therefore, it is possible to insert an appropriate formula in place of F and experimentally check whether the equality $F = ma$ is indeed true.

However, the problem is that the list of known forces (including electric and magnetic forces) is not exhaustive. It is possible that science has not yet discovered all the interactions that exist in the world. In that case, we have a problem with testing the second law. Imagine that we observe a certain body that suddenly, for unknown reasons, changes its speed (begins to accelerate or turns). None of the known forces are at work in the case under consideration. In such a situation, we have two options – either to accept that the second law is not universally applicable (it is false), or to assume the existence of a new unknown force (or forces), the value of which is exactly equal to the product of mass and the instantaneous acceleration of the body. The latter option protects the second law from being refuted – we will always be able to explain the observed behavior of the body (change in its velocity) by introducing additional forces. However, the price we have to pay for saving the second law of motion is to deprive it of its empirical content and reduce it to a terminological convention – force simply becomes definitionally identical with the magnitude of ma.

There is an intermediate solution, proposed by many philosophers, which preserves a certain empirical content of the second law and at the same time admits of the possibility of discovering new mechanical forces. The proposal is to treat the second law as a certain methodological rule, requiring the search for *simple* functions of force depending on time and

[2] In fact, neutralizing the force of gravity with another force is child's play – we place an object on the table, and the force of gravity acting on this object will be perfectly compensated by the reaction force of the table (thanks to the third law).

position, which would explain the observable behavior of bodies (their acceleration).[3] If we are able to reconstruct such functions on the basis of observation of the kinematic behavior of bodies, the second law is considered to be satisfied. It is possible, however, that the motion of the observed bodies will be so chaotic and unpredictable that the introduction of a "reasonable" force function will become practically impossible. In such a situation, we will have to revise our law linking acceleration with the acting force.

Let us now turn to the concept of mass. Newton's definition of this important concept leaves much to be desired – in his view, mass is a measure of the amount of matter, calculated as the product of density and volume. However, this determination of mass is unsatisfactory, because density in turn is defined as the mass of a body per unit volume. Some scholars (Ernst Mach, Henri Poincaré) proposed that Newton's third law should be used to define mass. Consider two isolated bodies A and B that interact with each other without contact with the environment. According to the third law, the force acting on body A is equal to and opposite to the force acting on body B: $F_A = -F_B$. Hence we have the following relationship between the accelerations and masses of the bodies:

$$\frac{m_A}{m_B} = -\frac{a_B}{a_A}.$$

This equation can be used to define the relative mass of one object with respect to another as the inverted ratio of their accelerations. The question arises again, what about the status of the third law of motion? Does it become an analytical sentence, true by definition? The answer is similar to that given in the case of the first law and the concept of equal time intervals. By selecting one standard body, we can determine the masses of all other bodies by comparing their accelerations when interacting with the standard. This definition ensures that the third principle will be fulfilled in the situation of the interaction of the standard body with any other body. However, it is not a foregone conclusion that during the interaction of two non-standard bodies, their mutual forces, calculated by means of previously determined masses and recorded accelerations, will balance each other. Thus the third law retains a well-defined empirical content.

The above analysis shows that Newton's laws of motion perform (or, more cautiously, can perform) a dual role. They can be used to characterize basic terms in mechanics, such as mass, force, and time uniformity, but this does not deprive them of their role as synthetic empirical generalizations that tell us something about the world. The concept of mass can be interpreted kinematically by means of the third law of motion applied to the interaction of a given body with a standard. But its empirical content will then be expressed in the statement that for each two interacting bodies the product of their predetermined mass and acceleration will be the same. On the other hand, we can characterize force in general as any "reasonably" behaving function F that satisfies the equation $F = ma$. The second law of motion will then state that there is such a reasonable function for every mechanical system. Finally, the first law can be used to define a good clock, and its empirical content will be reduced to the thesis that all free bodies, whose motion is measured by this clock, move uniformly.

---

[3] One of the possible conditions of simplicity, which we will discuss in detail later, is that the force function can be represented in the form of a rate of change with the distance of a certain scalar quantity (in the form of a derivative of this scalar, known as potential, with respect to position, or in general in the form of its gradient). The field of such forces is called conservative.

2.2 Determining the trajectory of motion

The main application of Newton's laws is to calculate the future evolution of bodies, i.e. to determine their trajectories. We will look more closely at how classical mechanics achieves this goal, starting with a simple example of a single body of negligible size (i.e., a so-called material point). For the sake of simplicity, let us also assume that the temporal evolution of a system proceeds in steps, i.e. discretely, starting from the zero moment, through the moment number one, and so on. We can assume, for example, that the body changes its parameters and position every second. At the zero moment t_0, the body has a certain position x_0 and the force acting on it is F_0. Is this information enough to predict the behavior of the body at the moment t_1? It is easy to see that it is not. According to the second law of motion, the applied force determines the change in velocity, but to calculate the velocity of the body at the next moment t_1, we need to know what its initial velocity was. Let us symbolize it with the letter v_0. The velocity v_0 will determine where the body will be in the next second, and the force F_0 will determine what its next velocity will be. This can be expressed symbolically as follows:

$$x_1 = x_0 + v_0 \Delta t$$

$$v_1 = v_0 + \frac{F_0 \Delta t}{m},$$

where $\Delta t = t_1 - t_0$ is the interval between two consecutive moments. Remember that in general, position, velocity and force are vectors, not numbers. For simplicity's sake, we will write our equations in numerical form, but compare Figure 2.2, which illustrates the procedure described above using vectors.

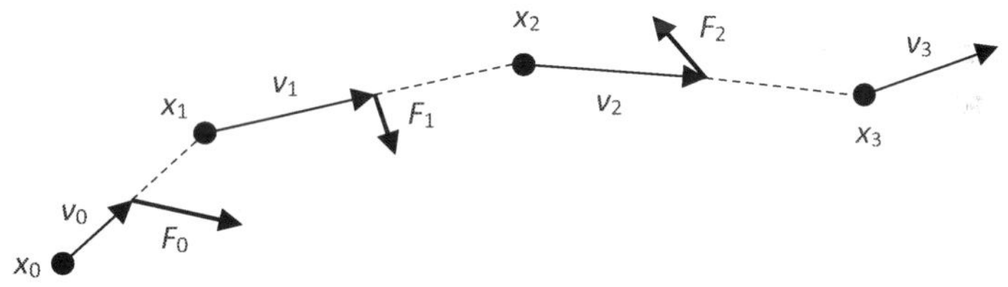

Fig. 2.2 Determining the trajectory using the numerical method. The position of a body at each successive moment is determined by its velocity at the preceding moment, while the new velocity arises as a result of adding the force vector to the previous velocity (we assume that the mass of the body, as well as the intervals between successive moments, are equal to one).

At the moment t_1, the situation repeats itself. Again, we have a set velocity of v_1, and a new force F_1. They will impact the evolution of the object in the next moment t_2 – the velocity will fix its position, and the force will determine its new velocity:

$$x_2 = x_1 + v_1 \Delta t$$

$$v_2 = v_1 + \frac{F_1 \Delta t}{m}.$$

Continuing this procedure for subsequent moments, we will get what we wanted, i.e. the trajectory of the object, given by successive positions x_0, x_1, x_2, ... The procedure presented above is called numerical calculation of a trajectory and is successfully used with the help of computers to determine the evolution of complex mechanical systems, consisting of, for example, celestial bodies – planets, asteroids, comets. It is worth noting that each subsequent step in the evolution of a body subjected to forces is unambiguously determined by the preceding step. This provides us with a clue that Newton's laws of mechanics may be deterministic in nature.

Step-by-step numerical trajectory calculation can be replaced under certain circumstances by a more elegant, "global" mathematical procedure. Let us take a closer look at this approach. Let $F(t, x)$ denote some mathematical function that expresses the value of the force acting at time t on the body located at place x. Newton's second law, as we remember, will take the form of an equation

$$F(t, x) = ma(t).$$

We now have to use the mathematically strict definition of acceleration. Until now, we have interpreted acceleration as the ratio of the increase in speed to the time in which this increase occurs. However, in this way, we will not determine the acceleration at a given time point t, but rather its average value in a given interval. In order to talk about the so-called instantaneous values, we need to calculate the limit of the relevant quotients with the value of the time interval converging to zero. As many of you probably know, the resulting number is called the derivative of the velocity function $v(t)$ at a point (this concept was first introduced by Newton and Leibniz). Using common symbols, we write the formula for acceleration as a function of time t as follows:

$$a(t) = \frac{dv(t)}{dt}.$$

Instantaneous velocity $v(t)$, on the other hand, is defined analogously as the derivative of the position function $x(t)$ with respect to time (see the entry in the box for further explanations). It follows that in order to calculate the acceleration, we need to differentiate twice the position function. The second derivative is written in a slightly strange way (even though this notation has a perfectly natural explanation in terms of ordinary algebraic operations, if we treat symbols d and dt as if they denoted usual numbers):

$$a(t) = \frac{d^2x(t)}{dt^2}.$$

Ultimately, Newton's equation takes the following form:

$$F(t, x) = m\frac{d^2x(t)}{dt^2}. \tag{2.1}$$

The unknown here is the form of the position function $x(t)$, i.e. trajectory.[4] Such equations are called differential equations, and their solutions are dealt with by a separate branch of mathematics, known as mathematical analysis. For many force functions $F(t, x)$ we find the solution relatively easily, while for others this is analytically impossible and then we have to

[4] Of course, it should be remembered that in general this equation has to be written in a vectorial form. In other words, we need to write three equations, each separately for one of the x, y, and z coordinates.

return to numerical methods. Those who have learned calculus and can compute derivatives of simple functions, such as power functions, can easily check that, for example, for a constant function $F(t, x) = F$ the solution is the quadratic function $x(t) = \frac{F}{2m}t^2$. This is a well-known formula for distance in uniformly accelerated motion, already anticipated by Galileo.

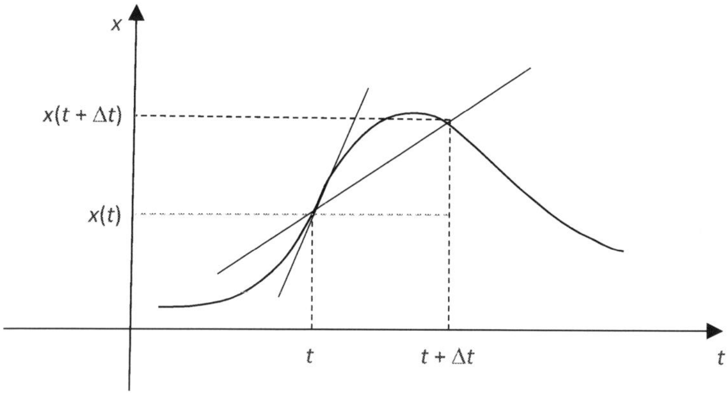

Fig. 2.3 A graphic representation of the derivative of a function

Probably many readers have come across the concept of derivative and its application to express the notion of instantaneous velocity. For those who have not had such an opportunity or would like to refresh their memory, I have prepared this short entry. Let $x(t)$ be a function that describes the position of a body in time. The average velocity in the time interval from t to $t + \Delta t$ will be the quotient of the distance traveled in this time and the interval:

$$\frac{x(t + \Delta t) - x(t)}{\Delta t}.$$

If we want to calculate the instantaneous velocity at a point t, we have to take smaller and smaller intervals Δt and calculate the appropriate quotients. Formally, we call it a transition in the limit of Δt approaching zero. The limit we are looking for will be the number near which the above quotients for smaller and smaller intervals "congregate". Of course, there is a formally correct definition of such a limit, but we will not use it here. This limit will be called the derivative of the function $x(t)$ at point t and denoted by $\frac{dx(t)}{dt}$. There are certain rules and formulas that show how to calculate derivatives of known functions. For example, the derivative of the power function t^n will be given by nt^{n-1}.

Let us now turn our attention to a simple geometric interpretation of the derivatives. Diagram 2.3 shows the graph of function $x(t)$. The quotient $\frac{x(t+\Delta t)-x(t)}{\Delta t}$ determines the angle of inclination (strictly, its tangent) of the chord connecting the points $x(t)$ and $x(t + \Delta t)$. By taking Δt closer and closer to zero, we get a line tangent to the graph at point $x(t)$. Thus, the derivative defines the slope of the curve $x(t)$ at a given point.

Let us now ask ourselves whether equation (2.1) has exactly one solution. It is easy to show that the answer is negative. For a given solution $x(t)$ of the equation there are infinitely many solutions of the following form:

$$\tilde{x}(t) = x(t) + \alpha t + \beta,$$

where α and β are arbitrary numbers. This is because differentiating the function $\alpha t + \beta$ twice yields zero, and therefore the second derivative of the function $\tilde{x}(t)$ is the same as that of $x(t)$. Assuming that function $x(t)$ and its derivative take the value 0 for $t = 0$, we get an interpretation of the constants α and β: the constant β is the position x_0 at the zero moment, and α is the velocity v_0. Therefore, the solutions differ in that they assume different initial values of position and velocity. By fixing the initial values of position and velocity, we can eliminate all but one of the functions of the above form.

There is a mathematical theorem which states that for every function of force F satisfying certain conditions, there is exactly one solution $x(t)$ of Newton's equation which takes a given value x_0 at the zero point, and its derivative at this point is equal to v_0. This is the mathematical basis for formulating the version of determinism in classical mechanics. In the case of one body, determinism of Newton's equation is based on the fact that by choosing the position and initial velocity, we thereby fix the entire unique trajectory in the future. For systems of many bodies, of course, we will need the position and initial velocities of each of them separately.

However, there is a minor problem here that is often overlooked in popular approaches to determinism. What about the force function? Do we not also need to fix all the forces acting on each body at the initial moment? And is such a determination of the initial force sufficient? In fact, we need much more – we need to know in advance what forces will act on our system in the future. Therefore, the doubt arises whether we can talk about determinism at all, since such an important element as the forces acting in the future must be added "by hand". The answer here is to make two additional assumptions, the second of which is usually ignored in philosophical formulations of Newtonian determinism. The first assumption is the thesis that the system under consideration is isolated from all external influences. Determinism in relation to a selected system of many bodies must be accompanied by this assumption, otherwise we will have to take into account the possibility of a completely unforeseen influence of the environment, not resulting from the initial state of the system. In the case of global determinism, i.e. the deterministic thesis concerning the entire universe, the assumption of isolation is unnecessary, because there is no external environment for the universe that could affect its evolution.

The second assumption is that in an isolated system consisting of n bodies, all the forces acting on individual bodies are functions of the positions and (possibly) velocities of these bodies. In other words, by knowing the instantaneous positions and velocities of the bodies, we can calculate all the forces acting at that moment. Under this assumption, the initial positions and velocities of the bodies play a dual role in determining the future behavior of the system. First, as we have already said, they are an indispensable element in choosing one of the infinitely many solutions to Newton's equation of motion. Second, they also determine the initial forces acting on bodies (whereas subsequent forces are functions of subsequent positions and velocities). Note that this additional assumption is by no means implied by Newton's laws of motion – it could happen that the forces in nature would come "out of nowhere" and the second law would still be satisfied.[5] The reduction of forces to the mutual positions and velocities of bodies is part of a larger doctrine known as mechanicism.

2.3 Two versions of determinism

We have not yet specified what the philosophical position of determinism says. We will have to take on this task in order to be able to answer the question of whether classical mechanics is a deterministic theory, as is commonly believed. As is usually the case with key terms in philosophy, there are many non-equivalent interpretations of the concept of determinism. Let us start with the most common approach, referring to the issue of predictability. We will call this version epistemological, or Laplacian, determinism, derived from the name of Pierre Simone de Laplace. Laplace was an outstanding French mathematician, physicist and philosopher living at the turn of the eighteenth and nineteenth centuries, his works significantly contributing to the development of theoretical mechanics. He was fascinated by the possibilities offered by Newtonian mechanics in determining the future behavior of celestial bodies – planets or comets. Extrapolating the successes of celestial mechanics to other branches of science, he announced that in principle there were no obstacles to applying its methods to predict all types of future events in the universe. The only condition for such a prediction is to know exactly the state of the world at a given moment and to perform appropriate calculations. Thus, epistemological determinism can be presented in the form of the following theorem:

> Knowledge of the initial state of a given physical system and the laws governing this system makes it possible to calculate all its future states.

The key concept used in the above wording is the notion of "possibility". How should we understand the possibility of calculating future states on the basis of present states? Laplace was aware that it might not be feasible for us to predict future states in practice. There are two reasons for this. First, we cannot know in the smallest detail the instantaneous initial states for complex physical systems. Second, even if we succeeded in this task, the calculations necessary to make predictions could exceed the capabilities of even the most powerful computing machines. Hence, Laplace, in his famous formulation of the principle of determinism, referred to a hypothetical "intelligence" whose perceptual and computational abilities are incomparably greater than human abilities. It is customary to refer to this intelligence as Laplace's demon, although the French scientist himself never used this term. However, it should be emphasized that the introduction of a "demon" with unspecified supernatural abilities deprives the principle of determinism of a clear empirical content. To reduce the problem to absurdity, we could speculate that a demon with superhuman abilities might simply have direct insight into future events, thus securing the perfect predictions. Either way, the epistemological interpretation of determinism entangles us in the difficult problem of computability. The concept of computability is intensively analyzed by logicians and mathematicians. In mathematics, interesting limitation theorems are proven, which show the impossibility of

---

[5] The introduction of forces that depend on other parameters besides the configurations of bodies is not at all uncommon in modern physics. For example, the interpretation of quantum mechanics, known as Bohmian mechanics, makes the trajectories of particles dependent on the force field, represented by the so-called wave function. We will mention this in the chapter on quantum mechanics.

solving certain mathematical questions using algorithmic computational procedures.[6] It is not excluded, although this has not been studied in detail, that similar constraints apply to predictions based on the equations of classical mechanics.

The weakest interpretation of the concept of possibility in the epistemological version of determinism is that there are some minimal conditions satisfied in the world that are necessary to make a correct prediction. Such minimal conditions necessary (although certainly not sufficient) for a successful prediction can take the form of determinism in the ontological version. It is quite obvious that if the world independent of the perceiving subject were not deterministic, it would be impossible to predict the future course of events for fundamental, not just practical, reasons. Determinism in relation to the world, and not to knowing subjects, can be formulated as follows:

> For every complete state of a physical system at the initial moment, there is exactly one sequence of future states of the system in accordance with this initial state and the laws of nature.

If the above condition were not met, i.e. if there were various alternative sequences of future events under the given initial conditions and laws, predictability would collapse. We will encounter such a situation later when discussing quantum mechanics, but for now let us stay on the firm ground of Newtonian physics. A temporally ordered sequence of states can be briefly called the evolution of a system, so the above formulation of determinism states that for a given initial state there is exactly one evolution predetermined by the laws of nature.

As we remember from the previous section, physical systems described by Newton's equations generally meet the condition of ontological determinism (given the additional assumptions we discussed earlier regarding the acting forces). This is a consequence of the theorem about the uniqueness of solutions to the equations of motion – the only function $x(t)$, which takes the initial values x_0 and v_0, unambiguously determines the future evolution of the system. Let us add that classical mechanics does not distinguish the future from the past in any way. Traditionally, the thesis of determinism is formulated in relation to the future – this is of course justified by practical reasons, since we are generally interested in what is yet to happen, not in what has already happened. However, Newtonian equations are completely symmetric with respect to time.[7] This means that determining the values of x_0 and v_0 at a certain moment unambiguously determines the trajectory of the object not only in the future, but also in the past.

The concept of the physical state of a system at a given moment (the instantaneous state) is an extremely important element of the deterministic doctrine. In general, the state of

[6] The most famous theorem of this type is Gödel's theorem, which states in the most general outlines that for every sufficiently complex mathematical theory there are theorems that cannot be formally derived from the axioms of this theory. A concrete example of mathematical non-computability is, for example, the problem of the existence of integer roots of polynomials with integer coefficients.

[7] Formally, this means that if function $x(t)$ is a solution of Newton's equation, then the function $x(-t)$ is also its solution. Of course, it is still true that for given values x_0 and v_0, only one function $x(t)$ is the solution. To obtain the "inverse" function, you need to flip the sign of the initial velocity v_0 thus replacing it with $-v_0$. Then the future evolution of the system from the zero moment will be an exact reversal of its past evolution to the zero moment, as in a movie run backward.

a system can be interpreted as any of its complete and non-redundant characteristics at a given moment. Completeness means that the information on the state at time *t* implies all the relevant facts about the system at *t*, while non-redundancy ensures that each element of the state is indispensable. As we remember, in classical mechanics the state of the system of *n* point particles is given by the exact values of their positions and instantaneous velocities. However, the question arises whether these characteristics are indeed instantaneous, or whether they contain hidden information about the behavior of the system before and after a given moment. This issue is quite important from the point of view of the deterministic thesis, because if it turned out that the "instantaneous" state contains such information in some form, determinism could be trivially satisfied (if, for example, the state of the system at the initial moment contains the information that in five seconds the state will be such and such, it is not surprising that the initial state unambiguously determines the state after five seconds). In this context, it may be problematic to use instantaneous velocities as elements of momentary states. Zeno of Elea in his famous arrow paradox already asked whether motion (and therefore a non-zero velocity) can be attributed to non-extended moments. Imagine the "freezing" of a moving body at a certain moment, as if in a single film frame. It is obvious that this body will still possess a well-defined position, but does it make sense to talk about its speed? Observationally, the body does not differ from the resting one, so in order to determine that we are dealing with movement, it is necessary to "play" the frozen film.

The mathematical concept of instantaneous velocity, as explained in the previous box, is defined by the derivative of the position function. This may give the impression that the speed depends on the position, so the state of the system would be redundant. However, this is not the case if we limit ourselves to one point. The value of the position function $x(t)$ at a given moment does not impose any restrictions on the value of the velocity at the moment. The derivative of the function is determined by the tangent to the graph (the slope), and thus to calculate it, we must consider the positions of the body at moments other than the selected moment t_0. This seems to imply that instantaneous velocity contains information about future states! However, the matter is not so simple. The information about the instantaneous velocity does not allow us to derive any specific value of position at any later moment *t*. For we can always take a function of position in a smaller interval around the moment t_0, which does not contain the selected moment *t*, and this is sufficient to determine the instantaneous velocity. Instantaneous velocity belongs to the category of quantities characterized infinitesimally – the position of the body at precisely the moment t_0 is not enough to determine it, but the position at any other specific moment is redundant.

What does the issue of predictability look like in the context of classical mechanics? It turns out that epistemological determinism falls into an even more serious quagmire than the one discussed at the beginning of this section. This is related to the phenomenon of so-called deterministic chaos, which has become the subject of intensive research in recent decades. At first glance, the concept of deterministic chaos seems internally incoherent – how can we talk about phenomena that are deterministic, i.e. orderly, and yet chaotic? However, this terminology is well justified. In the most general terms, a chaotic mechanical system is a system whose deterministic evolution is "sensitive" to minuscule differences in initial conditions. This means that a small, even imperceptible change in the initial conditions results in a huge difference in the subsequent evolution of the system. This effect is often referred to as the "butterfly effect": the flutter of a butterfly's wings in the Amazon jungle can later trigger a hurricane off the coast of North America. A typical example of a chaotic system is, of course, the Earth's atmosphere with all its complex weather phenomena. However, there are much simpler examples to illustrate this. For example, if we place a cylinder-shaped obstacle on a billiard table, as shown in the picture, it will turn out that the path of a single billiard ball will depend extremely strongly on small differences in the angle at which the ball was shot (Fig. 2.4). Therefore, chaotic systems do not have to be composed of many elements – they can consist of a single body subjected to standard mechanical forces.

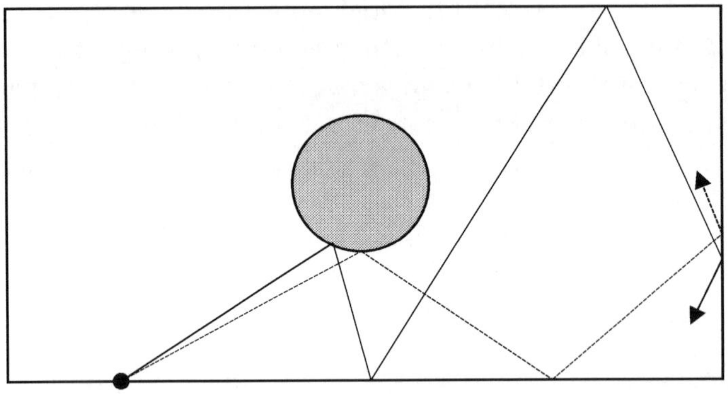

Fig. 2.4 Example of a chaotic system ("billiard table"). Two trajectories that differ minimally at the initial moment lead to completely different evolutions.

Predictability in chaotic systems breaks down for a simple reason: knowledge of the initial state is always limited by the accuracy of the measuring devices. Measurement errors are an indelible element of any experimental procedure. Therefore, if the initial state is known only within a certain margin of error, then it would be expected that future states would be similarly determined with a certain approximation only. However, chaotic systems expand this margin of error to an extent that makes future predictions all but impossible. Nevertheless, the evolution of a system is strictly deterministic, with each precisely defined initial state unambiguously determining its evolution.

Therefore, we are left with determinism in its ontological version, which seems to be guaranteed by the laws of classical mechanics. However, is it really impossible to break it by a mechanical system governed by Newton's laws? It turns out that such cases are known. Remember that the mathematical theorem about the uniqueness of the solution for the differential equation (2.1) was accompanied by the comment "for the functions of force satisfying a certain condition". It may seem that forces expressible by simple mathematical formulas should meet this condition without much problem. However, there are exceptions. There is a straightforward force function for which there are infinitely many solutions of the equations of motion with given initial conditions. This function looks like this:

$$F(x) = C\sqrt{x},$$

where C is a constant. It is relatively easy to prove that the solutions of Newton's equation for such a force are all functions of time of the form

$$x(t) = \begin{cases} 0 \text{ for } t < \tau \\ \dfrac{C^2}{144m^2}(t-\tau)^4 \text{for } t \geq \tau \end{cases}$$

where τ is any number and m is the mass of the object. I encourage readers familiar with the rules of differentiation of power functions to check that as a result of differentiating such a function twice, we will obtain exactly C times the root of this function divided by m, i.e. Newton's equation of motion $F = ma$ is satisfied. The physical meaning of the above function $x(t)$ is that it describes a body that remains motionless for a period τ, and then suddenly begins to move with increasing speed (traveling in proportion to the fourth power of time). It is important to keep in mind that the parameter τ can take any value, and therefore the time in which the body begins to move is not determined by the principles of Newtonian mechanics. This is an unmistakable example of indeterministic behavior.

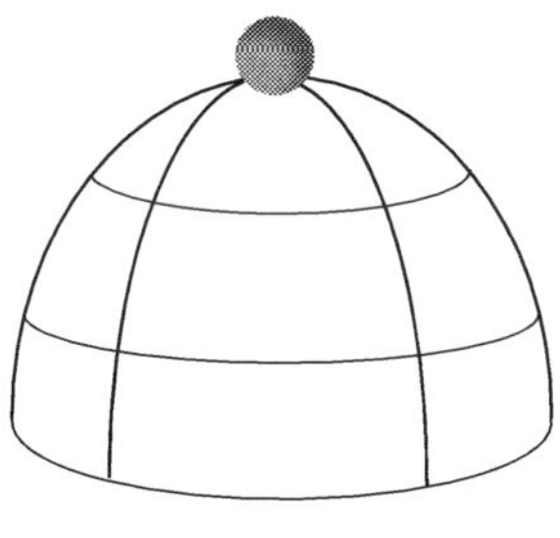

Fig. 2.5 Norton's dome

One may ask whether forces such as the above have a well-defined physical sense. John Norton, a contemporary philosopher of physics, answered this question in the affirmative. He showed that there is a simple physical model that implements the mathematical form of the force function given above. This model is the so-called Norton dome, i.e. a properly curved dome-shaped surface, placed in a gravitational field (we also assume that the dome is perfectly smooth, free of any friction). Its shape is described by a precisely selected function, which guarantees that the component of the gravitational force parallel to the surface will have exactly the value $C\sqrt{x}$, where x is the distance from the summit measured along the curvature (Fig. 2.5). Thus, the equations of classical mechanics imply that for a body placed on the top, there are an infinite number of possible evolutions. This body may remain at the starting point forever, but it may start to slide down at any time – in a minute or in a hundred years.[8]

Of course, in practice we are not able to construct a perfect Norton dome, because even small deviations from the mathematically defined curvature will cause the above reasoning

[8] You may ask yourself how it is possible that motion will occur at all, since at the initial moment the velocity of the body placed on top of the dome is zero, and the instantaneous force acting on it is also zero. There is no other answer to this question than that the equations "know better". Our intuitions based on common sense must give way to mathematical arguments – if there is a solution to Newton's equation describing an object that begins to slide off the dome, then apparently such a situation is possible.

to cease to apply. However, such a situation is not excluded by any general principle, and therefore the determinism of Newtonian theory is not a foregone conclusion. There are other examples of systems acceptable by Newton's theory whose behavior is non-deterministic. One of them is the so-called case of space invaders. Imagine a body subjected to a force that grows with time at such a rate that the speed of the accelerated body increases to infinity in a finite time interval (Fig. 2.6). This means that the graph of the path $x(t)$ as a function of time will asymptotically converge to the line with the time coordinate t_0 (a body accelerated in this way will "disappear" forever from the event horizon, never reaching the moment t_0). By reversing this process in time, which is always allowed in classical mechanics, we get a situation in which at the moment t_0 an object will fly in from infinity (e.g. a spaceship with alien invaders). Since the moment t_0 can be chosen completely arbitrarily, the arrival of the invaders is in no way determined by the past state of the universe.

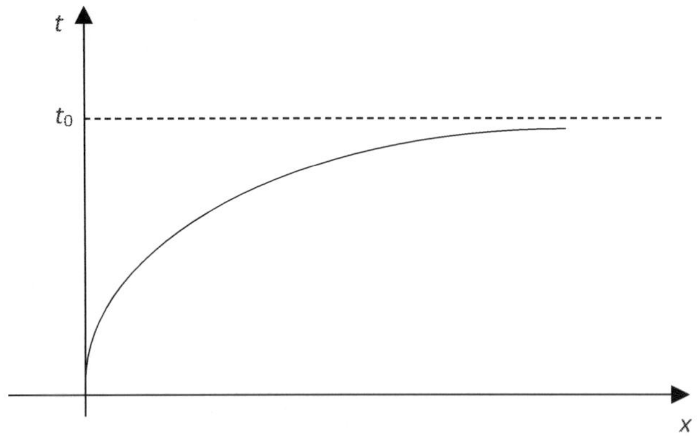

Fig. 2.6 The case of "alien invaders". By temporal reversal of the situation presented above, we get a non-deterministic process.

Let us conclude this review of nondeterministic situations in classical mechanics with another, more mundane example. Not all situations we encounter in mechanics can be described by Newton's equations with appropriately selected forces. A well-known phenomenon of this type is a perfectly elastic collision of bodies, requiring the application of infinite forces to describe the instantaneous change in momentum as a result of the collision. To predict the behavior of bodies after such a collision, we use other laws of mechanics, such as the principle of the conservation of momentum (or angular momentum) and of energy. However, it turns out that these rules do not allow us to unambiguously predict how three bodies (e.g. billiard balls) will behave after a perfectly symmetrical, simultaneous collision. Of course, we must emphasize that all the examples considered above are based on highly idealized assumptions that are not strictly speaking possible to meet. Thus, it can still be argued that the real world of Newtonian mechanics is *de facto* deterministic. However, if we want to apply the adjective "deterministic" to a theory, we must think twice before we say that Newton's theory is a deterministic theory. At best, we will have to add additional caveats or restrictions to such a claim.

2.4 Reference frames and Galilean transformation

Let us now move on to another important problem of classical mechanics, which is the issue of time and space. We know that the basic equations of Newton's theory contain two key parameters: position in space x and time location t. We will now discuss a more precise interpretation of these parameters, considering the issue first from the physical point of view, and then moving on to a more philosophical perspective. For a physicist, it is obvious that locations in time and space require a certain point of reference – we ask what the location of a given object is in relation to other objects, not "in itself". In order to avoid mentioning some concrete objects, the abstract concept of a frame of reference is called for. A frame of reference can generally be any set of objects that can enable us to measure spatial distances and time intervals. Usually, however, we imagine reference systems as analogous to Cartesian coordinate systems, in the form of three lines perpendicular to each other and hooked at a certain point. Such straight lines, i.e. coordinate axes x, y and z, allow you to spatially locate each instantaneous occurrence by specifying its coordinates. To this, of course, there must be added a clock, with which we can also determine the time coordinate of events.

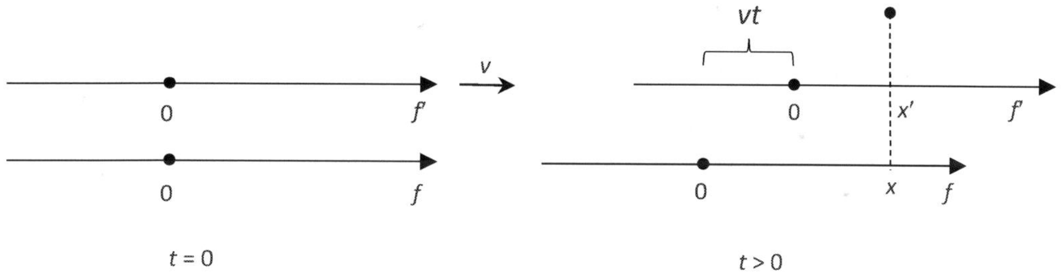

Fig. 2.7 Derivation of Galilean transformations

From the above definition, it is clear that there are many frames of reference in which we can describe the processes occurring in time and space. Some of these systems will differ in ways that are not important from the point of view of physics – for example, the axes of one system may be shifted relative to another or rotated. The most interesting, however, are systems that differ from each other in their state of motion, i.e. one moves relative to the other. For example, one system may be permanently linked to the Earth's surface and the other to a flying airplane. The question we will ask ourselves is the following: how do we relate the locations of objects specified in two frames of reference moving relative to each other? The answer is given in the form of so-called coordinate transformations, i.e. formulas that connect the coordinates x', y', z' and t' of the "new" frame with the coordinates x, y, z, t of the old frame. We will derive these transformations for the simplest case, when the reference frame f' moves relative to the system f with a constant velocity v along the x-axis (in the direction of increasing values of x). For the sake of simplicity, let us additionally assume that the direction of the x'-axis of the frame f' is the same as the x-direction of frame f, and that at the moment $t = 0$ the origins of both systems coincide. It is easy to see that under the adopted assumptions only the coordinate x' will differ from x (the other two spatial coordinates will remain unchanged). From the diagram above (Fig. 2.7), we conclude that the formula combining x' with x will be as follows:

$$x' = x - vt.$$

Add to this the intuitive assumption that time does not change when you move from one system to another (the clock readings remain the same whether you are on the surface of the Earth or on a flying airplane). The full set of equations including the above equation plus simple identities $y' = y$, $z' = z$, $t' = t$ is called the Galilean transformation.

With the Galilean transformation at our disposal, we can now introduce the fundamental concept of an *invariant*. An invariant of a given transformation is any concept that does not change during the transition from one coordinate system to another. The concept of invariant is important from the philosophical point of view, because it makes it possible to distinguish real properties of the world and physical objects from merely "apparent" properties. It is generally accepted that if a quantity changes when moving from one frame of reference to another, it means that it does not correspond to objectively existing reality. We treat different frames of reference in a similar way to different units expressing the same quantity, e.g. meters or feet for measuring distances. Objective facts about the world should not depend on what units we use, i.e. they must remain the same when changing units.

Let us consider two important examples illustrating the problem of invariance with respect to Galilean transformations. The first of these will be the velocity of an object. Is it an invariant of Galileo's transformation? It is easy to see that it is not. Let a body o move with a constant velocity v with respect to some frame of reference in the direction of the x-axis. This means that the velocity of this body will be expressed by the formula $v = \frac{x}{t}$. What will be the velocity of body o measured in the second system moving with respect to the first with velocity V? To calculate this, we need to apply the Galilean transformation to the coordinates x' and t'. As a result, we get the well-known formula for the new velocity v':

$$v' = \frac{x'}{t'} = \frac{x - Vt}{t} = v - V.$$

This is the formula for composing velocities: to calculate the new velocity of a body, the velocity of the second system relative to the first one must be subtracted from the old velocity of the body. So velocity is not an absolute quantity (Galileo noted this already).

Now let us discuss the second example, which will be acceleration. For this purpose, we will use the universal formula for instantaneous acceleration, which is given by the second derivative of the position function: $a = \frac{d^2x}{dt^2}$. In the "primed" system, it will of course be the same formula expressed by the new coordinates: $a' = \frac{d^2x'}{dt'^2}$. Inserting the expression $x - vt$ in place of x' (and, of course, t instead of t') we notice that the double differentiation of this function gives us exactly the second derivative of x (the second derivative of the function vt with respect to time is zero). Thus, acceleration does not change during the Galilean transformation. It is an invariant. This also implies an important consequence regarding Newton's second law. Since this law has the form $F = ma$, i.e. it uses acceleration and not, for example, velocity, its truth will not change when using the Galilean transformation, unless the acting forces change. Thus, the law of motion describes real regularity, occurring regardless of the system in which it is described.

However, it should be noted that the above conclusions are significantly limited. The Galilean transformation connects the coordinates of systems moving relative to each other in uniform and rectilinear fashion. But what if we consider any motion at all, such as rotational or rectilinear but accelerating? It turns out that things are getting complicated. Of course, it is possible to derive appropriate transformation formulas for any relative motion, but they will be much more complex than a simple Galilean transformation. However, more important than the degree of complexity is the issue of invariants. In the general case, acceleration will not be an invariant of transformations between arbitrary frames, and therefore Newton's second law will also cease to apply universally. To remedy this, the important concept of inertial frames is introduced. According to a popular approach, an inertial system is one that does not accelerate. However, this is not a correct definition, because acceleration by itself is essentially a relative concept (e.g. no system accelerates relative to itself). It is more correct, although somewhat abstract, to define inertial frames of reference as a certain distinguished class of coordinate systems, each of which remains in uniform and rectilinear motion relative to each other. Thus, Galilean transformations apply without exception to all inertial frames.

But how exactly do we distinguish the class of all inertial frames? The criterion of inertiality is based on the concept of apparent forces (also called, albeit somewhat confusingly, inertial forces), and more generally on the issue of the applicability of the second law of motion. In general, the idea is that if we take into account all the "normal" forces acting on given bodies and calculate the acceleration resulting from their action, and it turns out that all these bodies undergo some additional acceleration not resulting from the second law of motion, we attribute the additional acceleration to the action of apparent forces. Well known from everyday life, apparent forces occur in accelerating or braking vehicles, as well as in rotating systems (centrifugal forces and the Coriolis force mentioned in the previous chapter). Apparent forces give each body exactly the same acceleration, which distinguishes them from other types of forces (except gravity). Systems in which there are no apparent forces, and therefore Newton's law takes the simplest possible form in them, are called inertial.

Let us consider the simplest example of a non-inertial frame, which is a system moving with a constant acceleration a with respect to another inertial frame. In this case, the corresponding spatial coordinate transformation will look like this:

$$x' = x - \frac{at^2}{2}.$$

The acceleration of a given body with respect to the frame f is (after differentiating the above formula twice):

$$\frac{d^2x'}{dt'^2} = \frac{d^2x}{dt^2} - a\,,$$

Thus a new component of acceleration will appear equal in value and opposite to the acceleration of the entire system. To maintain the correctness of Newton's equation $F = m\frac{d^2x}{dt^2}$, we must add to the ordinary force F an apparent force equal to $-ma$.

2.5 Time and space in classical mechanics

The problem of time and space, which we will consider in this section, boils down to the question of whether time and space are absolute or relative. This question can be understood twofold. In the approach typical of physics, absoluteness is understood as invariance with respect to the choice of the frame of reference. Absolute time is the same in every frame of reference, while relative time varies depending on the frame in which it is measured. The same is true of space. On the other hand, the second understanding of the dispute of absolutism vs. relativism (also known as relationalism) is more philosophical in nature. This is a question about the ontological nature of time and space – whether they are separate entities, independent of other substances, or whether they are secondary in nature, depending, for example, on the existence and properties of material objects. The absolute character of time and space in the first sense does not yet guarantee absoluteness in the second sense. It may be that in every system both time and space look the same, but ontologically they are dependent on some other entities. On the other hand, the dependence of time and space on the choice of a frame of reference seems to deprive them of the character of ontologically separate entities.

It is relatively easy to answer the question whether according to classical mechanics time depends on the frame of reference. Due to the assumed Galilean transformation $t' = t$, time is measured equally in all systems and is therefore absolute in the physical sense. However, the status of space is not obvious. The Galilean transformation changes the value of spatial coordinates, but does this mean that space itself also changes from system to system? It depends on what exactly we mean by space. In the context of physical considerations, we can naturally identify space with a distance function defined on a set of points. A space is called invariant if the spatial distances between points or events are the same in every frame of reference.

Let us investigate this matter further. Let there be two events A_1 and A_2 with coordinates (x_1, t_1) and (x_2, t_2). (For simplicity, we omit the other two spatial coordinates y and z – you can always choose a coordinate system in which the y and z coordinates of both events are 0.) The spatial distance between A_1 and A_2 is, of course, $|x_1 - x_2|$. However, when we apply the Galilean transformation, this distance will typically change:

$$|x'_1 - x'_2| = |x_1 - x_2 - vt_1 + vt_2|.$$

The spatial distance between two events is not generally invariant with respect to the change of the frame of reference. This can be illustrated by a simple example – imagine a passenger sitting on a train and drumming his fingers on a table. Two consecutive finger taps in the train system have zero spatial distance, while from the point of view of an observer standing on the ground, their distance is equal to the distance traveled by the train in time between the finger taps.

Note, however, that the distance in the frame f' will be equal to the distance in f, as long as the time of the events A_1 and A_2 is the same: $t_1 = t_2$. This shows that the distance between simultaneous events is an invariant of the Galilean transformation (in fact, any coordinate transformation connecting two arbitrarily moving systems). Thus, space can be considered invariant with respect to the choice of reference frame, if space is defined by all spatial distances taken at the very same moment. The set of all events (points) occurring at the same moment and connected by the distance function is called instantaneous space (snapshot space). From our analysis, we conclude that instantaneous space is an invariant of the Galilean transformation, and therefore it is absolute in the physical sense. On the other hand, the temporarily extended "global" space, including objects lasting in time, is not absolute, because its characteristics by means of the distance function depend on the choice of the reference frame.

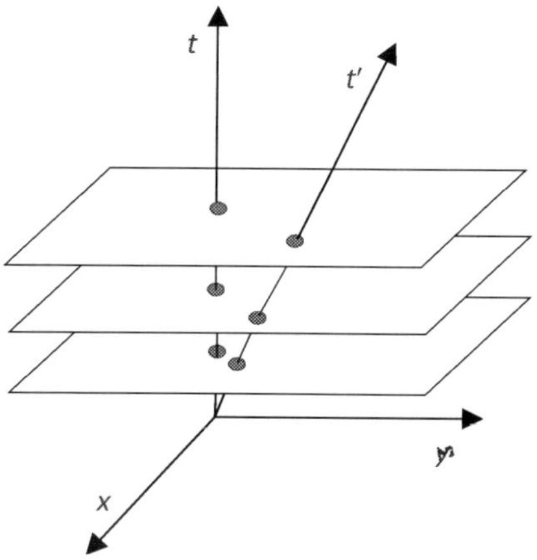

Fig. 2.8 Two reference frames and planes of simultaneity (instantaneous spaces)

The above considerations can be illustrated with the help of space-time diagrams. Such diagrams take into account spatial axes and one temporal axis on which we mark the time coordinate of a given object or event. For obvious reasons, we cannot graphically represent the general case of three spatial axes and one temporal axis, so we usually reduce the number of spatial axes to two, and sometimes even to one x-axis. The diagram (Fig. 2.8) shows an oblique line that represents the path of a body moving at a constant velocity relative to a given system. This line can also represent the temporal axis of a new reference frame, which is in motion with respect to the original frame. The time axis connects events that in a given system have spatial coordinates equal to zero (they occur at the origin of the system), and the origin of the system f' moves relative to the system f. Horizontal planes represent instantaneous spaces that look the same in any frame of reference. On the other hand, the lack of global space is revealed in the fact that it is not possible to absolutely connect points belonging to different planes by the relation of being in the same place. Two non-simultaneous events that occur in the same place with respect to one frame of reference do not occur in the same place according to another frame. In other words, the concept of the same place is relativized to the frame of reference. The interrelated concepts of motion, rest, and same location turn out to be only relative, not absolute.

Let us now move on to the philosophical issue. What is time or space, really? Should we imagine space as a kind of empty container filled with places ready to receive physical objects? Is time an extension made up of entities—moments—that follow one another, even when nothing is happening in them? In order to present this issue more clearly, we can make use of a concept taken from Aristotle – namely that of a substance. In Aristotle's approach a substance is a being capable of independent existence, i.e. one that is not dependent on other beings in its existence. For example, specific physical objects – stones or trees – are substances, but their properties are not. We may therefore ask whether time and space are substances in the above-mentioned sense. Would space exist if there were no objects filling it? Can time exist without events occurring in it?

An affirmative answer to the above questions is given by the position of absolutism (also called, for obvious reasons, substantivalism). Newton himself was an ardent supporter of philosophical absolutism in relation to time and space. Widely known and quoted excerpts from his *Principia* leave no doubt about it:

> Absolute, true, and mathematical time, of itself, and from its own nature, flows equably without relation to anything external, and by another name is called duration.
>
> Absolute space, in its own nature, without relation to anything external, remains always similar and immovable.
>
> Place is a part of space which a body takes up, and is according to the space, either absolute or relative.
>
> Absolute motion is the translation of a body from one absolute place into another.

Note that Newton links the position of absolutism in the philosophical sense with its physical interpretation, maintaining that there is an absolute concept of place, motion, and rest independent of the state of motion of the observer (Newton did not yet use the concept of a frame of reference). The founder of classical mechanics presented theological arguments for his position, but his most famous argument refers to a physical experiment. This is the famous bucket argument.

Fig. 2.9 Two stages of the experiment with a bucket and water. The observed rotational movements take place relative to the marked fixed stars.

Imagine a bucket filled with water and hung on a rope. The rope is twisted and then released freely, resulting in the bucket starting to rotate. In the first phase of movement, the bucket rotates, while the water remains at rest due to its inertia. When observing the surface of the water, we do not notice any effect – the surface is flat. The water then begins to take over the rotational motion of the bucket, resulting in a concave surface. In the final stage, the bucket comes to an abrupt stop, while the water continues to spin and its surface remains concave until the rotation stops. Comparing the first and third stages of the experiment, we can see that in both cases the relative motion of the water and the bucket is the same – the water rotates relative to the bucket or the bucket relative to the water, which is the same (Fig. 2.9). However, there is clearly an observational difference between the two situations – in one case the surface of the water is flat, and in the other it is concave. Newton concluded that only in one of the two situations does the water "really" rotate, and in the other it really (i.e. absolutely) rests. The criterion of absolute rotation is, of course, the occurrence of apparent forces, in this case the centrifugal force, which changes the shape of the water's surface.

Relative to what does true rotational motion occur? Newton had no doubt that the only possible answer to this question is: relative to absolute space. The water molecules participating in the motion change their position in relation to absolute places in space, and this absolute, not relative, movement of water manifests itself in a real physical effect. However, Newton was aware that not every type of motion has observable effects. Uniform motion, as Galileo had already noted, does not produce the effect of deformation of the moving body. Why does absolute space manifest itself only when the body accelerates relative to it, and not when it simply changes its position? This question remains unanswered. Consequently, for fundamental reasons, we cannot empirically verify the existence and identity of absolute positions in space. We assert that the accelerating body must change its position in absolute space, but we do not know from where to where. It may be that a given molecule of water in a bucket moves absolutely from point A to point B on the circumference of a circle, but this motion may also be combined with an unobservable uniform motion of the entire universe at an incredible speed in one direction, and thus the absolute position of the particle under consideration after a second may be many thousands of kilometers away. But we have to rule out the possibility (if we accept Newton's point of view) that a particle of a spinning water sample could be in the same absolute place all the time. The presence of apparent forces does not give us an unambiguous criterion for identifying absolute places, but the assumption of the existence of these places is the best explanation for the observable behavior of bodies subjected to acceleration.

However, the assumption that the existence of absolute space is the only (or, more cautiously, the best) possible explanation for inertial effects has been challenged. Ernst Mach, an Austrian physicist and philosopher of the late nineteenth century, noted that Newton completely ignored the existence of bodies other than the bucket and the water, as if the bucket of water was the only thing in the universe. However, this is not the case: the bucket and the water are spinning relative to the rest of the universe. The first and third stages of the bucket experiment differ empirically: in the first, it is the bucket, not the water, that rotates relative to the Earth, the Sun, and the fixed stars; in the third, the bucket rests and the water begins to swirl in relation to the surroundings. In order to obtain the full symmetry necessary to derive Newton's conclusion, the following two situations should be compared: one in which the water is spinning relative to the bucket and the rest of the universe, and the other in which the water is resting and the bucket and the entire universe are turning around it. However, this is impossible for obvious reasons – we cannot "turn" the entire universe.

Mach's argument against Newton's conception of absolute space can be understood in two ways. In the weaker interpretation, Mach simply shows a gap in Newton's reasoning that blocks the way to its conclusion. However, many commentators interpret Mach as authoritatively stating that the rotation of the entire universe relative to the water would have exactly the same effect as the rotation of water itself—a centrifugal force would have occurred to change the shape of the surface of the water. Mach himself was quite cautious in formulating "what if" hypotheses, however his name is associated with a specific doctrine. This doctrine, known as Mach's principle, states that inertial effects must be explained by the physical influence of the entire universe on a given object. In particular, inertial mass is the result of the influence of the rest of the universe on each body. A solitary body in an empty universe would not have any inertial properties. For obvious reasons, this is speculation that we cannot directly verify, but certain consequences of Mach's principle can be tested. For example, you can study the effect of the rotation of a super-heavy hollow cylinder on the mass of the objects at its center. Experiments of this type have been carried out, although their results are not conclusive. In any case, Mach's ideas had a great influence on the young Albert Einstein and his first attempts to create a general theory of relativity, in which there would be no room for absolute space, and all motions without exception would be relative.

Newton's argument, if we find it convincing, rehabilitates the notion of global space that we have rejected above in favor of momentary space. For this reason, Newton's theory of time and space is sometimes called neo-Aristotelian because it reinstates the Aristotelian concept of absolute motion and rest, although Newtonian dynamics remains non-Aristotelian in the sense that forces don't cause motion but change in motion. On the other hand, the theory of time and space without absolute global space, but only with instantaneous spaces, is called Galilean theory. Thus, Newtonian physics can be fundamentally connected with two concepts of time and space.

A staunch opponent of the absolutist concept of time and space was Leibniz. He found it absurd that space could exist without the objects that fill it. Geometric relationships, such as distance, occur primarily between physical bodies, and only derivatively between spatial points. Spatial and temporal points and places owe their identity and existence to the objects and events located in them. Leibniz's conception of space and time is called relationalism, as it emphasizes the relational nature of the temporal and spatial properties of objects. It makes sense to say that body A at the moment t is 100 meters away from the body B, but it cannot be assumed that body A is absolutely located at a certain place, 100 meters away from the location of the body B. Leibniz formulated a famous argument against the absolutist concept of space, known as the shift argument. Let us assume, as the absolutists would have it, that space exists objectively and independently of the physical objects that occupy it. In such a situation, one can consider moving all material bodies in the universe in a certain direction by a constant distance, e.g. five meters, without changing their relative positions. The situations before and after the shift would be different, because the objects would occupy numerically different places in them. But this difference would in no way reveal itself empirically or qualitatively.

Two states of affairs, differing only in the position of the whole world with respect to absolute space, would be an example of pure numerical difference without any difference in qualitative properties. Leibniz was a supporter (and *de facto* the author) of the metaphysical principle of the identity of indiscernibles. It states that absolute qualitative indiscernibility (by means of properties) implies identity (being the same object) – or, in a logically equivalent form, that every two numerically different objects or states of affairs must differ by at least one property. Leibniz justified his principle by referring to an even more abstract metaphysical rule, which he called the principle of sufficient reason. Putting this principle in theological terms, we can say that God, when taking any action, always has a rational reason to choose one of many possible options. According to absolutists, God first created space and time, and then placed the physical world in them. The problem is that the Creator could have no reason to choose one of the infinitely many possible locations of the physical world, and thus the principle of sufficient reason would have been broken. The solution to this theological/metaphysical problem, according to Leibniz, was the assumption that it was only the creation of the world with the space-time relations between its elements that allowed time and space to exist as secondary entities

The principle of the identity of indiscernibles can be formulated as the following implication: if objects *a* and *b* have exactly the same properties, then *a* is identical with *b* (*a* is the same object as *b*). However, the meaning of this principle, as well as its satisfaction, depend on the adopted interpretation of the term "property". In the most liberal sense, every true proposition about a certain object corresponds to a certain property. With such an interpretation, among the properties of a given object there will be the property of being exactly this object and nothing else (such a property is often referred to by the scholastic term *haecceitas*, anglicized to "haecceity"). Then, of course, the Leibnizian principle will be trivially satisfied, since each two objects differ in their *haecceities*. However, most philosophers believe that in the context of the principle of the identity of indiscernibles, only the so-called pure (qualitative) properties should be used, which do not contain reference to any specific objects.

Another important conceptual distinction is the division into intrinsic and extrinsic (relational) properties. An intrinsic property of an object is one whose possession does not depend on the presence of other objects, while extrinsic properties also take into account the outside of the object. For example, the feature of being two meters away from a birch tree is extrinsic, because by cutting down the birch tree we deprive the object placed next to it of this property. In the stronger version, the principle of the identity of indiscernibles states that numerically distinct objects must differ by at least one intrinsic feature, while the weaker version allows for their intrinsic indistinguishability, as long as they have different extrinsic features. According to the weaker rule, two objects can be distinguished by the relationships between them and other objects (for example, when one of them is closer to a certain birch than the other).

For naturalistically oriented philosophers, the principle of the identity of indiscernibles may have a more empirical justification. Assuming that only qualitative facts based on the possession of certain properties can be empirically available to us, we must conclude that purely numerical differences, based on numerical individuality, cannot be subjected to empirical verification. When it comes to the location of the entire world in absolute space, it is clear that we cannot distinguish between the available options. Facts about the absolute spatial location of the world (as well as the location in absolute time) have no empirical consequences, and therefore can be treated by us as unnecessary baggage. Leibniz, a rationalist philosopher, would not be satisfied with such an argument, but the principle of the identity of indiscernibles is usually justified in this way by contemporary philosophers of science educated in the spirit of logical empiricism.

Leibniz's shift argument is widely regarded as an extremely serious challenge to absolutism with respect to time and space, which balances and even outweighs Newton's argument from the presence of apparent forces. Ultimately, however, the dispute between absolutism and relationalism does not find an unambiguous solution on the basis of classical mechanics. In the following parts of the book, we will see how this dispute developed with the arrival of more recent theories of time and space – special and general relativity. There will be completely new approaches and arguments, such as a modern version of Leibniz's shift argument, known as the hole argument. Despite the considerable technical advancement of the new arguments, the debate between the two positions is still ongoing, and its resolution is as distant as in the times of Newton and Leibniz.

2.6 Newton's theory of gravity

Classical mechanics in the Newtonian version provides us with an extremely useful tool for describing and predicting the mechanical behavior of bodies, including the motion of the planets. However, in order to use this tool, one more key element is needed. We must describe in a mathematically precise way the forces that act on moving celestial bodies. From a school course in physics, we know that these are the forces of gravitational attraction between bodies endowed with mass, and in particular the force of attraction between the Sun and other celestial bodies. These forces are described by the well-known law of universal gravitation. However, few people know how Newton arrived at the mathematical formulation of his law (that the gravitational force between two bodies is inversely proportional to the square of their distance). The anecdote about the apple that fell on Newton's head to inspire him to formulate the law of gravity has no historical confirmation, and it obscures the true source of his discovery. Newton would not have been able to derive his formula from the study of the free fall of bodies in the Earth's gravitational field, because bodies located near the Earth's surface fall with approximately the same acceleration, as Galileo had already noticed. The dependence of acceleration and gravitational force on distance is revealed only on a planetary scale. The actual source of the discovered formula was Kepler's laws, which synthetically captured the observed regularities of planetary motion.

Newton set himself the task of finding a suitable force that would explain the fact that the planets follow elliptical orbits around the Sun, and also substantiate the known quantitative facts about this motion. His reasoning can be divided into several stages. In the first stage, he showed that it is possible to draw the conclusion from Kepler's second law that the force acting on each planet is directed centrally towards the Sun. In the second step, Newton derived the formula for centripetal acceleration in circular motion. The final stage of Newton's reasoning was to apply Kepler's third law to calculate the value of the central force that would give the corresponding centripetal acceleration of the planet. All three steps were based on extremely ingenious geometric arguments, the qualities of which should be appreciated even several hundred years after they were formulated. So let us take a closer look at this reasoning.

Newton successfully used approximations of "smooth" geometric figures, such as circles or ellipses, using polygons. In his first argument, he assumed that the motion of the planet around the Sun occurs in steps, as in the case of the numerical solution to the equations of motion presented in the second section of this chapter. So let us suppose that in the first second the planet moves a certain distance, and in the next second covers another segment, turning a little more towards the Sun. From the second law of motion, we know that the change in velocity, which is the vector formed by subtracting the velocity in the first second from the velocity in the next second, is in the direction of the applied force. Considering two triangles representing the areas drawn by the planet's leading radius at equal one-second intervals, we conclude from Kepler's second law that the areas of these triangles must be equal. In addition, the first triangle has the same area as the triangle formed by moving the first vector to the anchor point of the second velocity vector (cf. Fig. 2.10 with detailed explanations). From this it follows that the difference of the two successive velocity vectors will be parallel to the line connecting the planet with the Sun, and therefore the force changing the speed of the planet must be directed towards the Sun. It is important to note that Newton's argument can be repeated for arbitrarily short time intervals (one millisecond, one nanosecond, etc.), and thus in the limit the conclusion remains valid for a smooth trajectory of a planet (ellipse or circle).

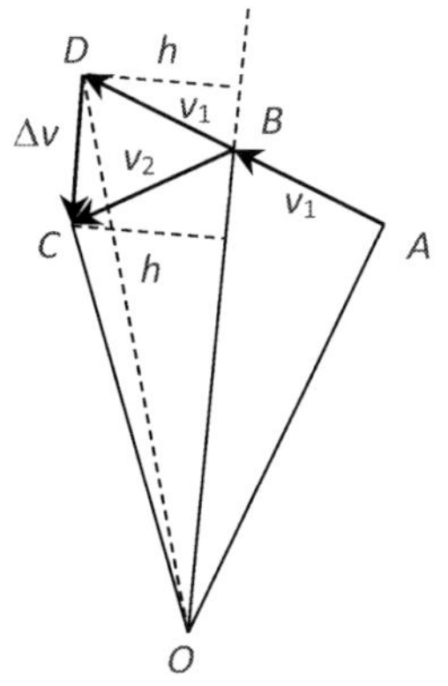

Fig. 2.10 Newton's argument for the centrality of the gravitational force. The planet moves from point A through B to C (velocity vectors are equal to displacements, because time intervals are equal to unity). The triangles OAB and OBC have the same areas from Kepler's second law. The triangles OAB and OBD also have equal areas (equal bases AB and BD and the altitude falling on these bases. Thus, the OBC and OBD triangles must have the same areas. Since they have a common base of OB, the altitudes falling on this base must be equal. This means that the velocity change vector Δv (DC segment) is parallel to the line OB, and therefore the force acting at the moment when the planet is at point B is directed centrally to O.

A similar method of approximation was used by Newton to derive the formula for centripetal acceleration in circular motion. The first approximation of such a motion is the movement along a square inscribed in a circle – this is a very inaccurate approximation, but as it turns out, it can be easily extended to any regular polygon. For a circle with a radius r, the side of the inscribed square is $\sqrt{2}r$ (Fig. 2.11). If you denote by v the value of the velocity of the body (length of the velocity vector), then at the moment of passing through the vertex of the square, the change in velocity is $\sqrt{2}v$. Within one cycle period T the change of the velocity vector takes place four times, so on average the change of velocity per unit of time (i.e. the average acceleration) is $\frac{4\sqrt{2}v}{T}$. The period of revolution T is linked to the speed v by a simple formula (the length of the circumference of the square divided by v): $T = \frac{4\sqrt{2}r}{v}$, from which we easily derive the formula for the average acceleration in the motion on a square:

$$a = \frac{v^2}{r}.$$

The above reasoning can be easily repeated for a pentagon, hexagon and generally any regular n-sided polygon, which shows that in the limit $n \rightarrow \infty$ the centripetal acceleration in circular motion is given by the above formula.

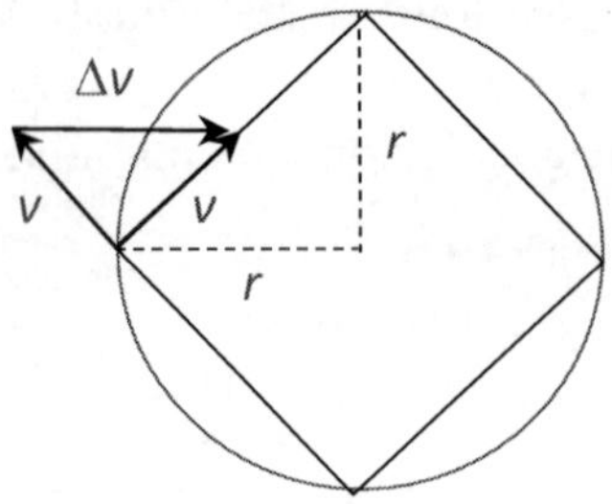

Fig. 2.11 Calculation of centripetal acceleration in a circular motion, approximated by a square. Notice that in each of the four vertices, the change in velocity Δv (and thus also the acceleration) is directed towards the center of the circle.

All that remains is to calculate the amount of the force that must act on a given planet to give it the acceleration needed to keep it moving in a circle. Let us express this force using the period of the revolution of the planet T instead of the velocity, inserting in the above formula in place of v the expression $\frac{2\pi r}{T}$:

$$F = ma = m\frac{4\pi^2 r}{T^2}.$$

From Kepler's third law we know that the ratio of the radius cube to the square of the orbital period is the same for all planets: $\frac{r^3}{T^2} = C$, where C is a constant. Multiplying the numerator and denominator of the above formula by r^2 and applying Kepler's law, we get:

$$F = \frac{4m\pi^2 C}{r^2}.$$

Thus, the force that keeps the planets in motion around the Sun must be inversely proportional to the square of the distance of the planet from the Sun. Newton put his formula to the test by applying it to the motion of the Moon around the Earth. The Moon is not subject to Kepler's laws, as it does not revolve around the Sun, and therefore the above reasoning does not apply directly here. Knowing the distance of the Moon from Earth and its orbital period (about 28 days), Newton calculated its centripetal acceleration and then compared it with the known acceleration of free fall at the Earth's surface. The ratio of these two accelerations was found to be equal to the ratio of the squares of the radius of the Earth and the distance from the Earth to the Moon, in accordance with the inverse-square law. Thus, the gravitational force coming from the Earth is enough to keep the Moon in orbit around our planet.

Generalizing the above considerations, Newton came to the conclusion that each two bodies attract each other with a force inversely proportional to the square of their distance. Using the principle of action and reaction (the third law of motion), it can be argued that this force must also be proportional to both the mass of one of the bodies and the mass of the other. The result is the well-known law of gravitation:

$$F = G\frac{m_1 m_2}{r^2},$$

where G is the gravitational constant, the exact value of which was determined much later by Henry Cavendish in experiments with a so-called torsion balance.[9]

Let us note an important aspect of deriving the law of universal gravitation from Kepler's laws of planetary motion, which is not always adequately emphasized. Kepler's third law implies that a planet at a given distance r from the Sun will be subjected to the same centripetal acceleration, regardless of its mass, size, etc. This is because the determination of the radius r unambiguously determines the period T of the orbit and therefore also the centripetal acceleration. As a consequence, the formula for the gravitational force includes the inertial mass from the second law of motion, and thus mass begins to play a dual role: as a quantity characterizing the inertia of a body subjected to forces and as a measure of gravitational interaction. However, it does not have to be this way. For example, in Chapter 4, we will discuss Coulomb's law of electrostatics, according to which the force of interaction is proportional to the new quantity that characterizes the interacting particles – their electric charge. If celestial bodies interacted with each other electrostatically, the period of orbit of a planet with a given orbital radius would additionally depend on its charge (or more precisely, its relation to the mass of the body). This peculiarity of gravitational interaction, the earlier manifestation of which we have already seen on the occasion of Galileo's law of free fall, will only be explained within the general theory of relativity.

Newton's success was met with mixed reactions from his contemporaries. On the one hand, the mathematical description of the forces needed to keep the planets in their orbits was a huge breakthrough in science, but on the other hand, the nature of gravitational forces remained mysterious to many. How does the Sun affect the motion of planets millions of kilometers away? Does it send out invisible threads that bind the planets and draw them towards itself? Let us remember that in Newton's time, the prevailing paradigm of mechanical force was interaction by contact. To force a resting body to move, you need to physically push or pull it. A popular concept of planetary motion, propagated by Descartes' supporters, was the so-called vortex theory. It assumed that interplanetary space was filled with an invisible substance in a state of spinning around the Sun. According to the Cartesians, it is this substance that drags the planets with it like pieces of cork on the surface of a whirlpool. For the Cartesians, it was inconceivable that in empty space (the existence of which they did not recognize, following their master) forces acting on objects could appear. Something must be carrying these forces – some physical, material substance.

When questioned about the nature of gravitational forces, Newton replied with his famous and characteristically arrogant statement "I don't invent hypotheses" (*Hypotheses non fingo*). He probably meant that the basic goal of a scientist should be to capture facts and empirical data in mathematical formulas, and that the metaphysical interpretation of these formulas should be left to – well, metaphysicians. We do not need to model gravitational interactions based on known mechanical interactions—we just need to be able to describe

[9] Of course, we still assume that we are dealing with point bodies. In the case of extended bodies, the matter becomes more complicated, unless they have a perfectly symmetrical shape, e.g. a sphere. Then Newton's law still applies if the distance r is calculated from the centers of both spheres. In the general case, in order to determine the total force, one must add up the contributions from small, almost point-like parts of the body, using the mathematical method of integration.

them precisely. Such an instrumentalist approach to science, which we have already talked about on the occasion of the analysis of the Copernican revolution, resonates in later statements of physicists, for example in the famous exclamation "Shut up and calculate", used when asking questions about the interpretation of the formalism of quantum mechanics. Although the instrumentalist attitude implies the rejection of many philosophical questions as non-scientific, it is worth noting that its acceptance is also a philosophical decision in itself, not resulting from empirical data alone.

Closely related to the problem of the hidden nature of gravitational forces is the issue of so-called action at a distance. The gravitational force, it seems, is able to cover huge distances without any delay. Newton assumed that the force of gravity coming from a distant body (e.g. the Sun) is felt at any point in space immediately. This violates the intuitive principle of locality, which we will talk about later in the chapters devoted to the theory of electromagnetism and quantum mechanics. Scientists who demanded that Newton provide the mechanism of gravitational forces had in mind precisely the problem of non-locality or action at a distance. Intuitively, it seems clear that the action of the Sun on the planets must somehow propagate through space in the form of a perturbation of finite velocity. However, Newton, as we know, rejected all divagations on this subject. Again, we will have to wait until the arrival of the general theory of relativity to solve this interpretative problem.

Newton's law of gravitation, combined with the second law of motion, allows us to accurately calculate the trajectory of a single body in the gravitational field of another body (the two-body problem). Depending on the initial velocity of a given body, its trajectory will be one of the conic sections: a circle or ellipse for lower velocities, and when the limiting velocity is reached and exceeded, a parabola or hyperbola. These last two trajectories are open – a body moving along these trajectories will never return to the starting point. This is a path of motion typical of bodies outside the Solar System, captured by the Sun and "launched" back into space. On the other hand, the problem of three bodies is already causing major technical difficulties. They are usually solved on the assumption that one of the bodies is much smaller or much more distant, and therefore the motion of the other bodies can be described as a combination of the problem of two bodies and a slight disturbance (perturbation) coming from the third body. A typical effect of a slight perturbation is the slow rotation of the orbit of one body around another body (this rotation is called the precession of the *perihelium*, i.e. the point in the orbit closest to the central body). However, in general, the three-body problem is not analytically solvable, which reminds us of the serious limitations of the epistemological version of determinism that we discussed earlier.

A spectacular success of Newton's theory of gravity was the correct explanation of the phenomenon of tides, which caused a lot of trouble for scientists. Galileo put forward the bold hypothesis that the tides are caused by the combined action of the centrifugal force from the rotation of the Earth around its axis and the centrifugal force from orbiting the Sun. A point on the Earth's surface that is on the opposite side of the planet from the Sun "feels" the maximal centrifugal force, which is the sum of the "rotational" and "orbital" forces, and the point directed towards the Sun is minimal (the difference of two forces), which should give the effect of raising and lowering the surface of the water. However, this hypothesis gives incorrect predictions of the frequency of high tides: according to Galileo, we should have one high tide and one low tide per day (and for this they must be correlated with the position of the Sun in the sky). In fact, we observe two high tides and two low tides a day, with the tides occurring when the Moon is either near the zenith or in the nadir (on the opposite side of the celestial sphere).

Newton's law of gravitation, on the other hand, implies that the side of the Earth facing the Moon will be attracted more strongly than the opposing side. This creates an effective force that stretches our planet and the oceans on it along the axis connecting the Earth with the Moon, resulting in two ocean "bulges" (high tides) and two places with reduced water levels (low tides) – see Fig. 2.12. It can be calculated that tidal effects from the Moon are much stronger than the analogous forces from the Sun – not because the gravity of the Sun felt on Earth is smaller than that of the Moon (in fact, it is much greater), but because the diameter of the Earth is negligible in relation to the distance from the Sun (but not in relation to the distance from the Moon), and therefore the difference between the acceleration of the closer and the farther part from the Sun is negligible.

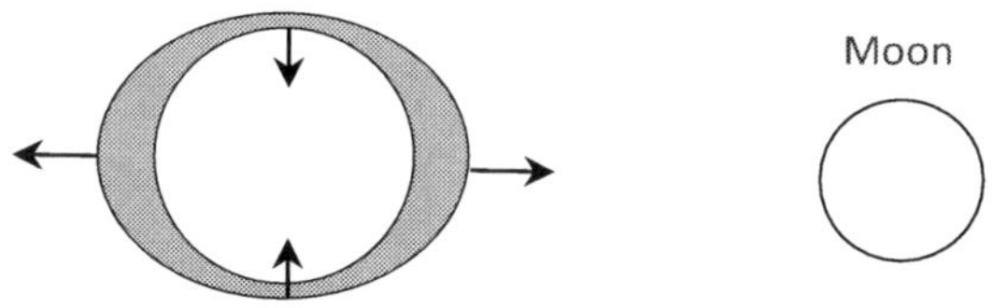

Fig. 2.12 Explanation of the origin of tides in Newton's theory of gravitation. The gray area represents the mass of water on the Earth's surface.

2.7 Classical mechanics after Newton

For the next two hundred years after Newton's death, classical mechanics developed at an incredible pace. Progress was made both in terms of expanding the scope of its applicability and developing completely new mathematical techniques. A whole panoply of brilliant physicists and mathematicians contributed to these advancements: Jean-Baptiste d'Alembert, Leonhard Euler, Joseph Louis Lagrange, Pierre Simone Laplace, Wiliam Hamilton and many others whose names have been immortalized in the mathematical and physical terminology still used today. The first natural extension of the scope of applicability of mechanics was to take into account the sizes of bodies subjected to forces, which led to the creation of so-called rigid body mechanics. Rigid solids (i.e. bodies that do not change their shape) can participate both in translational motion, described by the ordinary laws of motion applied to their center of mass, and in rotational motion. We know from school that to describe rotational motion, new concepts are needed, such as a moment of force, also known as torque (force times the distance from a fixed point), angular velocity and acceleration, angular momentum, and moment of inertia. The moment of inertia, as a generalization of the concept of mass, contains information about both the total mass of the object and its shape and mass distribution, which affects inertia in rotational motion. Newton's second law of motion applied to rotational motion with respect to a fixed axis will be in the form of the following equality: moment of acting force = moment of inertia times angular acceleration.[10] A further generalization of the motion of rigid solids will be the dynamics of incompressible fluids, taking into account the possibility of relative motion of body fragments (e.g. a stream of water), but without changing its volume. It is an extremely mathematically complex part of mechanics, which has very important practical applications, e.g. in aerodynamics.

A key addition to the laws of motion are the conservation principles. Already in the Middle Ages, it was hypothesized that the amount of motion of a given body (its so-called *impetus*) should not change. In Newtonian mechanics, this idea took the form of the conservation of momentum, which in the case of a system of two bodies not subjected to external forces can be easily derived from the third law of motion. The equality of action and reaction forces can be written as follows (p_1 and p_2 denote the instantaneous momenta of the first and second bodies):

$$\frac{dp_1}{dt} = -\frac{dp_2}{dt},$$

from where we quickly get that

$$\frac{d(p_1 + p_2)}{dt} = 0,$$

i.e. the total momentum of an isolated system does not change (its rate of change equals zero). The angular momentum of a rotating isolated system is also conserved.

The next important conserved quantity is energy. Kinetic energy, associated with motion, is expressed in the form of the well-known formula $\frac{mv^2}{2}$. This quantity is obviously not conserved in the presence of acting forces that cause the change in velocity. However, isolated systems retain their kinetic energy (in the case of a single body, this is quite a trivial observation, because Newton's first law immediately implies that its velocity must be constant over time). On the other hand, in the case of forces of a special type (the so-called conservative forces), a certain preserved quantity can be found, of which kinetic energy will be a part. This quantity is the total energy of the system, the second component of which is the potential energy dependent on the force field. A well-known example is the potential energy in a homogeneous gravitational field (e.g. at the Earth's surface), depending only on the elevation of the body relative to a certain arbitrary level.

A very interesting fact, proven at the beginning of the twentieth century by the mathematician Emmy Noether, is the relationship between the laws of conservation in mechanics and symmetries. It turns out that every law of conservation of a certain quantity corresponds to a special kind of symmetry of the equations of motion. By symmetry we mean a transformation that does not change the corresponding mathematical formulas. (For example, the Galilean transformation is a symmetry of Newton's equations.) The principle of the conservation of energy results from the symmetry with respect to shifts (translations) in time, and the principle of the conservation of momentum is a consequence of the symmetry with respect to shifts in space. Conservation of angular momentum, on the

[10] If we allow the axes of rotation to change in time, such as the axis of a spinning top, which also performs the so-called precessional ("nodding") motion, then the equations of rotational motion will take the form of so-called Euler's equations, containing moments of inertia, forces and angular velocities with respect to three perpendicular directions in space.

other hand, is a manifestation of rotational symmetry (isotropy). More details about Noether's theorem can be found in the section with an asterisk.

The most spectacular aspect of the development of classical mechanics after Newton was the introduction of completely new formulations of the basic laws of mechanics and mathematical methods of their application. Generally speaking, in post-Newtonian mechanics we can distinguish two basic trends, which are sometimes called Lagrangian mechanics and Hamiltonian mechanics. Both of these approaches, although equivalent to traditional Newtonian mechanics under certain assumptions, can be applied to situations far removed from classical mechanical problems. The impulse for the development of these new techniques, and in particular Lagrangian mechanics, was the problem of how to describe the behavior of systems subjected to restrictions of motion, i.e. the so-called constraints. An example of a constrained system can be, for example, a bead sliding without friction on a bent wire in a gravitational field. Other types of constraints, in turn, are present in the case of a wheel that rolls without sliding. Constraints of the first kind (also called holonomic) can be mathematically represented in the form of an equation connecting the coordinates of a body, while equations defining the second type of constraints (non-holonomic) also include speeds. In the case of a rolling wheel, the constraint equation links rotational velocity to the linear velocity of the center of the wheel.

It is extremely difficult to apply Newton's equations directly to the analysis of constrained motion. For example, in the case of the movement of a bead, we would have to take into account at each moment the reaction forces from the wire, which depend on its shape and the speed of the bead. Similar difficulties are encountered in an attempt to describe the motion of a compound pendulum, made up of two or more parts connected in a way that allows their relative motion. Problems such as the above were considered by eighteenth-century scholars – Jacob Bernoulli, Jean-Baptiste d'Alembert and Joseph Louis Lagrange. They drew inspiration from statics, specifically from the well-known law of the lever, which states that in the equilibrium state, the products of the forces acting on the object times their effort arms should add up to zero. This eventually led to Lagrange's formulation of the so-called principle of virtual work, also known as d'Alembert's principle. Let us consider in general the arrangement of bodies connected with each other by certain constraints limiting their movements. For each body separately, let us calculate the force acting on it and subtract from it the product of mass and instantaneous acceleration during a small (virtual) motion. To the forces calculated in this way, we then apply the principle of static equilibrium due to small shifts in the direction of the given force. This principle requires that the sum of the products of all forces and their virtual displacements be zero.

Lagrange then showed that the principle of virtual work formulated above is mathematically equivalent to a differential equation called the Lagrange equation (other terminology is the Lagrange equation of the second kind or the Euler-Lagrange equation). This equation contains a function $\mathcal{L}$ referred to as the Lagrange function or in short Lagrangian. The Lagrangian of a given system is represented as a function of the so-called generalized coordinates. Generalized coordinates are derived from "ordinary" coordinates (e.g., Cartesian coordinates) by applying constraint equations to them. Due to the existence of constraints, the coordinates of a given body are not independent of each other – e.g. if we know the "vertical" coordinate of a bead sliding on a wire, we can calculate the horizontal coordinate. Generalized coordinates, on the other hand, are independent, so their number must be less than the number of ordinary coordinates. To describe the movement of a bead, it is enough to have one generalized coordinate in the form of, for example, the distance from the starting point, calculated along the wire.

The Lagrangian also contains reference to velocities expressed in generalized coordinates (i.e. derivatives of generalized coordinates with respect to time). The Lagrange equation includes the so-called partial derivatives of the Lagrangian with respect to the generalized coordinates and their velocities (the exact form of the equation is given in the "asterisked" section). In the event that the forces acting on a given object can be presented as the gradient of the potential, the Lagrangian of the system is identical to the difference between the kinetic energy and the potential energy of the system. The dynamic behavior of a given system can therefore be described by knowing the dependence of its kinetic and potential energy on generalized coordinates and generalized velocities. Note that with this approach, we do not have to refer to the concept of force at all. This is good news for those who, like d'Alembert, have serious doubts about the ontological nature of the forces (their "mysterious" or "occult" nature). On the other hand, however, it can be argued that the status of the energy is also not entirely clear, since the energy of the system is not given to us in direct observation. In any case, the transition from classical Newtonian to Lagrangian mechanics involves a paradigm shift in mechanics – from the concept of a "pushing" force to the total energy that determines the future behavior of the system.

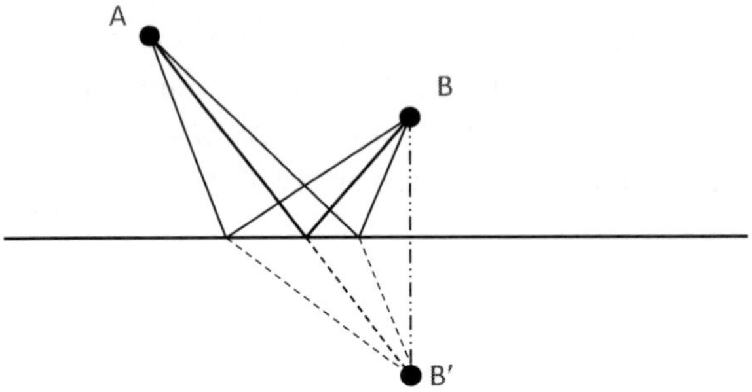

Fig. 2.13 Minimizing the path between point A, the "river" and point B. Reflecting point B with respect to the line, we can see that the problem can be reformulated as a search for the shortest path between A and B′, which is of course the straight line AB′.

An even more radical departure from the worldview presupposed in classical Newtonian mechanics can be seen in the approach to mechanics based on the so-called principle of least action. Its sources can be found in the "minimization" problems, known since ancient times. One such problem is the question of what geometric shape allows a given surface to be closed with the shortest circumference (the answer is a circle, of course). Another example, more akin to physical problems, is the following practical problem: what route should be taken from point A (which can be, for example, a house) to a certain straight line (a river), and then to point B (a bed that should be watered with water taken from the river) so that the total path is as short as possible (Fig. 2.13). Incidentally, this problem can be easily solved by using properly applied symmetry (reflection with respect to the line). The solution is a path whose parts form the same angles with the selected line.

The latter example brings to mind associations with the law of reflection, known from geometric optics. This law can be derived from the so-called Fermat principle, which states that light moves in such a way as to minimize the time it takes to reach a given point. In the case of light propagating in a homogeneous medium at a constant speed, minimizing time is obviously equivalent to minimizing the distance, and thus we get an analogous law of reflection as in the above river example. On the other hand, a more interesting case is the phenomenon of refraction during the transition from one medium to another, when the speed is different in different media. It turns out that assuming the inverse proportionality of the speed of light in a given medium to the refractive index of this medium, Fermat's principle can be used to derive Snell's law of refraction known from optics (the ratio of the sine of the angle of incidence to the sine of the angle of refraction is constant for two mediums).

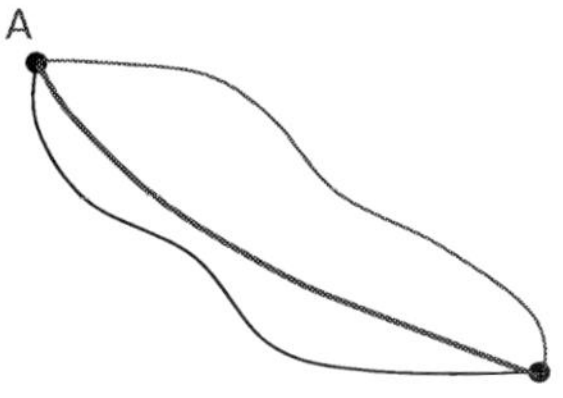

Fig. 2.14 Illustration of the minimization problem in mechanics. For each possible trajectory from A to B, we calculate the action and select the trajectory for which this action is minimal (bold line).

The general procedure used in Fermat's principle is as follows: we choose the starting point A and the end point B, then consider all possible trajectories from A to B and calculate a certain quantity for each of them (e.g. arrival time). The trajectory for which this quantity takes a minimal value is the trajectory actually chosen in the real phenomenon (Figure 2.14). The application of this scheme to the movement of material objects was taken up by Pierre Louis Maupertuis in Leonhard Euler. They called the quantity to be minimized "action", so the principle of least action was the basis of the proposed approach. For Maupertuis, action was determined by the product of momentum and the distance traveled by the body. More precisely, to calculate the total action for a given trajectory of a body, the actions calculated on small sections of the path where the change in momentum (velocity) is negligible should be summed up. In the limit, when the selected segments tend to zero, the corresponding sum passes into integration operation (the "inverse" of differentiation – see box on p. 92 for detailed explanations). The principle of least action requires us to choose from all possible trajectories of a body of a given energy the one for which the integral described above gives the minimum value. It should be emphasized that in Maupertuis's view, only those trajectories for which the total energy of the body is as in reality should be considered.

The brilliant mathematician Euler formalized the above procedure, giving it a mathematically precise form of the so-called variational calculus. The calculus of variations can be seen as a generalization of Newtonian and Leibnizian differential calculus. In calculus, we compute the derivative of a given function of a real variable $f(x)$ in order to determine a local minimum or maximum of this function, that is, such an argument x_0 for which the value of the function is locally the largest or smallest. As some of you may know, the criterion of minimization or maximization is the zeroing of the first derivative at this point (the possibility of the existence of an inflection point should also be excluded – for this purpose the second derivative is calculated). On the other hand, in the case of the principle of least action, we want to find in general not a number, but a function (trajectory) for which the corresponding value of action is minimal. A mapping that assigns appropriate values to functions (rather than numbers) is called a "functional". Euler showed that the condition of minimizing (or maximizing) a given functional can be expressed by the so-called functional variation, which in this case must be zero.

The principle of least action in the most general form was given by Lagrange.[11] It turns out that an operation whose minimization (or in some cases maximization) selects a real trajectory from the set of all possibilities can take the form of an integral from the previously introduced Lagrange function. It is an integral whose integration parameter is time, but it contains "hidden" information about the selected trajectory, because for each moment the value of the Lagrangian is calculated assuming that the system is on the selected trajectory. With this general approach, we no longer have to limit ourselves to trajectories describing the motion of an object with a fixed energy – we can take into account all mathematically acceptable trajectories. Lagrange proved that the assumption of zeroing variations from such a defined operation (i.e. the condition of maximizing or minimizing the functional) implies the satisfaction of the Euler-Lagrange equation, the derivation of which from the principle of virtual work we talked about before. Thus it turns out that the Euler-Lagrange equation can be arrived at in two different ways.

The principle of least action is extremely interesting from a philosophical point of view. Let us compare it with the standard method of calculating a trajectory using Newton's second law of motion, discussed in section 2.2. To calculate the future trajectory of a body or system of bodies, the current state of the system (position and velocity) and the forces currently acting must be considered. Thus the arrow of prediction is directed from the past to the future, in accordance with the intuitive belief that causes precede their effects in time. However, the situation is different in the case of the approach based on minimizing (or maximizing) the action. In order to calculate the trajectory, we must analyze the future options, i.e. take into account the total action on each of the possible paths of the system's evolution and choose the path with the least value. However, the full result is known only at the end of the completed movement, not at the beginning. It seems, therefore, that at least part of the cause of the future motion of a body (or bodies) is to be found in an even more distant future, at the completion of the motion itself.

In philosophy, explanations of an event referring to a later state of affairs are well known. These are explanations based on the so-called final cause ("teleological"), already considered by Aristotle. We apply them primarily to human actions and behaviors, and more broadly to all beings endowed with consciousness. It is natural to explain the behavior of an archer aiming his bow and arrow at a target by his future goal, i.e. hitting the target. Hitting the target is considered to be the intentional cause of the earlier bow shot. However, doubts arise here as to the fundamental nature of such an explanation. Many believe that the actual cause of the shot is not so much a later hit, but an earlier thought about the hit (the desire to hit or the pursuit of it). That is why it is so important in a teleological explanation to limit ourselves to conscious beings capable of making decisions and plans.

---

[11] Due to the fact that in some cases the correct trajectory of the system is selected by maximizing the action, we should talk about the principle of extremal (minimal or maximal), also referred to as stationary action. However, for historical reasons, the term "principle of least action" is still in use.

Despite the above reservations, final causes appear in sciences, mainly biological sciences. The complex structures of organs, such as the heart or lungs, are often explained by the function of these organs (pumping blood, exchanging oxygen and carbon dioxide). Similarly, evolutionary biology explains the development of appropriate characteristics of organisms (phenotype) by the fact that their presence increases the chances of survival of the organism. However, the partial and vicarious nature of such an explanation is usually emphasized. The "real" cause of the development of a new feature of organisms (e.g. the appearance of the opposable thumb in primates) are complex physico-chemical processes at the cellular level participating in the formation of daughter organisms from the cells of parent organisms, which are described by genetics. Because of the extraordinary complexity and breadth of these processes, spanning countless generations, it is easier to apply a simplifying explanation based on the purpose of the change. However, the fundamental science of physics should not refer to this type of simplification.

The role of extremal principles in mechanics undermines such an optimistic belief. The explanation of the mechanical behavior of bodies suggested by these principles is based on the future "goal" of minimizing (maximizing) a certain quantity, although we cannot of course attribute consciousness, willingness, and desires to physical bodies. Note that there is a fundamental difference between the use of goal-oriented explanations in biology and the method based on extremal principles in physics. Teleological causes in biology do not allow us to make detailed predictions. We are not able to precisely predict what new traits may appear in organisms in order to increase their chance of survival. On the other hand, the method of extremization of action in physics gives us an unambiguous answer to the question of what the future trajectory of the body will be. For this reason, we cannot treat the method based on "finality" as a mere rough approximation of the exact Newtonian method, using efficient causes. Of course, under certain assumptions, both methods are equivalent, so it can be argued that the ontological world view of classical Newtonian formalism is more fundamental. But is this attitude not the result of a historical accident that classical mechanics was first formulated in the language of efficient forces and causes? What about the argument that extremal principles and the Euler-Lagrange equation based on them have wide applications outside the realm of strictly mechanical phenomena? Let us leave this matter to the reader to ponder.

The last milestone in the development of classical mechanics, which we will discuss in this chapter, will be so-called Hamiltonian mechanics. This approach arose in the nineteenth century as a result of mathematical research on the already existing Lagrangian formalism. Thus, the development of Hamiltonian formalism was not inspired by a desire to solve some new problems or to generalize an existing theory. In principle, the framework of Lagrange mechanics, with the fundamental Euler-Lagrange equation at the forefront, was quite sufficient. However, the novelty of Hamilton's approach became apparent much later, when it was successfully applied to new theories such as quantum mechanics and general relativity. The Hamiltonian formalism also plays a very important role in statistical mechanics.

As we remember, Lagrangian mechanics introduced the concept of generalized coordinates, i.e. basically arbitrary, independent parameters, characterizing the position of an object at a given moment. In Hamiltonian mechanics in addition, for each such coordinate, a correlated concept of generalized momentum is defined (such momentum is also called conjugate to a given coordinate). As a result of mathematical analyses, it turned out that for typical coordinates, such as linear Cartesian coordinates, differentiating the Lagrangian with respect to the generalized velocity for a given coordinate simply gives ordinary momentum (mass times velocity) along that coordinate. This suggested that such a partial derivative of the Lagrangian with respect to the generalized velocity should be called generalized (conjugate) momentum for all possible coordinates. For example, if the generalized coordinate is the angle that determines the position of the rotating body, then the generalized momentum coupled to it will be the angular momentum of that body.

The basic equations of dynamics in Hamiltonian mechanics take into account a new object, called Hamilton's function or Hamiltonian for short. The Hamiltonian of a given system is a function of its generalized coordinates and the generalized momentums coupled to them. It is defined in a rather complicated way by means of a mathematical operation on the Lagrangian[12], but in most typical cases, the Hamiltonian turns out to be equal to the total energy of the system (kinetic plus potential energy). Hamiltonian mechanics is based on two very simple equations. The first of them equates the rate of change of a given generalized coordinate (and thus the generalized velocity) with the derivative of the Hamiltonian with respect to the momentum coupled to this coordinate. The second equation is almost a "mirror image" of the first: the rate of change of the generalized momentum is equal to the derivative of the Hamiltonian with respect to the conjugate coordinate, but with a minus sign. For typical situations, Hamilton's second equation is in fact equivalent to Newton's second law of motion (the derivative of energy with respect to a given coordinate is a measure of the force acting in the direction of that coordinate). The first equation, in turn, is something like a definition of velocity, or better yet, the relationship between velocity and momentum.

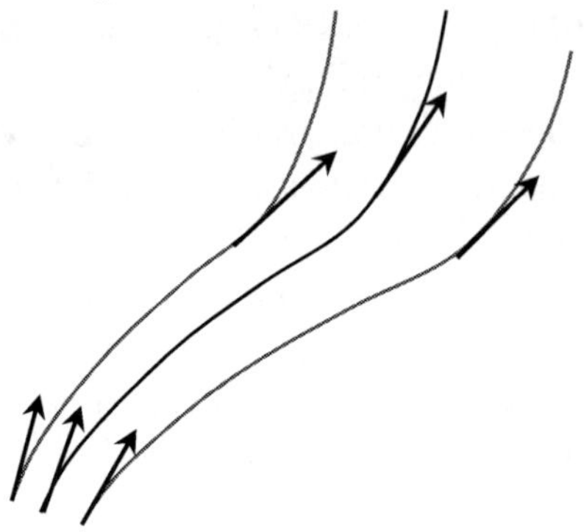

Fig. 2.15 Phase space (of momenta and positions) with three possible trajectories marked. The arrows represent "abstract" velocities (rates of change of positions and momentums), calculated using Hamilton's equations.

Hamiltonian mechanics uses an important concept of phase space. It is an abstract mathematical space in which the coordinates of points are given by the coordinates and momenta of a particular system. For example, a point body with no constraints requires six parameters to determine its position and momentum – three parameters representing three spatial coordinates and three numbers characterizing the momentum components in the directions of the corresponding coordinates. Therefore, the phase space for such a body must have six dimensions. On the other hand, when describing N free particles, the dimension of the phase space increases to $6N$. A single point in phase space represents the instantaneous state of a given

[12] The operation that transforms the Lagrangian of a given system into a Hamiltonian is called the Legendre transformation.

physical system (as we remember, positions and momenta uniquely determine the state of material particles in Newtonian mechanics). On the other hand, trajectories (curved lines) represent possible evolutions of the system over time, i.e. how the positions and momenta of the bodies in the system change (Fig. 2.15). Hamilton's equations determine the rate of change of position and momentum coordinates, so it can be said that they characterize an abstract "velocity" in phase space. For a given Hamiltonian, we can attach a vector representing the rate of change of each coordinate at each point in the phase space. This vector determines the trajectory for a given point. Thus, in general, all possible evolutions of the system are represented in phase space as analogous to the flow of liquid. This analogy is even more complete due to some mathematical facts, but we will return to this in the chapter devoted to statistical mechanics.

2.8* Elements of analytical mechanics

In this section, intended for more advanced readers, we will present slightly more precisely the mathematical foundations of analytical mechanics, and in particular the approaches of Lagrange and Hamilton. Let us begin by introducing the concepts of potential and kinetic energy and showing that the second law of motion implies the conservation of the sum of both energies for a single body. The formula for kinetic energy of a given body is well known from the school course in physics:

$$T = \frac{1}{2}mv^2,$$

where v is the velocity of the body (more precisely, the length of the velocity vector). Potential energy can be introduced with an additional assumption regarding forces. Namely, we assume that the net force acting on the body in a given direction x can be represented as a derivative of a certain function of position x, which we will mark as $V(x)$:

$$F_x = -\frac{dV(x)}{dx}.$$

Function $V(x)$ is called potential energy.[13] The above assumption means that the force can be treated as the slope of a certain curve, in some analogy to the slope of a hill. Energy V in general, is a function of three spatial coordinates x, y and z. In such a situation, the derivative with respect to a given coordinate is formally called a partial derivative and is denoted slightly differently than the "ordinary" derivative. However, in essence it is the same mathematical operation, only it is performed under the assumption that the other variables are constant numbers:[14]

---

[13] The minus sign follows from the fact that if V is an increasing function of the x coordinate (i.e., the potential energy is increasing in the direction of positive x's, then the acting force will be directed oppositely, towards decreasing energies.

[14] However, it should be remembered that in general, the partial derivative of a function of many variables with respect to one variable will not be equal to the total derivative with respect to this variable. The total derivative also takes into account that other variables can be dependent on the differentiation variable. A common example of this is a function that depends on a time variable and spatial variables. The partial derivative of the function f with respect to variable t takes into account only the

$$F_x = -\frac{\partial V(x, y, z)}{\partial x}.$$[15]

Let us write down the formula for the total energy (the sum of kinetic and potential energy) for the one-dimensional case:

$$E(x) = \frac{1}{2} m \left(\frac{dx}{dt}\right)^2 + V(x)$$

Function $E(x)$ does not change when its time-derivative is zero everywhere. Let us differentiate the above expression with respect to time using the chain rule, specifying how to differentiate the composition of two functions $f[g(x)]$. The principle is simple: the derivative of the whole is the product of the derivative of the first function f and the second function g (in our case f is the quadratic function of velocity $\frac{1}{2}mv^2$, and g is just velocity v as a function of time t; of course we keep in mind that $v = \frac{dx}{dt}$):

$$\frac{dE(x)}{dt} = m \left(\frac{dx}{dt}\right)\left(\frac{d^2x}{dt^2}\right) + \frac{dV(x)}{dt}.$$

We assume that the dependence of energy V on time comes only from the changing position x of the body. In such a situation, the derivative of potential energy with respect to time can be calculated by the chain rule as the derivative of energy with respect to x multiplied by the derivative of x with respect to time:

$$\frac{dE(x)}{dt} = m \left(\frac{dx}{dt}\right)\left(\frac{d^2x}{dt^2}\right) + \frac{dV(x)}{dx}\frac{dx}{dt}.$$

So we can see that equating the above formula with zero means that:

$$m\left(\frac{d^2x}{dt^2}\right) = -\frac{dV(x)}{dx}.$$

The right side is nothing but the acting force, and the left side is the product of mass and acceleration. This is the well-known Newtonian equation of motion (second law of motion) $F = ma$. Thus, the second law of motion results in the principle of the conservation of total energy (as long as such energy is well defined, i.e., there is a potential energy function that is not explicitly dependent on time.)

We will now introduce the Lagrange function, marking it with $\mathcal{L}$:

$$\mathcal{L} = T - V.$$

Since kinetic energy is a function of velocity, and potential energy is a function of position, the Lagrangian will be a function of both of these variables. In the one-dimensional case, the Lagrangian variables will be: the coordinate x and the velocity v_x in this direction, i.e. the

---

explicit dependence of f on time; the total derivative also takes into account the dependence of spatial coordinates on time.

[15] The operation of taking partial derivatives with respect to the x, y, and z coordinates from some scalar function is called the gradient of this function. More precisely, the gradient of a function V is a vector whose components are of the form $\frac{\partial V}{\partial x}, \frac{\partial V}{\partial y}, \frac{\partial V}{\partial z}$. It is abbreviated as ∇V.

time derivative of x. To simplify notation, we will follow the common practice to symbolize the derivative of a function with respect to time by a dot above it. Therefore, the velocity in the direction of x is given by $\dot{x}$, and we will present the Lagrangian as $\mathcal{L}(x, \dot{x})$. With the help of the Lagrangian, we define the total action of $\mathcal{A}$ on a certain path traveled by the body in the interval from the moment of t_1 to t_2:

$$\mathcal{A} = \int_{t_1}^{t_2} \mathcal{L}(x, \dot{x}) dt.$$

The above integral can be treated approximately as the sum of the values of the Lagrangian in small time intervals Δt times the interval: $\sum \mathcal{L}\Delta t$, where $\sum \Delta t = t_2 - t_1$. It is important to note that the value of an action depends on the adopted trajectory of the body, i.e. the function $x(t)$. The dependence of the Lagrangian on time is the result of the "composition" of two relationships: the position and velocity of the body on time (i.e. its trajectory) and the dependence of the Lagrangian on these two parameters. An action can therefore be looked at as a function that assigns to a given trajectory $x(t)$ a certain number. Such a function is called a functional.

We can then think about what will happen to our functional with small changes in trajectory. This process is called calculating the variation of a given functional. If for a given trajectory, the variation of the functional is zero, i.e. small (infinitesimal) deviations from this trajectory do not cause a change in the value of the functional, this means that the functional for this trajectory takes an extreme value (minimum or maximum). Thus, the condition of extremalization takes the form of the equation:

$$\delta \mathcal{A} = 0.$$

where δ denotes a variation. The condition of zeroing of the variation of a given functional can be equivalently stated as a condition expressed by means of "ordinary" derivatives of the corresponding functions included in this functional. We won't show it, but it turns out that in the case of the action defined in the above way by the Lagrangian, zeroing out its variation is equivalent to the following equation, called the Euler-Lagrange equation (for the one-dimensional case)[16]:

$$\frac{d}{dt}\left(\frac{\partial \mathcal{L}}{\partial \dot{x}}\right) - \frac{\partial \mathcal{L}}{\partial x} = 0. \tag{2.2}$$

It is a differential equation, i.e. it characterizes our system at every moment of its evolution. It follows that if we calculate at a given moment how our Lagrangian changes with the change of one argument x and the second argument $\dot{x}$, then the rate of change in time of one parameter of variation (relative to $\dot{x}$) will be equal to the other of them (relative to x). It is worth noting that in this way we shift the perspective from "global" (minimizing the action on the entire path from t_1 to t_2) to "local" (interdependence of certain instantaneous parameters). We therefore return to the conception of mechanical phenomena in terms of efficient causes acting at a given moment, instead of final causes involving future trajectories. We will not decide which of these philosophical perspectives is more fundamental.

[16] In the three-dimensional case, the Euler-Lagrange equation looks the same for each coordinate separately.

We can see from the definition of the Lagrangian that equation (2.2) is in fact Newton's second law of motion "in disguise". We will show this by calculating the partial derivatives of the Lagrangian with respect to the variables x and $\dot{x}$. Note that for a single body in a force field, the explicit dependence of its Lagrangian on velocity $\dot{x}$ comes only from its kinetic energy, and the dependence on x is contained in the potential energy. So we have the following equations:

$$\frac{\partial \mathcal{L}}{\partial \dot{x}} = m\dot{x}$$

$$\frac{\partial \mathcal{L}}{\partial x} = \frac{\partial V}{\partial x}.$$

Inserting the above into the Euler-Lagrange equation, we get

$$m\ddot{x} = -\frac{\partial V}{\partial x},$$

so of course the well-known equation $F = ma$ (two dots mean the second derivative with respect to time). Incidentally, we have just found out that the derivative of the Lagrangian with respect to $\dot{x}$ is nothing more than the momentum of the body mv. This leads directly to the definition of generalized momentum conjugate to the x coordinate. In the general case, the Lagrangian can be treated as a function of any coordinates q (and their rate of change), and then the momentum p coupled to a given coordinate will be defined exactly as $\frac{\partial \mathcal{L}}{\partial \dot{q}}$.

This brings us to the Hamiltonian formalism. Consider the general case of a physical system that is characterized by a certain number of generalized coordinates q_i. As before, we define a generalized momentum p_i for each coordinate q_i:

$$p_i = \frac{\partial \mathcal{L}(q_i, \dot{q}_i)}{\partial \dot{q}_i}.$$

We will now introduce a new function $H(p_i, q_i)$ of coordinates and momenta, called the Hamiltonian:

$$H = \sum_i p_i \dot{q}_i - \mathcal{L}. \tag{2.3}$$

The above definition may seem a bit strange and unintuitive, but there is an important observation in favor of it. If the total derivative of the expression H with respect to time is calculated, then after making appropriate transformations it will turn out that the result will be exactly equal to the partial derivative of the Lagrangian with respect to time (with a minus sign). The partial time-derivative characterizes the explicit dependence of a given parameter on time. The Lagrangian is usually dependent on time due to its dependence on coordinates, which in turn change over time during the evolution of the system. However, typical Lagrangians do not contain an explicit dependence on time – in other words, they are invariant with respect to shifts in time. Thus, the Hamiltonian of a given system will be a function that does not change at all in time, as long as the Lagrangian does not explicitly depend on time.

It is not difficult to check that in typical situations, the Hamiltonian is simply identical to the total energy of the system, which, as we already know, should be conserved. Let us examine this again using the example of a particle moving in one dimension x in a field of force. In this case, the Lagrangian has the form (written using the momentum p_x):

$$\mathcal{L} = \frac{{p_x}^2}{2m} - V(x).$$

Of course $p_x\dot{x} = \frac{p_x^2}{m}$, so inserting this into (2.3) we get the total energy:

$$H = \frac{{p_x}^2}{2m} + V(x). \tag{2.4}$$

Hamilton's equations have the following general form:

$$\begin{aligned} \dot{q}_i &= \frac{\partial H}{\partial p_i} \\ \dot{p}_i &= -\frac{\partial H}{\partial q_i}. \end{aligned} \tag{2.5}$$

Again, substituting the linear coordinate x for q_i, the momentum p_x coupled with it for p_i, and inserting formula (2.4) in place of the Hamiltonian, we get from Hamilton's equations:

$$\begin{aligned} \dot{x} &= \frac{p_x}{m} \\ \dot{p}_x &= -\frac{\partial V(x)}{\partial x}. \end{aligned}$$

The first of the above equations is the well-known relation momentum = mass times velocity. The second, of course, is Newton's law of motion: the rate of change in momentum equals the applied force. Let us emphasize once again that although Hamilton's equations are equivalent to Newtonian laws of motion in special cases, they are also used where Newtonian mechanics ceases to work.

At the end of the chapter on classical mechanics, let us take a closer look at the aforementioned relationship between symmetries and conservation laws (Noether's theorem). A symmetry of a given object is a transformation that does not change the object (this object is an invariant of this transformation). We remember from sec. 2.4 that Galilean transformations are symmetries for Newton's equations. We will now consider the symmetries of the specific Lagrange functions. For example, it may happen that a given Lagrangian does not vary with spatial shifts (translations) or with rotations. This means that the interactions to which a given system is subject do not distinguish any place in space or any direction.

Formally, spatial transformations are described in the form of coordinate transformations. For example, translation in a given direction can be represented as the transformation $x \to x + d$, where d is a number. Translations and rotations belong to the category of continuous transformations, for which we can specify the so-called infinitesimal generators. For example, an infinitesimal shift along the x-axis would be the transformation $x \to x + \delta$, where δ is an arbitrarily small magnitude. In general, if we consider any geometric transformation acting on spatial coordinates q_i, then its infinitesimal generator will be of the form

$$\delta q_i = f(q)\delta,$$

where f is some function and q symbolically denotes all q_j variables. The function f may therefore depend on coordinates other than the transformed coordinate q_i (this is the case, for example, in the case of rotations). For translations, however, function $f(q)$ is identical to 1.

What will the infinitesimal change (variation) of the Lagrangian subjected to the above coordinate transformation look like? The answer is provided by calculus, and it is slightly complicated. In the simplest case, when we have a function of one variable $F(x)$, the change of this function during the infinitesimal change of the value of x is given by the following formula:

$$\delta F = \frac{dF}{dx}\delta x.$$

One can imagine that dx in the denominator and δx are "reduced", and in this case d and δ denote the same infinitesimal changes of F. However, in the general case, the Lagrangian's variation under the influence of the corresponding infinitesimal transformation will be more complex due to the fact that the Lagrangian depends on many coordinates q_i and $\dot{q}_i$. The general formula, which is a generalization of the above equation, is as follows:

$$\delta \mathcal{L} = \sum_i \left(\frac{\partial \mathcal{L}}{\partial q_i}\delta q_i + \frac{\delta \mathcal{L}}{\partial \dot{q}_i}\delta \dot{q}_i \right).$$

Noting that $\frac{\delta \mathcal{L}}{\partial \dot{q}_i}$ is just the generalized momentum p_i, and applying the Euler-Lagrange equation to the first component of the sum, we can get the following formula for the infinitesimal variation of the Lagrangian:

$$\delta \mathcal{L} = \sum_i (\dot{p}_i \delta q_i + p_i \delta \dot{q}_i).$$

Readers familiar with the basic rules of differentiation know that the derivative of the product of functions FG is given in the form of the derivative of F times G plus F times the derivative of G. The expression under the sum has a similar structure (remember that the dot means time-differentiation), so we can somehow pull the time derivative out of the brackets:

$$\delta \mathcal{L} = \frac{d}{dt}\sum_i (p_i \delta q_i).$$

Now we can insert an expression for the infinitesimal change of the coordinate q_i, thus obtaining:

$$\delta \mathcal{L} = \frac{d}{dt}\sum_i (p_i f(q))\delta.$$

If the transformation under consideration is a symmetry of the Lagrangian, its variation under the influence of this transformation should be zero. This means that the time derivative of the corresponding expression is also zero, and thus the expression does not change with time. In other words, it is preserved. Thus, we have obtained the result that the existence of spatial symmetries is always connected with a certain principle of conservation. If the symmetry under consideration is a translation, then since in this case $f(q) = 1$, the conserved quantity will be the sum $\sum_i p_i$, i.e. the total momentum of the system. Thus, the principle of the conservation of momentum results from invariance with respect to spatial translations.

On the other hand, it can be shown (we will not do this) that the angular momentum will be conserved in the case of the rotational symmetry of the Lagrangian.

It is worth noting at this point that not every Lagrangian meets the condition of symmetry with respect to spatial translations. For example, the Lagrangian for a single particle in the force field is not symmetric unless the function $V(x)$ is constant. It is quite obvious that in such a situation, the momentum of the particle is not conserved, as it will accelerate due to the force. On the other hand, if we consider two interacting particles whose potential energy depends only on their relative distance $V(x_1 - x_2)$, then such a Lagrangian is already symmetric with respect to spatial translations, and therefore the total momentum will be conserved.

There is also a connection between symmetry in time and the conservation of a certain quantity. As we noted above, the Hamiltonian function has been defined by the Lagrangian in such a way that the following equality occurs:[17]

$$\frac{dH}{dt} = -\frac{\partial \mathcal{L}}{\partial t},$$

So the total rate of change of the Hamiltonian with respect to time is equal to the partial rate of change of the Lagrangian with a minus sign. This means that if there is no explicit time variable in the Lagrangian, then the Hamiltonian of the entire system does not change with time, and is therefore conserved. But the Hamiltonian is nothing but the energy of the system. Thus, the existence of symmetry with respect to time translations implies that the energy of the system will be conserved.

Questions and problems

1. Can Newton's laws of motion be interpreted as implicit definitions of some of the terms they contain? Which terms will it be? Does the acceptance of appropriate semantic postulates reduce Newton's laws to *a priori* analytic statements?

2. Is it possible to definitively refute Newton's first and second laws of motion?

3. Show that the definition of the mass of an object in the form of the postulate "Each two objects interact with one another in such a way that the product of their masses is equal to the inverse of the product of accelerations" is non-conservative. Hint: consider three bodies *a, b,* and *c,* and apply the above postulate to their interactions in pairs: *a* with *b*, *b* with *c,* and *a* with *c*. Prove that the adopted postulate results in a certain relationship that does not include the masses of bodies *a, b* and *c* at all.

4. Explain how Newton's second law of motion allows the trajectory of a given body subjected to forces to be calculated. What additional information about this body is needed here?

5. Present two versions of the thesis of determinism: ontological and epistemological. What are the logical relationships between them? What crucial concept impacts the meaning of epistemological determinism and its validity?

---

[17] This can be shown by explicitly calculating the total derivative of the Lagrangian with respect to time. The following formula will then be obtained: $\frac{d\mathcal{L}}{dt} = \frac{\partial \mathcal{L}}{\partial t} + \sum_i \left(\frac{\partial \mathcal{L}}{\partial q_i}\dot{q}_i + \frac{\partial \mathcal{L}}{\partial \dot{q}_i}\ddot{q}_i\right)$. Next, from the Euler-Lagrange equation, we need to convert the partial derivative of the Lagrangian with respect to q_i into the time derivative of its rate of change with velocity, and use the definition of the generalized momentum $\dot{q}_i p_i$, which will give us the following sum $\sum_i(\dot{p}_i\dot{q}_i + p_i\ddot{q}_i) = \frac{d}{dt}\sum_i p_i\dot{q}_i$.

6. How can the concept of the state of a physical system at a given moment be characterized in general? Why should the instantaneous state of the system not contain any information about the behavior of the system after this moment? Does the state of the system described by the velocity function (interpreted as a derivative of the position function) satisfy this condition?

7. Discuss examples suggesting that classical mechanics is not a fully deterministic theory.

8. Does deterministic chaos undermine the validity of ontological or epistemological determinism?

9. What is the mathematical form of the Galilean transformation and its physical meaning? Give an example of an invariant of this transformation.

10. Present two understandings of the dispute about the absoluteness of time and space: physical and philosophical (metaphysical).

11. What is momentary (instantaneous) space? Is it an invariant of the Galilean transformation? Consider the same question in relation to a time-extended (temporally persistent) space.

12. Analyze Newton's bucket argument for the existence of absolute space. Does this argument give us a method for identifying the absolute points of space?

13. What loophole in Newton's bucket argument did Ernst Mach reveal? Present a weak and strong interpretation of Mach's argument.

14. Present Leibniz's shift argument against absolutism (for relationalism) in the dispute over the status of space. What role do metaphysical assumptions play in this argument (the principle of the identity of indiscernibles, the principle of sufficient reason)?

15. Leibniz's argument for relationalism can be formulated in a "kinematic" version, in which instead of the absolute position of the universe, its velocity relative to absolute space is discussed. Try to reconstruct such an argument and compare it with the "static" version. Does the kinematic argument have to be based on the principle of the identity of indiscernibles, or is it sufficient that we should not assume the existence of any facts that cannot be empirically verified?

16. Analyze how Newton justified that the planets must be held in orbit by a force directed to the Sun whose magnitude is inversely proportional to the square of the distance. Indicate the role of Kepler's laws in the subsequent steps of Newton's argumentation. Pay attention to the method of approximating smooth trajectories (along a circle or ellipse) with polygons.

17. Discuss the problem of "action at a distance" in the case of gravitational forces. What was Newton's reaction to this problem?

18. Present the general idea of determining the trajectory of motion using the principle of least action. What consequences does this approach have for the causal explanation of motion?

19. Present in broad outline the essence of Hamiltonian mechanics, using the concept of phase space. What is the dimension of the phase space describing the motion of N point bodies?

Further reading

The problem of the methodological status of Newton's laws of motion and the issue of time and space in classical mechanics are discussed in a well-known work: E. Nagel, *The Structure of Science: Problems in the Logic of Scientific Explanations*, Harcourt, Brace & Worlds 1961, chapters 7 and 8.

The question of the determinism of classical mechanics is thoroughly analyzed in a monumental philosophical work: J. Earman, *A Primer on Determinism*, D. Reidel, Dordrecht 1983, chapters II and III.

A recommendable introduction to chaos theory for philosophers: P. Smith, *Explaining Chaos*, Cambridge University Press, Cambridge 1998.

Details of Newton's derivation of the law of gravitation and many other interesting facts about astronomy and celestial mechanics are given in the book: J. Cushing, *Philosophical Concepts in Physics*, Cambridge University Press, Cambridge 1998, parts II and III.

A discussion of the historical development of classical mechanics from antiquity to modern times, with particular emphasis on analytical mechanics after Newton, can be found in the book: L. Sklar, *Philosophy and the Foundations of Dynamics*, Cambridge University Press, Cambridge 2013.

For those interested in a more thorough analysis of the mathematical apparatus of classical mechanics, a recently published book by a renown physicist will be a valuable item: L. Susskind, G. Hrabowsky, *Theoretical Minimum. What You Need to Know to Start Doing Physics*. Basic Books, New York 2014.

CHAPTER 3. THE SCIENCE OF HEAT

Mechanical phenomena such as motions, collisions of bodies, transfer of energy and momentum, etc., form the backbone of Newtonian physics. However, there are types of processes in the physical world that seem fundamentally different from simple mechanical interactions. These phenomena include, above all, various types of thermal processes – heating, cooling, combustion, melting, solidification, evaporation, condensation. Of particular interest is the behavior of liquids and gases that do not have a well-defined shape like solids, but take the shape of a container. Already in antiquity, attention was paid to the specific behavior of fluids, which resulted in formulation of interesting laws of hydrostatics, e.g. well-known Archimedes' law of buoyancy. Phenomena related to gases, on the other hand, began to be studied in more detail much later – in the seventeenth and eighteenth centuries, when technology made it possible to produce sealed vessels necessary to study changes occurring in a specific gas sample. It was quickly noticed, for example, that heating a gas sample increases the pressure exerted on the walls of the vessel. If the pressure is kept constant by using a movable piston, the supply of heat will increase the volume of the gas, and the piston will do some work at the same time. On the other hand, the compression of a gas by providing it with a certain amount of mechanical energy results in an increase in its temperature. Similar observations were used to formulate quantitative laws correlating macroscopic parameters of pressure, temperature and volume, which in turn opened the way to the creation of a new theory linking thermal phenomena with macroscopic mechanical processes (mechanical work), called thermodynamics.

Archimedes' law of buoyancy states that a body submerged in water loses exactly as much weight as the weight of the water displaced by this body. This principle has interesting and non-obvious consequences. For example, consider the following issue. Let us imagine that a certain vessel has been filled to the brim with water in which a piece of ice is floating. What happens when this ice melts? Will the water overflow the edges of the vessel or will its level drop slightly? It seems that without reference to experience we will not be able to settle this. And yet Archimedes' law gives us an immediate answer: the water level will not change (you can check it yourself if you don't believe the famous physicist from ancient Syracuse). The reason is simple: because a piece of ice floats on water, it displaces exactly as much water as it weighs, and therefore, after melting, the water formed from the ice will occupy precisely the volume previously displaced by the ice. By the way, we can see the reason why floating pieces of ice stick out slightly above the surface – this is the result of the difference between the density of water and the density of ice. Water with a volume equal to the volume of the submerged portion of the ice must weigh the same as a whole piece of ice, and thus ice has a lower density than water.

Why do thermal phenomena deserve the attention of a philosopher? There are many reasons. An ontological question comes to the fore: do heat and the processes associated with it involve a separate category of properties of material objects, or are they only manifestations of hidden processes of a strictly mechanical nature, such as motions and collisions? In other words, we ask about the ontological nature of heat and its relationship to other known properties of matter. A natural answer to this question, adopted at the beginning of the development of heat science, is the thesis about the ontological independence of thermal phenomena from other physical phenomena. A specific implementation of this doctrine was the theory of the so-called caloric, i.e. an invisible fluid that was to be responsible for all thermal processes – heating, cooling, etc. For example, the process of heat transfer from a hot body to a cold body was presented as the flow of a certain amount of caloric contained in the first body to the second body until the equilibrium level was reached. However, the development of science dealt a blow to this doctrine. The close relationship between heat and mechanical energy, which we will discuss in more detail later in this chapter, has suggested a possible reduction of thermal phenomena to purely mechanical processes occurring at the microscopic molecular level. This leads to the philosophical and methodological question of what it means to reduce one phenomenon, property or description to other phenomena (properties, descriptions).

We can come across the concept of reduction in many areas of philosophy and methodology. Typical philosophical reductions concern certain ontological categories – for example, some try to reduce the category of concrete objects to properties and relations; Others, on the other hand, propose inverse reductions, in which abstract objects are reduced to certain constructs based on the concrete. At the interface between philosophy and science, we encounter the extremely important question of the possibility of reducing mental processes and properties (such as thoughts, desires, and emotions) to purely physical or neurological phenomena occurring in the brain. This touches on the fundamental dispute about the nature of mind between dualism and various forms of physical monism. Analysis of thermodynamic phenomena in Newtonian mechanical terms provides a model example of a successful reduction between theories, so it is worth taking a closer look at the structure and features of this reduction procedure.

The main reason for the interest that philosophers take in thermodynamics is its surprising and subtle relationship with the problems of the philosophy of time. Common experience teaches us that time has an unambiguously determined direction. There seems to be an objective difference between the past and the future – the past is fixed, closed, and the future is open. On the other hand, the fundamental physical theory, which is classical mechanics, does not confirm this asymmetry of time. Any process allowed by the laws of Newtonian physics may well occur backward in time. In this respect, thermodynamics differs fundamentally from classical mechanics. Processes involving heat are clearly directional and irreversible – for example, heat always flows from a warmer body to a colder body, never the other way around.[1] This temporal irreversibility finds expression in a fundamental principle called the second law of thermodynamics. The question arises, therefore, how to reconcile the temporal asymmetry of the laws of thermodynamics with the symmetry of the laws of classical mechanics, in the light of the above-stated fact that thermal phenomena are nothing more than mechanical motion and the interaction of a huge number of molecules. Many physicists and philosophers have attempted to solve this mystery, and the key here is the statistical (probabilistic) nature of the description at the molecular level. We will look more closely at the ways in which the asymmetric laws of thermodynamics can be derived from the symmetric assumptions of classical mechanics combined with some statistical assumptions. As it turns out, in order to obtain the desired result, we will have to make additional assumptions that go beyond both theories and concern certain global properties of the entire universe.

3.1 Heat and temperature

But let us start with the basic question: what is heat anyway? How is it measured? In everyday life, we have no problem using the concept of heat – we know that fire is hot and ice is cold. With the help of the sense of touch, we can compare bodies in terms of their heat (of course, being careful not to burn ourselves in the process). Heat in the colloquial sense is sometimes identified with temperature, but in physics these two concepts should be kept apart. Heat in the physical sense belongs to the category of quantities called *extensive*, while temperature is an *intensive* quantity. Extensive quantities characterizing a given object depend on its size, while intensive quantities do not (it can be said informally that extensive quantities are global, while intensive quantities are local, changing from point to point). The heat contained in a bathtub with water at 35 degrees Celsius is much greater than the heat of a glass with water at the same temperature. The combined heat of two bodies is the sum of the heat of each body separately, whereas temperature does not add up in this way.

The difference between heat and temperature can be elegantly explained using the fundamentally incorrect but useful caloric model. Suppose that each body contains a certain amount of invisible caloric fluid, which can assume different degrees of density. The higher the temperature of a given body, the higher the density of the caloric, which gives us a greater amount of fluid within the constant volume of the body. Heat is a measure of the total amount of caloric in a body, while its density relates to the temperature. The transfer of part of the caloric from one body to another causes it to dilute in the first body and congest in the second, i.e. the temperature of the first body decreases and the second body increases. Let us also

[1] An attentive reader may protest at this point: after all, it may seem that in classical mechanics we are also dealing with a lot of processes that cannot be reversed. For example, a ball rolls down a hill in a gravitational field, but it never rolls back to the top by itself. However, we must avoid jumping to conclusions. Classical mechanics describes the process by which a body with some potential energy (at the top of a hill) begins to lose this energy in favor of kinetic energy, increasing its speed. At the end of the rolling process, the ball has the highest speed. The process thus described is absolutely reversible if only the necessary reversal of the direction of the final velocity of the ball is applied: we are then dealing with a ball that has been given an initial velocity, enabling it to roll to the top of the hill. One may respond that we know from experience that the ball will eventually stop at the foot of the mountain – however, this goes beyond Newtonian mechanics. The kinetic energy of the ball has been converted into heat as a result of friction, and this is an irreversible process, described by thermodynamics.

note that the total amount of caloric in an isolated system cannot change – as much caloric leaves one body, that much must appear in the other body.

How to quantify a measure of body temperature? We are accustomed to using thermometers to determine the temperature of, for example, the human body. But how does the thermometer work and what does it really measure? Consider a simple thermometer based on the phenomenon of thermal expansion of a certain liquid contained in a thin tube. Such a liquid can be mercury (today withdrawn from common use for environmental reasons) or alcohol. The basis for the operation of such a thermometer is the assumption that the thermal expansion of a given material is proportional to the difference between the final and initial temperature. With this assumption, we can scale the thermometer by marking equal segments and assuming that they correspond to equal temperature jumps. Of course, there remains some doubt whether the assumption of proportional expansion of the working liquid does not entangle us in a regress, because in order to determine that the thermometer works correctly, we must use a new thermometer to check the correctness of our assumption. We have already encountered a similar problem when analyzing Newton's laws, so our solution will also be similar. We have to refer to a certain convention, but it will have a non-negligible empirical component in the form of the statement that the readings of different thermometers should agree with each other.

Temperature measurements are burdened with an additional problem that does not affect the time measurements discussed earlier. The point is that any temperature measurement involving physical contact of the object being measured with the thermometer disturbs the measured value. The thermometer receives a certain amount of heat from the object being measured, thereby reducing its temperature.[2] Of course, in practice, this effect can be made negligible by using small thermometers with near-zero heat capacity. However, it should be remembered that as a rule, the thermometer does not measure the initial temperature of the object, but the final temperature of the system of the object plus the thermometer.

The second limitation of the measurement procedure described above is that it only allows us to quantify the difference between temperatures, not its absolute value. The expansion of the liquid is (ideally) proportional to the difference between the initial temperature of the thermometer and the temperature of the body being measured. However, in practice, we use absolute values. This is guaranteed by the adoption of a conventional value of zero for a particular type of phenomena. In the case of the Celsius scale, as we know, this conventional value is the melting point of ice. Of course, there are new problems associated with this – how do we know that ice always melts at the same temperature? In fact, we know that this is not the case – the melting point of ice depends on many additional parameters, such as air pressure or chemical contamination of the ice. Scientists have put a lot of effort into "purifying" the definition of zero on the Celsius scale of factors that could affect its correctness. As a result, a temperature scale was obtained in which an unambiguous numerical value can be assigned to each material object with well-defined thermodynamic characteristics. However, it should be kept in mind that temperature can be freely rescaled using a linear transformation ($y = ax + b$); then we will get another, equally good scale (e.g. Fahrenheit). Temperature belongs to a type called interval quantities, in which only the differences make physical sense. The situation is different in the case of quantities such as mass, where the zero value is ob-

[2] Or it gives off a certain amount of heat if the initial temperature of the object was lower.

jectively "fixed" and cannot be moved. Such quantities are called ratio variables, and the only acceptable scaling transformations are multiplications by a number.[3]

Now let us deal with the mathematical relationship between heat and temperature. We know that the supply of heat to a given body results in an increase in its temperature (unless additional mechanical work has been done). What amount of heat ΔQ is needed to raise the temperature of e.g. a certain amount of water by a given interval ΔT? Experience teaches us that this amount depends on the size of the body – more precisely, on its mass (and not its volume). We also know that different substances have different abilities to "accumulate" heat. The same amount of heat (e.g. coming from exactly the same source) can raise the temperature of one gram of one substance by one degree Celsius, and of another by one and a half. Taking this fact into account requires an introduction of the concept of specific heat – an appropriate conversion factor different for different substances. To avoid problems with a vicious circle, we must arbitrarily assume the value of such a specific heat for a given substance, and then compare the behavior of other substances during the supply of the same heat and thus empirically determine their specific heats. The discussed dependence of heat on temperature is expressed by the well-known formula:

$$\Delta Q = cm\Delta T,$$

where c is the specific heat and m is the mass.

With the help of this simple formula, and using the assumption that the total heat of a closed system cannot change, we can solve easy calorimetric problems, such as determining the final temperature of a mixture of two substances with given initial temperatures and specific heats. However, the situation becomes much more interesting when we introduce mechanical work (energy) into the description.

3.2 Heat and mechanical work

We know from experience that heat can be generated by mechanical work. Rubbing your hands on a cold morning will warm you up a bit, and prolonged use of the brakes in the car will certainly increase the temperature of the brake pads, which can lead to a dangerous overheating of the brake fluid. There is also the possibility of reverse conversion, when heat is transformed into mechanical work. Consider a portion of gas enclosed by a moving piston. By heating up this gas, we will cause the piston to slide out due to the increased pressure, and the gas will take up more and more volume. It is worth noting that the end result of such a process is twofold – on the one hand, the piston has done work that can be used for practical purposes, but on the other hand, the gas has taken on an expanded form that makes it impossible to further use the heat supplied. This is a very important fact, which we will later use to formulate the second law of thermodynamics.

The problem of the relationship between heat and mechanical work was dealt with in more detail by James Joule, who was not only a skilled experimenter, but also a successful brewer. Already in school we learn that Joule proved the equivalence of heat and mechanical

---

[3] It should be added here that there is a so-called absolute scale of temperature, in which zero is also fixed. This is related to the reductive definition of temperature as a measure of the average kinetic energy of molecules, which we will talk about later. Zero temperature corresponds to zero energy (no motion), which is absolute.

energy. However, this is a slight overstatement. Joule's experiments made it possible to compare two seemingly unrelated quantities: the mechanical energy used and the heat generated as a result of the application of this energy. In Joule's device, a weight suspended on a rope was connected to a rotor, the blades of which rotated in a tightly sealed vessel with liquid (Fig. 3.1). The installed thermometer measured the increase in the temperature of the liquid. Joule calculated the energy obtained by the weight falling from a certain height, and then the amount of heat released in the liquid. On this basis, he determined a conversion rate that allows the conversion of energy measured in units of mechanical work (e.g. newtons times meters, called joules) into heat measured, e.g. in calories. However, the experiment conducted in this way does not yet prove the thesis of equivalence of energy and heat. No result of the experiment can refute our hypothesis (unless no heat is released in the liquid). Any non-zero value of the heat released is recorded, and it only affects the numerical value of the conversion rate.

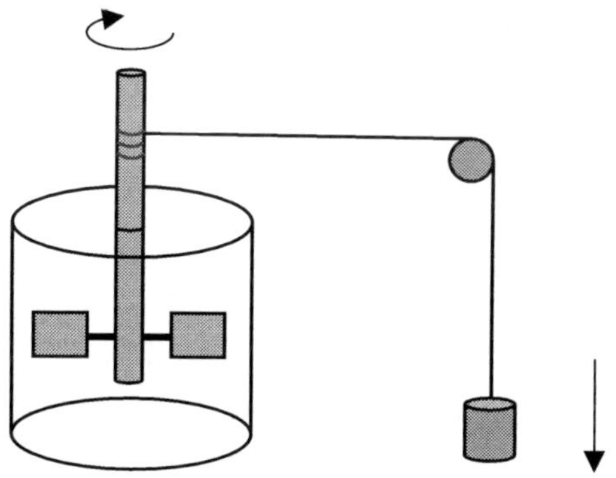

Fig. 3.1 Diagram of the Joule experiment

Does the thesis about the equivalence of work and heat have no empirical content? Fortunately, this is not the case. The empirical content of this principle is revealed when we repeat Joule's experiment using different substances. There is no *a priori* reason to believe that the amount of heat released for these substances will be the same in each case (with the same amount of weight drop). In other words, it could happen that the energy-to-heat conversion factor would be different for each substance. Then the thesis about equivalence would collapse. However, it turns out that regardless of the type of substance, the material from which the rotor blades are made, the time of day, etc., a given amount of mechanical energy is always converted into the same amount of heat. This gives us a good inductive justification for the close correlation (proportionality) between the two quantities, and since the choice of units is a matter of convention, we can assume that our two quantities – heat and mechanical work (energy) – are equal.

Mathematical equality does not mean the ability to freely transform one form of energy into another at will. Converting mechanical energy into heat is easy – you can use a Joule's device for this purpose. However, let's try to recover the mechanical energy expended in the form of the heat released in the vessel. How to do it? It is difficult to imagine a situation in which the temperature of the substance in the vessel begins to fall, the rotor blades suddenly start rotating in the opposite direction, and the weight rises to its initial height. We do not observe such processes in reality.

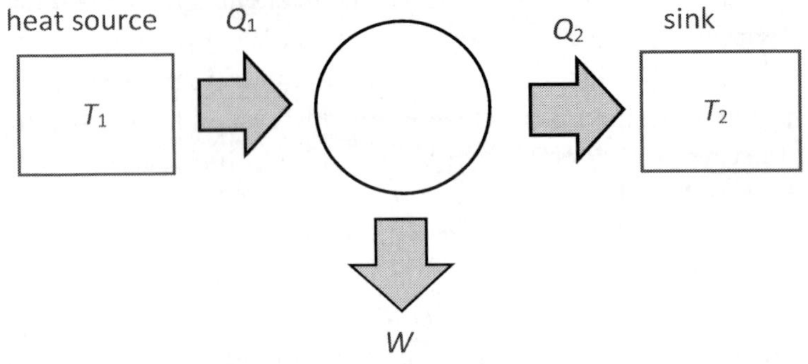

Fig. 3.2 Diagram of the operation of a heat engine

However, human ingenuity knows no bounds. The industrial revolution of the eighteenth and nineteenth centuries made it possible to create a machine with the help of which heat could be converted into useful work in a continuous fashion. The first successful design of this type was, of course, the well-known steam engine, used for both production and transport. In parallel with the improvement of the technical details of steam engines, work was carried out to create the theoretical foundations for their operation. The result is a general scheme of operation of any device capable of continuously and cyclically converting heat into mechanical work. This diagram is presented above (Fig. 3.2). The basic elements of a machine (called a heat engine) are: the working element (water in the case of steam engines), the heat source (boiler) and the sink (environment). During one cycle, the working element extracts a certain amount of heat Q_1 from the source, then performs work W and gives off a small amount of heat Q_2 to the sink/cooler (e.g. in the form of hot water vapor escaping into the environment). Then the cycle is repeated. We can now write a simple equation that is a consequence of Joule's equivalence of work and heat:

$$Q_1 = W + Q_2.$$

The percentage of heat Q_1 effectively converted into work is given by the equation:

$$\frac{W}{Q_1} = \frac{Q_1 - Q_2}{Q_1}.$$

This is the so-called heat engine efficiency. As a result of many observations and theoretical considerations, it has been concluded that the efficiency of the engine can never reach 100%; This would only be possible if no amount of heat was transferred to the sink ($Q_2 = 0$). In fact, the maximum value of the efficiency that can be achieved is given by:

$$\frac{T_1 - T_2}{T_1},$$

where T_1 is the temperature of the heat source and T_2 is the temperature of the sink (on an absolute scale). In other words, it is impossible to completely convert heat into mechanical work in one work cycle, in which the workpiece returns to the starting point and is ready to absorb the next portion of heat. The loss of some heat to the cooler is inevitable. At the same time, it should be remembered that this is not a loss that can be made arbitrarily small. For example, we know from experience that as a result of resistance or friction, every movement on earth slows down. However, this is an effect that can be made near to zero, e.g. by using lubricants. In contrast to that, there is an insurmountable barrier to the improvements of the heat engine's efficiency. It is exactly the above temperature ratio. It can be said that the practical imperfections of the machine reduce its efficiency below this value and we can fight them, but we will never achieve a result better than this ratio. We are prohibited to do so by some fundamental rule characterizing physical reality, which we will discuss in more detail in the next section.

3.3 Two laws of thermodynamics

Thermodynamics is a branch of physics that synthetically captures all the phenomena described above (and much more). Like any fundamental theory, it formulates several basic laws. The first law of thermodynamics is nothing more than the generalized law of the conservation of energy, which takes into account the fact, discovered by Joule, that heat is also a form of energy. This principle can be represented in the form of the following simple equation:

$$Q + W = U,$$

where Q is the heat supplied to a given body (isolated from the environment), W is the work done on this body, and U is the change in its internal energy. Moreover, we adopt the convention that the negative values of work and heat mean the work done by the body and the heat given off by it, respectively. The above equation can be applied to various possible situations. For example, when $U = 0$, i.e. the internal energy of the body does not change, all the work done on the body is given away by the body in the form of heat: $Q = -W$. On the other hand, when U is negative, the principle implies that the body can give off a certain amount of heat and do work at the expense of internal energy. Note that the first law of thermodynamics allows for a situation in which all the heat supplied to a body will be converted into mechanical work without changing the internal energy: $W = -Q$. But such a process, as we shall see in a moment, is subject to the limitations formulated in the second law of thermodynamics.

The second law of thermodynamics has several versions which appear to be independent, but turn out to be mutually equivalent. All of them are given in the form of a rule that prohibits a certain process from occurring. The formulation known as the Kelvin principle is based on the fact stated above that there are no 100% efficient heat engines:

> There is no physical process whose sole outcome is a complete conversion of a certain amount of heat into mechanical work.

First of all, we must highlight the expression "sole outcome" in the above formulation. Without this addition, the formulated thesis would be patently false, as the example from the previous section with the expanding gas shows. It is possible to convert all the heat supplied to the gas sample into work, provided that the final state of the gas is different from the initial one – the gas expands. Heat engines, on the other hand, are devices returning exactly to the initial state after work. In such a situation, the second law of thermodynamics forbids the total use of heat – a certain part of it must be dissipated into the environment.

Clausius' formulation of the second law refers to the process of heat transfer between bodies with different temperatures. We know from experience that bodies with a higher temperature can transfer heat to bodies with a lower temperature. Is the opposite situation possible? Clausius' version confirms what we know well from everyday experience:

> There is no physical process whose sole outcome is the transfer of heat from a body with a lower temperature to a body with a higher temperature.

Again, it is extremely important to emphasize that we are talking here about a process whose only result is this heat transfer. Otherwise, Clausius' principle would exclude the existence of refrigerators, which, after all, collect heat from cold objects and transfer it to a hotter environment. However, this happens at the expense of the energy supplied, which must have its source somewhere. The second principle in Clausius' version excludes only the "spontaneous" flow of heat from cold to hot bodies, without external intervention.

The Kelvin and Clausius versions look different, but it can be proven that they are equivalent. To illustrate, we will show how to derive the first version from the second in a fairly simple way. So let us assume that the first version is false, i.e. there is a way to convert all heat into mechanical work without any additional change in the physical system. It is easy to see that this would allow heat to be transferred from a cold body to a hot body. Let's take a colder body and apply a hypothetical 100% efficient heat engine to it. It will take away a certain amount of heat from this body and convert it without loss into mechanical work, e.g. the sliding motion of a piston. The complete conversion of the obtained amount of mechanical work into heat is a simple matter – it is enough, for example, to use a suitably modified Joule device for this purpose. In this way, we obtained the final effect of cooling the cold body and warming the hot body without any additional changes in the environment. Clausius' formulation of the second law of thermodynamics turns out to be false, which shows that Clausius' version implies Kelvin's version. It can also be shown that the assumption that the principle in Clausius's version is false entails the falsity of the Kelvin formulation, but it is a little more complicated. In any case, the two principles, although they sound different, are equivalent.

3.4 The second law of thermodynamics and entropy

The above formulations of the second law of thermodynamics are not widely known, although many readers have probably encountered them before. However, I am sure that everyone has heard about the third, most famous phrase:

> The entropy of an isolated system never decreases (increases or remains unchanged).

This version introduces the concept of entropy, which we will have to take a closer look at. It is often said that entropy is a measure of disorder, and thus the second law of thermodynamics states that the disorder of an isolated system increases or at most remains the same (when it reaches its maximum value). This is true, of course, but disorder is a statistical concept, not a thermodynamic one. We will deal with the statistical definition of entropy later, but for now we will introduce a slightly less known concept of thermodynamic entropy.

Let us go back to our analysis of heat engines for a moment. The above considerations show that the amount of heat supplied to a given engine is not the only parameter determining the mechanical work obtained. The efficiency of the engine in converting heat into work depends on the temperature, and more specifically on the temperature difference between the heat source and the sink. If the difference is small, then only a tiny fraction of the heat supplied can be used effectively in the form of mechanical work. This means that heat can come in different "forms", more or less useful. The amount of heat (or, more correctly, internal energy) contained in a pot of boiling water is much more useful than the same amount of heat in a bathtub of lukewarm water. Entropy in the thermodynamic sense is exactly the quantity that is related to the usefulness of a given system in converting heat into work. More precisely, entropy is the inverse of utility – the greater the entropy, the less useful the system and the heat it contains.

It should therefore come as no surprise that the formula expressing thermodynamic entropy should take into account temperature in such a way that a higher temperature implies less entropy. The exact definition of entropy is a bit complicated, but it is worth putting some effort into understanding it. At the outset, let us emphasize that we will not define entropy as such, but only the difference in entropy between two states of a given system. Thus, thermodynamic entropy is an interval quantity, just like temperature. Determining, for example, the value of zero entropy is a matter of convention – we can do it, but we do not have to. Let us therefore choose two states A and B of a given physical body (e.g. gas). The states A and B may differ in the value of macroscopic parameters – pressure, temperature, volume, density. There are infinitely many ways to get from state A to state B. From this multitude of ways, let us select those that represent a certain reversible process, i.e. one that can occur in both directions. How do you know if a process is reversible? A useful rule of thumb here is that the process must be slow. For example, removing a wall in a vessel containing pressurized gas will cause a rapid expansion of the gas, which is certainly not reversible – the gas will not spontaneously "push" itself into the vessel in which it was previously enclosed. On the other hand, if we connect a piston to the vessel and pull it out slowly, such a process can be easily reversed by applying the reverse force on the piston and compressing the gas.

Let us therefore consider such a slow, controlled, and reversible process that begins in state A and ends in state B. Let us divide this process into small fragments in such a way that the body temperature is approximately constant in each of them. For each such part, let us calculate the quotient of the heat ΔQ absorbed by the system and the body temperature T: $\frac{\Delta Q}{T}$. Adding up these quotients over all parts of the way from A to B, we get the entropy difference between the final state B and the initial state A. We denote it by S_{AB}. So we have the following formula:

$$S_{AB} = \sum_i \frac{\Delta Q_i}{T_i},$$

where index i numbers individual fragments into which we have divided the whole way from A to B. By choosing smaller and smaller fragments, we will get the exact value of entropy within the limit, which is expressed by the mathematical operation of integration:

$$S_{AB} = \int_A^B \frac{dQ}{T}.$$

Although the definition of entropy refers to a specific path (process) from state A to B, it can be shown that the result is independent of the details of this path, as long as it is reversible.

In other words, the magnitude of S_{AB} is the same for every reversible path from A to B. We can therefore assume that S_{AB} characterizes the initial and final states.[4]

It may be useful for some readers to explain in greater detail an intuitive interpretation of the important mathematical operation of integration, which we will encounter over and over again in this book. Suppose that we have a function *f*(*x*) of real numbers defined on an interval from *a* to *b*, whose graph is drawn in Fig. 3.3. Our goal is to calculate the area below the curve. This can be done as follows: we can divide up the interval (*a*, *b*) into small sections, each of the length Δx. For each such section we consider the rectangle whose height is the average value of the function *f* within the section. In other words, we replace a smooth continuous graph with a bar graph, well known from statistics. Then the first approximation of the area under the curve will be the sum of the areas of the selected rectangles (bars). This can be written algebraically as $f_1(x)\Delta x_1 + f_2(x)\Delta x_2 + \cdots + f_n(x)\Delta x_n$, where $f_1(x)$ is the height of the first bar, and Δx_1 its width. Or, we can write this sum in an abbreviated form as $\sum_{i=1}^{n} f_i(x)\Delta x_i$. Now, the smaller the sections Δx_i are, the better the considered approximation is. The integral of the function *f*(*x*) in the interval from *a* to *b* is the limit of the sequence of the sums $\sum_{i=1}^{n} f_i(x)\Delta x_i$ when *n* goes to infinity (and thus sections Δx_i must tend to zero). The notation used to express this operation is very intuitive: we replace the sign of the (discrete) sum Σ with a "stretched" sigma symbol $\int$, and we substitute the infinitesimal (that is, arbitrarily close to zero) interval *dx* for Δx_i, while dropping the unnecessary index *i*. The result is the definite integral $\int_a^b f(x)dx$ from *a* to *b*, which we can still think of as a long, infinite sum of appropriate infinitesimal fragments.

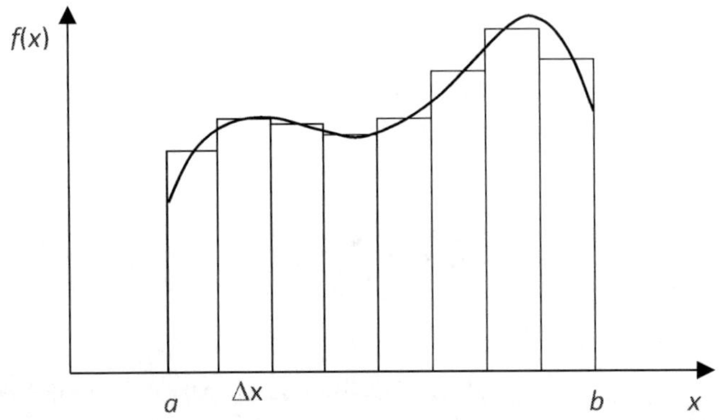

Fig. 3.3 Integration of a given function

Let us use a simple example to illustrate the concept of thermodynamic entropy that has been introduced. Consider a gas enclosed in one part of a vessel with the other part empty. When the barrier is removed, the gas will spontaneously expand, occupying the entire volume of the vessel (Fig. 3.4). What will be the entropy difference between the final and initial state? On the surface, it may seem that the entropy change is zero, as no heat is supplied or extracted.

[4] Of course, we must also assume that for every two states there is a reversible process connecting them.

(The temperature of the gas also does not change – this is due to the simple fact that removing the barrier has no mechanical effect on the movement of the molecules, and thus their average kinetic energy remains unchanged.) However, this is a wrong conclusion. As we have already noted, spontaneous gas expansion is not a reversible process. To correctly calculate the entropy change, we need to find another, reversible way from the initial to the final state.

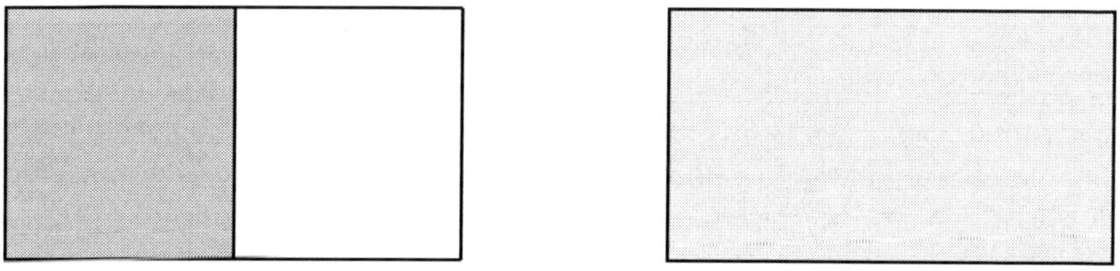

Fig. 3.4 Gas enclosed in one half of the container (left) has less entropy than gas expanded in the entire volume (right).

One of the possibilities is as follows. Instead of a barrier, let us introduce a movable piston and start pulling it out slowly, thus increasing the volume of gas. We know from experience that such a process results in a cooling of the gas (this can be justified using the molecular theory of gas – gas molecules collide with the piston moving away from them and therefore lose parts of their momenta). In order to maintain the stability of the temperature in the final state, we must put the container into contact with a heat source that will make up for the lost heat. However, this means that during the entire process, the amount of heat supplied to the system will be non-zero (and also positive, since the heat is absorbed by the gas). Dividing this number by the constant value of temperature T, we will get a positive value of the entropy difference S_{AB}. Thus, as a result of the spontaneous expansion of the gas, its entropy increases.[5]

How do we know that the second law of thermodynamics in the entropic version is equivalent to the other two formulations (Kelvin and Clausius)? Again, we will not prove this generally, but we will show that the principle of increasing entropy implies the Clausius version (and thus also the Kelvin version, since we have previously proved that "Clausius" implies "Kelvin"). So let's assume, contrary to what Clausius' version says, that it is possible to transfer heat from a colder body to a hot body without any additional changes. In fact, this implies that it is possible for two systems with the same temperature to spontaneously change their temperature – the temperature of one of them will increase and the other will decrease.[6] What is the entropy difference between the state of a system in "thermal equilibrium" and a state in which one part of the system is at a higher temperature than the other? It is not difficult to show that the entropy of the first state (equilibrium) will be higher than the entropy of the second state (non-equilibrium).

---

[5] Let us remember that strictly speaking entropy characterizes states, not connecting pathways or processes (the use of reversible processes of arrival in the definition is only a tool that can be rejected in the end, like Wittgenstein's ladder). However, informally we often talk about the entropy of a process.

[6] Try to prove yourself that if Clausius' principle is false, such a process is perfectly possible.

To prove this, let us consider two bodies with temperatures T_1 and T_2 respectively, where $T_1 < T_2$. When they come into contact, the final temperature will be equal to T_3, with $T_1 < T_3 < T_2$. How can this temperature equalization process be "simulated" with a reversible procedure? In order to cool a body with a higher temperature T_2 in a controlled manner, it can be brought into contact with a gas enclosed in a container with a piston, and the piston can be pulled out slowly. The temperature of the gas will fall, so it will start to draw heat from our body until it reaches the desired temperature of T_3. We will do the same with the cooler body – we will place it in contact with the gas in the vessel, but this time we will start pushing the piston. The heat released in this way will heat our body to the temperature of T_3. The change in entropy in the case of the colder body will be positive, because the heat is absorbed, while the change in entropy in the hot body is negative – the heat is released to the environment. The entropy of the entire system is assumed to be the sum of the entropy of individual parts. But what will be the sign of the sum of a positive number and a negative number? This, of course, depends on their absolute values. Note that when calculating, for example, the change in entropy of the hotter body, at each step we have to divide the heat supplied by a number in the range from T_3 to T_2. In the case of the cold body, the denominator is the number between T_1 and T_3. Since in both cases the total amount of heat absorbed and given off is the same, the absolute value of the entropy change for the hotter body will be lower than the absolute value of the entropy change for the cooler body. Thus, in the end, the change in entropy during temperature equalization is a positive quantity – the thermal equilibrium state is characterized by a higher thermodynamic entropy than the non-equilibrium state.[7]

However, as we have already indicated, if the second law in Clausius' version were broken, a spontaneous process would be possible leading from a state of equilibrium with higher entropy to a state of non-equilibrium with lower entropy. In such a case, the principle of the non-increase of entropy would also be violated. This proves that the entropy-based principle implies the principle in Clausius' formulation.[8] The second law of thermodynamics in the entropy version is clearly asymmetric in time – processes permissible in one direction are excluded when the time's arrow is reversed. We will leave a more detailed analysis of this phenomenon of nature for later, but for now we will take a closer look at the interpretation of thermodynamics in terms of the molecular theory of matter.

On the occasion of the analysis of the meaning of the second law, we can ask the following question of a more philosophical nature: would a world in which it would hold that entropy decreases or remains unchanged in all isolated physical systems be fundamentally different from ours? On the surface, it seems so. However, the concepts of "increasing" and "decreasing" *implicitly* assume that time already has a certain direction from smaller to larger values. A given function F of time (e.g. the entropy function) is increasing when, for any time values $t_1 < t_2$, $F(t_1) < F(t_2)$. But how do we know that *time* t_1 is "earlier" than t_2 and what does it even mean? There is a hypothesis according to which the temporal order of moments is secondary to the entropy order: moment t_1 will be considered

[7] It is interesting that in this case we would come to a similar conclusion even if we considered the fundamentally incorrect irreversible way from the initial to the final state. However, the numerical value of the "entropy" calculated in this way would differ from the true value.

[8] Note that we have already proven two implications: Entropy → Clausius and Clausius → Kelvin. To prove the equivalence of the three formulations, it is enough to show that there is a third implication that "closes" the whole cycle: Kelvin → Entropy.

earlier than t_2 when the entropy of the universe at t_1 is less than the entropy at t_2. In this case, however, it is incorrect to describe a world in which entropy is always decreasing: entropy by definition must increase. If we think that entropy is decreasing, it is only because we have reversed the order of time. Of course, here comes our good friend – the problem of the empirical meaning of the second law of thermodynamics. Does defining the temporal order by means of the universal increase in entropy reduce the second principle to the status of a mere convention (an *a priori* analytic proposition)? The matter is not clear-cut. A world in which this principle would be broken seems certainly possible. This would be a world in which some isolated processes would increase entropy, and some would decrease it. On the other hand, the world in which entropy is always decreasing is simply our world, only wrongly described, in which we have exchanged the future with the past.

3.5 Reduction of thermodynamics to mechanics

The idea that all matter is made up of small, invisible, and indivisible components sprouted in the minds of humans very early. The ancient Greek and Roman atomists (Democritus, Leucippus) based their hypothesis on observations of pollen floating in the air, but the essential part of their atomistic doctrine was pure speculation without empirical confirmation. It should be emphasized, however, that even in its naïve version, the atomistic theory made it possible to explain a wide range of phenomena, from changes in the states of matter of bodies to acts of perception. This ability of atomism to synthetically capture and explain the multitude of observed phenomena was a strong impulse to develop its modern version. As we will see in a moment, the atomistic model of the structure of bodies gives us an opportunity to see thermodynamic processes in a completely new light.

The basic assumption of the atomistic (or better molecular) theory of physical bodies is, of course, ontological (substantial) in nature. It states something about the structure of material objects, namely that they cannot be divided indefinitely. At the base of every physical body lies a vast number of identical elemental components—molecules or individual atoms. Therefore, we are dealing here with an ontological reduction through the identification of material objects with a collection of particles. The next step in the reduction procedure may be an attempt to characterize the observable or measurable properties of bodies in terms of the properties of the groups of molecules that make up these bodies. It should be noted, however, that the mere assumption that bodies are composed of molecules does not guarantee the success of such a reduction. It is conceivable that the right combination of the basic elements will result in a structure whose characteristics cannot be reduced to a suitable combination of the properties of these elements. Such irreducible properties of complexes are called emergent properties. In the history of science, for example, the idea that life processes cannot be completely reduced to the physicochemical properties of molecules that make up a given organism has achieved some popularity. With the increasing complexity of physical objects, from simple atoms, molecules (e.g. DNA), cells, to whole organisms, new characteristics appear, the existence of which cannot be deduced from the properties of the constituent elements.

However, the reductionist trend in physics has always been very strong. A spectacular example of this is the science of heat. The assumption of the molecular structure of matter was the starting point for scientists to interpret thermodynamic properties and processes in terms of mechanical interactions between molecules that make up a given substance. With the rejection of the substantial concept of heat in the form of the caloric, the idea of the emergent character of thermal properties of bodies was also abandoned. Efforts have been focused on redefining known thermodynamic concepts in terms of collections of molecules and their mechanical behavior. Let us start with the fundamental concepts of heat and temperature. As is well known, both of these quantities are interpreted in terms of the kinetic energy of particle motion, which is standard in classical mechanics. One of the fundamental assumptions of the molecular theory of matter is that the molecules that make up a given substance are in constant motion, and therefore their kinetic energy is non-zero. The total kinetic energy of all molecules is a measure of heat – or, more correctly, the heat taken up by a given body is equal to the increase in the total kinetic energy of the movement of the molecules. Body temperature, on the other hand, is defined as the average kinetic energy. Due to the chaotic nature of the internal motion of particles, different particles will have different kinetic energy (in this case, we are talking about the statistical distribution of velocity, and therefore also kinetic energy). Averaging this distribution, we get the absolute value of the temperature of a given body. Thanks to the adopted definition, we can non-arbitrarily determine the zero value of temperature, which corresponds to the cessation of motion (due to quantum effects, such a cessation is physically impossible, but we can get close to it).

Let us pay attention to one important aspect of the proposed definition of temperature – namely its statistical nature. As it turns out, reducing thermodynamic concepts to purely mechanical ones is not feasible for quite obvious reasons – we are not able to trace the motion of every individual molecule. Therefore, we must rely on statistics, which uses the concepts of averages, distributions, probability, etc. The theory to which we will reduce the science of heat is called statistical mechanics. This is in fact a new theory in relation to Newtonian mechanics. Of course, it contains well-known elements of the latter, but it also introduces completely new concepts. Moreover, statistical mechanics makes some additional assumptions, often in a hidden way, about which one may doubt whether they are justified. We will return to this matter later in the chapter.

The reductive definitions of temperature and heat easily explain the relationship between these two concepts, described in the first section of this chapter. Since the mass of a given body is proportional to the number of its constitutive particles, the product of the mass of the body and the average kinetic energy of its molecules gives us a measure of the total energy, i.e. heat (remember that the average energy is the sum of all the energies of the particles divided by their number). This simple example illustrates another important step in any successful reduction of one theory to another. After defining the basic concepts of the reduced theory in terms of the more fundamental theory, the next step is to derive the laws of the reduced theory from the laws of the reducing theory, using the definitions introduced earlier. Ideally, a given law of the reduced theory, after replacing all the terms in it with their definitional equivalents in the reducing theory, should turn into a law of this basic theory, or at least into a logical consequence of such laws. In practice, the matter is sometimes complicated, but the main goal is as I have described.

Following the way outlined above, we can also reduce the first law of thermodynamics to a well-known law of classical mechanics, which is the principle of the conservation of energy. Since the transferred heat is the kinetic energy supplied to the body, it follows automatically from the conservation of energy that the kinetic energy plus the work done on the body must result in an increase in the internal energy of the body. This reduction does not require the use of any strictly statistical concepts. On the other hand, in order to derive the other laws of thermodynamics, such as the laws governing the behavior of ideal gases (the equation of state of an ideal gas, combining temperature, volume and pressure), we need to reduce another concept, which is pressure. The pressure exerted by the gas on the walls of the vessel is the result of the bombardment of the walls by gas molecules. As a result of the collision with the wall, each particle changes its momentum, transferring a part of it. The change in momentum divided by the duration of the interaction gives us the force with which the particle acts on the wall (we know this from Newton's second and third laws of motion). On the other hand, force per unit area is called pressure. Of course, the total pressure will be the sum of all the changes in momentum of the molecules hitting the walls per unit time and unit area. From the additional assumption that no direction of motion of particles is distinguished (i.e. the average number of particles and their velocity in each direction is the same), Pascal's law of hydrostatics follows at once, stating that the pressure exerted on the walls of the vessel is equal in all directions.

Using statistical definitions of temperature and pressure, we can derive the ideal gas law, which is well known from thermodynamics. You may remember this derivation from a school course in physics, so I will only present a general sketch to capture the essence of the statistical approach. We consider a gas enclosed in a cube-shaped container whose side has length l (this is an assumption that simplifies the reasoning). We assume that the collisions of molecules with the walls of the vessel are perfectly elastic, i.e. a molecule with a velocity component v_x perpendicular to a given wall will change this component to the opposite one of the same value. This gives a momentum change of $2mv_x$. Assuming that v_x is an averaged value, we can conclude that the average time between two collisions with a wall is $\frac{2l}{v_x}$, i.e. the number of collisions per time unit is $\frac{v_x}{2l}$. Thus, the total change in momentum in the x-direction per the unit of time is equal to the change in momentum in one collision times the number of collisions, so it is $\frac{mv_x^2}{l}$. But the change in momentum per time unit is nothing but the force (from the second law of dynamics) that divided by the surface area will give us the pressure exerted by one particle on one wall of the vessel. If we multiply this by the number of particles N, we get the expression for pressure: $\frac{Nmv_x^2}{l^3}$. Taking into account the formula for kinetic energy and assuming that the average velocity in each direction is the same, while replacing l^3 with the volume of the vessel V, we get:

$$pV = \frac{2}{3}N\langle E_{\text{kin}}\rangle,$$

where $\langle E_{\text{kin}}\rangle$ is the average kinetic energy. Since average energy is nothing more than a measure of temperature, we finally get the functional relationship between pressure, temperature, and volume known from experiment.

It can be seen, then, that in order to derive the macroscopic law of ideal gases, we needed both the principles of classical mechanics (the second law of motion, the dependence of energy on velocity, etc.) and statistical assumptions (the same average velocity of particles in each direction). As we have already said, we are not able to determine the velocities and positions of individual particles for a simple reason – we cannot even identify individual particles due to their too small size and unimaginable large number. However, we can introduce a description of the particles that will contain more information than the average values of the given quantities (energy, velocity). Such a description may include the statistical distribution of a given quantity in the entire population, i.e. the information about the relative number of particles that have a given value of the quantity under consideration. Statistical distributions are commonly used in the analysis of human populations – e.g. we can talk about the distribution of individual income (how many people earn income in a given range). In statistical mechanics, the issue of velocity (and thus energy) distribution comes to the fore. The question that the researchers asked themselves was as follows. Consider a gas in a thermodynamic equilibrium, i.e. one whose parameters (pressure, temperature, volume) do not change. What function will correlate the number of particles with their speed?

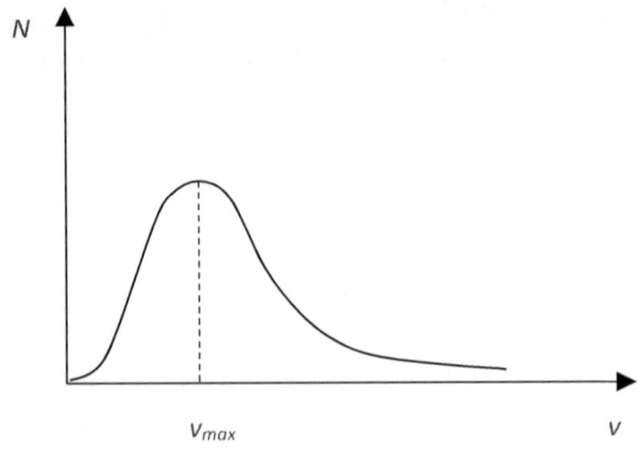

Fig. 3.5. Maxwell's velocity distribution. This distribution determines the number of particles having a given velocity v in any direction

This issue was solved independently by James Clerk Maxwell and Ludwig Boltzmann. Using only general assumptions, such as the anisotropy of a gas sample (the average velocities in each direction are the same), they derived a formula to determine what fraction of the total number of particles will be characterized by velocities in a small (infinitesimal) range from the value of v into $v + dv$. This formula is called the Maxwell (or Maxwell-Boltzmann) distribution. Function $f(v)$ of the Maxwell distribution takes a value of zero for the velocity of $v = 0$, then it increases to the maximum value for a certain velocity v_{max} and begins to decrease to zero at infinity (Fig. 3.5). It can be concluded that in a given gas sample, the number of particles with a velocity close to zero converges to zero (but compare the entry in the box). Speeds far exceeding the value v_{max} can occur, but they are statistically very unlikely.

The problem of statistical velocity distribution becomes slightly more complicated when we take into account that velocity is a vector, not a scalar. To consider this issue a little more precisely, let us introduce an abstract three-dimensional velocity space, in which each point represents a velocity vector with the origin placed at the origin of the coordinate system (Fig. 3.6). The statistical distribution of velocities can be illustrated by shading this space in a way that reflects the number of particles of a given velocity (the greater the shade, the more particles are in a given area). Let us now ask the question about the density of the area near a selected velocity vector with three components v_x, v_y, v_z – this area can be represented in the form of a small cube attached to the tip of the vector. The assumption of anisotropy implies that the function representing this density should depend only on the length of the vector, and not on the specific values of its components (i.e. its direction). On the basis of general considerations referring to the assumed symmetries, it is possible to derive a formula characterizing the density of the distribution of particles with velocities (vectors) in the vicinity of a given vector. The distribution function takes the form of a decreasing exponential function from the square of the length of the vector $v = \sqrt{v_x^2 + v_y^2 + v_z^2}$ (this is the so-called Gaussian distribution, or normal distribution):

$$g(v_x, v_y, v_z) = \alpha e^{-\beta v^2}.$$

The maximum of this distribution is at $v = 0$, and then its values fall (the so-called bell curve – see Fig. 3.7).

In order to determine the distribution of particles with given velocities treated as scalars, considered in the main text, it is necessary to calculate the number of particles in a spherical shell with a thickness *dv* and radius *v* (this shell contains all velocity vectors with a length between *v* and *v* + *dv*). We will do this by multiplying the above function *g* by the area of the sphere $4\pi v^2$. As a result, we obtain the Maxwell distribution, whose maximum is for some non-zero value of *v*, and whose value at *v* = 0 is zero:

$$f(v) = 4\pi\alpha v^2 e^{-\beta v^2}.$$

It is important to emphasize that the functions *g* and *f*, although they both depend on *v*, have different physical meaning. The first one determines the probability (more precisely: the probability density) of finding a particle moving with velocity *v in a given direction*, and the second one determines the probability that a given particle will have velocity *v* in *any* direction.[9] But where does the paradoxical discrepancy between the values of the functions *g* and *f* in zero come from? What is our conclusion about the number of particles moving at very low speeds? The non-zero value of function *g* at *v* = 0 is, in a sense, a mathematical artifact. It results from the fact that for very low velocities bordering on 0, their direction becomes irrelevant – a small cube placed infinitesimally close to zero in practice covers all directions of motion. On the other hand, for large *v*, function *g* chooses only one of an uncountable number of possible directions, so it is not surprising that the value of the function decreases with increasing *v*. In the limit $v \to 0$, the direction of the vector ceases to make physical sense, and this is the reason for the "anomalous" counting of particles with infinitesimally low velocity by the function *g*. Physically, it remains true that the number of particles with velocities closer and closer to zero decreases to zero, as described by the Maxwell function *f*.

---

[9] A mathematically attentive reader may ask why the functions *g* and *f* have different dimensions, since it is clear that in the latter we multiply the entire expression by the square of the velocity of dimension $\frac{m^2}{s^2}$? This is because these functions determine probability densities in different areas. In order to obtain a dimensionless probability from the function *g*, it must be multiplied by an element of the volume of $\Delta v_x \Delta v_y \Delta v_z$ of dimension $\frac{m^3}{s^3}$, while the function *f* represents the density in the volume Δv, whose dimension is $\frac{m}{s}$.

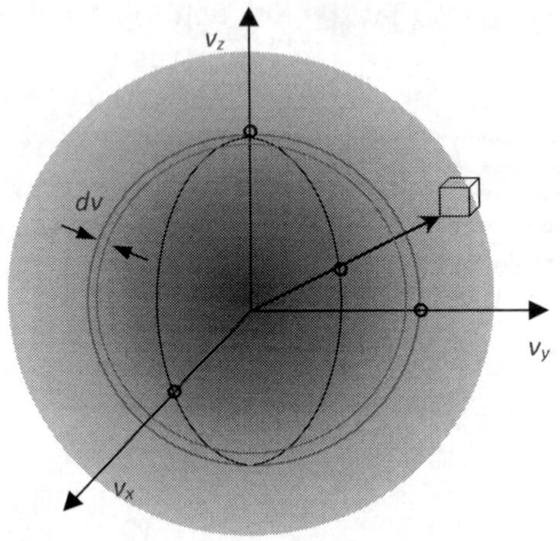

Fig. 3.6. Velocity space with an area of a given vector velocity (cube) and given scalar velocity in any direction (sphere shell)

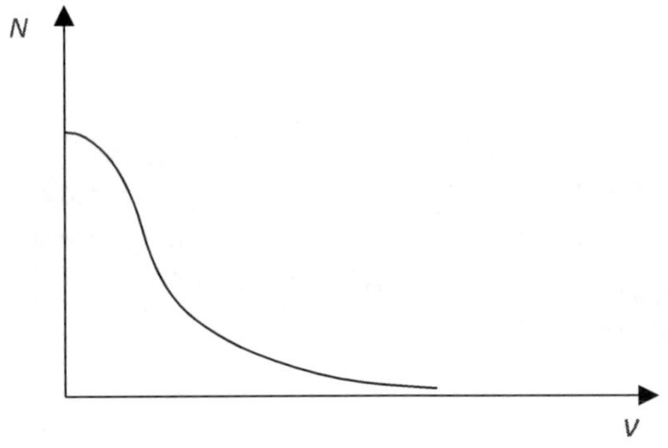

Fig. 3.7. The distribution of velocity in a specific selected direction. Although the distribution function depends only on the length of the velocity vector, its meaning is different from that of the Maxwell distribution

Let us summarize the conditions for a successful reduction of theory T_1 to another theory T_2, which we have been able to determine on the basis of the case of thermodynamics and statistical mechanics. We can collect them in the form of points:

1. *Object reduction*. In the first step, we identify the objects described by the reduced theory T_1 as the corresponding constructs from the objects of the theory T_2. In the case of heat science, we identify macroscopic physical bodies (gases, liquids, solids) with complexes consisting of a huge number of molecules.

2. *Reduction of properties*. The basic properties described by theory T_1 must be redefined with the help of the terms used in theory T_2. Thermodynamic concepts such as pressure, temperature, and heat are defined in terms of classical mechanics (and statistical terms) – mean kinetic energy, mean change in momentum, etc.

3. *Reduction of laws*. The laws L_1 of the reduced theory T_1 should be derivable from the laws L_2 of the reducing theory T_2, after replacing the terms used in L_1 by the terms of theory T_2 as described in point 2. It is permissible to use additional assumptions containing both some terms of theory T_1 and some terms of the theory T_2 (e.g. assumptions of a statistical nature). Such assumptions are often referred to as "bridge laws". An example of a reduction of laws is the derivation of the first law of thermodynamics from the principle of conservation of energy, or the law of ideal gases from Newton's laws of motion and statistical assumptions.

We still have to discuss how to derive the second law of thermodynamics from the laws of statistical mechanics. For this purpose, however, we will need an additional conceptual apparatus, which we will introduce in the next section. In it, we will also encounter the fundamental problem of temporal irreversibility.

3.6 The second law of thermodynamics in statistical mechanics

We will start this section by introducing the extremely important concepts of macrostate, microstate and phase space. By the macrostate of a given physical system (e.g. a gas sample) we understand a complete set of macroscopic values of measurable quantities characterizing this system. Examples of such quantities are well known: pressure, temperature, volume, heat (given off or absorbed), density, and also entropy. On the other hand, the microstate characterizes the molecules that are parts of the selected system. We treat these particles as impenetrable point bodies that interact with each other only mechanically, and thus can be completely characterized by giving their positions and velocities (or momenta). The microstate of the system assigns to each individual particle its exact position and momentum.

Let us gather some fundamental facts about microstates and their relationship to macrostates. Firstly, microstates – unlike macrostates – are not available to us in practice. We have no way of observing individual particles due to their small size and their huge, unimaginable number. Thus, the microstate is in a sense a useful theoretical construct that has no direct experiential basis. Secondly, a given microstate unambiguously determines the appropriate macrostate. Due to the assumption that macroscopic parameters are reducible to the properties of molecules, we must assume that the exact values of the positions and velocities of the particles uniquely fix everything that can be measured in relation to a given sample. However, an analogous relationship does not hold in the opposite direction. For a given macrostate, there is an enormous number of microstates implementing it. This is because macrostates operate with averaged quantities, and the average values may remain the same for different actual distributions. The pressure or temperature of the gas may not change if, for example, we modify the individual values of the velocity of the molecules accordingly.

Thus, the relationship between microstates and macrostates is many to one. Microstates and macrostates can be presented in a convenient way using the concept of phase space introduced by us in the previous chapter. The phase space for the group of N particles is a $6N$-dimensional space in which each point is characterized by coordinates representing numerically all the components of the position and the momentums for all the particles. Since each individual particle requires six numbers to describe its state (three position coordinates and three momentum components), we get the number of $6N$ independent parameters.[10] Thus, points in phase space are simply microstates. And what about macrostates? Since many microstates can correspond to the same macrostate, it is natural to represent macrostates in phase space as areas containing a sufficient number of adjacent points. In other words, the phase space breaks down into a large number of cells, each corresponding to a specific set of macroscopic parameter values. An important question arises: how do the sizes of individual cells relate to each other – are they more or less the same size or are they different from each other? We'll leave this question unanswered for now, but we'll come back to it in the next section.

As we remember from the previous chapter, classical mechanics in the Hamiltonian version represents the evolution of a system of many particles in the form of a trajectory (curve) in phase space. At each point in phase space, an abstract vector is attached, the length and direction of which is given by Hamilton's equations, and which sets the direction for the evolution of the system whose state occupies this point. Due to the determinism of classical mechanics, exactly one trajectory comes out of each point. This implies at once that trajectories in phase space cannot intersect with each other. By considering all the trajectories coming out of a given macrostate (a given cell), we can trace (of course only theoretically) the macroscopic evolution of a physical system. In classical mechanics, an important Liouville's theorem applies, which states that the volume of a given region subjected to a Hamiltonian evolution does not change. Therefore, a given macrostate must be transformed as a result of the evolution of the system into an area of the same size. However, it does not have to be an area of the same "shape" – it can be, for example, very fragmented.

Now we can present the problem related to the irreversibility of the second law of thermodynamics in a slightly more precise way. As we remember, this principle states that there is a certain macroscopic parameter – entropy – that cannot decrease over time for an isolated system. This means that if we plot trajectories coming out of a cell representing the initial macrostate of the system, they should only pass through cells corresponding to higher or equal entropy.[11] However, we know from classical mechanics that any trajectory representing the evolution of a given system can be reversed in the sense of agreement with Newton's laws of motion. If a trajectory in phase space connects two points A and B, this implies that both evolutions from A to B and from B to A are physically admissible. Consider a trajectory leading from some point in a low-entropy macrostate cell to a point in a higher entropy macrostate. Classical mechanics allows for the reversed process, which, however, would lead from a state with higher entropy to a state with lower entropy, and this is forbidden. (The physical realization of such a process would consist in a momentary reversal of all instantaneous velocities of the particles, in which case the particles would retrace their past paths to the starting point.) So we have a serious problem that calls into question the entire program of reducing thermodynamics to classical mechanics.

---

[10] As a matter of fact, in the case of thermodynamic systems this number may be smaller, since we have to impose certain additional constraints on the available microstates, such as the condition that the total energy remains constant. Since such a condition is given in the form of an equation connecting all positions and velocities of the particles, the number of independent parameters will be less than $6N$.

[11] As we shall see, this is not exactly fulfilled in reality. This condition would be satisfied only if the second law of thermodynamics were a non-exceptional law, which is not the case. For now, however, we assume that the laws of thermodynamics apply strictly.

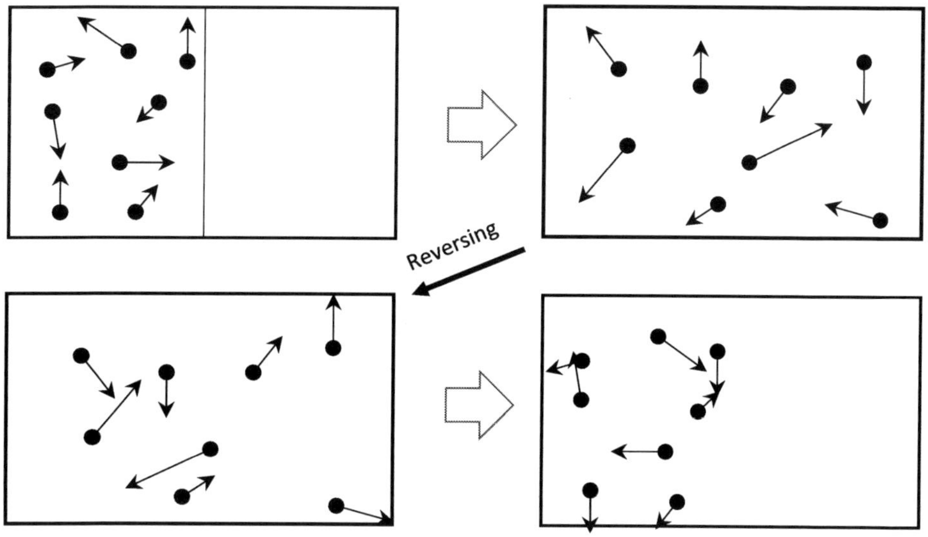

Fig. 3.8. Loschmidt's argument from the reversal of velocities

The problem of the reversibility of microscopic processes and its incompatibility with the second law of thermodynamics was noticed at an early stage of the development of statistical mechanics. The argument for the theoretical possibility of reversing thermodynamic processes by "flipping" velocities is called Loschmidt's argument after the name of its creator. Imagine a situation in which a gas initially enclosed in a certain part of the vessel begins to spontaneously expand after the barrier is removed. At some moment when the gas occupies the entire volume, each molecule has an appropriately directed velocity vector. Let us now consider a state in which every particle has the same velocity in value, but pointing in the opposite direction. It is true that we cannot practically create such a state, but it seems that its existence is absolutely possible according to the fundamental laws of nature. In this situation, the molecules will begin to "retract" to their previous state, and in the end the entire gas sample will occupy half of the vessel without external intervention (Fig. 3.8). This, however, violates the second law of thermodynamics.

Another spectacular counterexample to the universal validity of the second law was presented in an argument known as Maxwell's demon (with a rather obvious reference to Laplace's demon). Consider a box consisting of two parts separated by a partition (Fig. 3.9). In one part there is a gas at a higher temperature and in the other part a gas at a lower temperature. An opening was made in the partition covered with a shutter. The shutter opening mechanism is designed in such a way that if a slow-moving particle approaches from the colder gas, the shutter remains closed. On the other hand, if the particle has a sufficiently high speed, the aperture opens, allowing the high-energy particle to pass through[12] (Due to the obvious technical difficulties involved in making such a mechanism, its operation is somewhat jokingly entrusted to a demon with superhuman abilities.) As a result of the setup described above, the hotter gas will acquire a number of high-energy particles, which will increase its

[12]Note that according to the Maxwell distribution (Fig. 3.5), even in a cold gas there will be particles of very high velocity, although their number will be small compared to the number in a hotter gas.

temperature, while the colder gas will cool even more. This, however, is inconsistent with the second law of thermodynamics in Clausius' version.

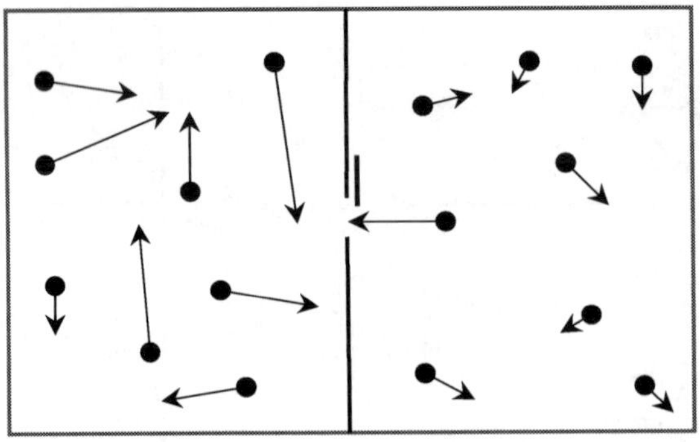

Fig. 3.9. Maxwell's demon. The shutter is opened when a high-energy particle approaches from the colder portion of gas (right side)

Maxwell's argument is based on the hypothesis that the demon's action described above does not violate any fundamental principle of classical mechanics. There is an even more convincing argument against the second law, which shows that breaking it is not only possible, but certain. This argument uses the so-called Poincaré recurrence theorem. Henri Poincaré proved in a mathematically exact way that every trajectory in phase space undergoing Newtonian evolution must necessarily pass arbitrarily close to the starting point at some future moment. The time it takes for such an infinitesimally close return is of course unimaginably long, but it is mathematically certain that it must happen at some point. Therefore, in order to show that the entropy of a system can spontaneously decrease, we do not have to reverse anything – it is sufficient to wait long enough for the system to return from a high-entropy state to the original macroscopic state with lower entropy.

3.7 Boltzmann's statistical argument and the arrow of time

Physicists realized that they were facing a serious problem when trying to interpret the second law of thermodynamics. The only way that seems possible in the light of the arguments for reversibility is to assume that the second law is not a universal, non-exceptional principle, but only a statistical regularity. Simply put, processes that reduce entropy on the time scale available to the human observer are extremely unlikely. But why should it be so? How can we derive from the principles of classical mechanics the thesis about a very high probability (bordering on certainty) of an increase in entropy? This task was undertaken by Boltzmann in his famous statistical (combinatorial) argument, the sketch of which we will present below. But before we do that, we need to introduce one more kind of abstract space (as you may have already figured out, physicists love all kinds of spaces). This is basically a variant of phase space – we call it a μ-space. In phase space, one point contains complete information about the state of N particles. However, in the case of μ-space we need N points, each representing the state of a single particle. So μ-space is "only" six-dimensional – three dimensions to determine the position and three dimensions of momentum. Of course, the two spaces are closely related – each point in the phase space corresponds to N points in μ-space.

Now let us divide the μ-space into a large number of cells of equal shapes and sizes. Each cell represents a certain range of possible values of momentum and position in which the momentum and position of each particle can be located. In this case, we will simply say that a given particle is in a specific cell. Assigning each particle to a specific cell is called an arrangement. On the other hand, determining the number of particles per cell for each cell without specifying which particles are in which cell is called a distribution. Of course, there are many arrangements for one distribution in general. For example, the distribution in which there are two particles in cell number one and three particles in cell number two, can be realized by arranging: particles number 1 and 2 in cell 1, particles number 3, 4 and 5 in cell 2, but equally well by the arrangement: particles number 3 and 5 in cell 1, particles number 1, 2 and 4 in cell 2 (Fig. 3.10). The differences between the arrangements for a given distribution are not physically significant, as we presuppose that all particles are identical. Boltzmann assumed that it is possible to choose the size of cells in such a way that each distribution will correspond to exactly one macrostate. Therefore, one can ask the question of how many arrangements realize a given macrostate (distribution).

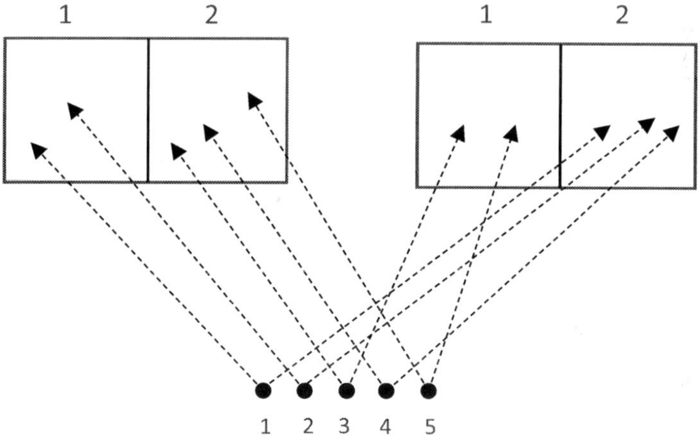

Fig. 3.10. Two arrangements implementing the same distribution

To answer this question, let us first of all note that the number of arrangements depends very strongly on the uniformity of a given distribution. Uneven distributions have few arrangements that implement them, and even distributions – a lot. To illustrate, consider an uneven distribution in which all five particles are in one of the five cells (Figure 3.11). In such a situation, we have only one arrangement that realizes this macrostate. On the other hand, a perfectly uniform distribution allocating one molecule to each of the five cells has 5! = 120 arrangements. However, the number of arrangements for a given distribution is strictly correlated with the size of the corresponding area in the phase space.[13] A distribution that has

[13] This is so because each arrangement of a given distribution corresponds to an area in the phase space of exactly the same size. Try to see it for yourself with a simplified example. Suppose we have only two cells in μ-space and two particles, each of which is characterized by only one parameter whose value ranges from –1 to 1. One cell includes values from –1 to 0, and the other from 0 to 1. The phase space in this case will be two-dimensional (we put the parameter characterizing the first particle on the x axis, and the parameter of the second particle on the y axis). Please draw the areas in the phase space that correspond to the four possible arrangements, and see that they are squares of the same size.

120 arrangements corresponds to an area one hundred and twenty times larger than a distribution with one arrangement. It can be concluded that different macrostates will correspond to areas in phase space of very different magnitudes and that macrostates with relatively evenly distributed distributions will be the largest. For realistic numbers of particles and cells, the size differences between macrostate cells will be incredibly large, beyond our imagination. In fact, the phase space for a typical macroscopic physical system will look like this: almost its entire volume is occupied by one giant cell of one macrostate, and the remaining cells will be tiny specks on its periphery, invisible even under a microscope (Fig. 3.12). This huge macrostate is a state of equilibrium in which no parameters of the system change.

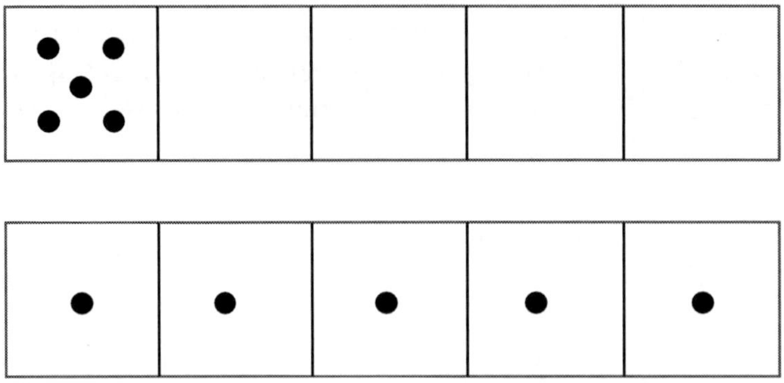

Fig. 3.11. Two possible distributions of five particles among five cells – uneven (top) and uniform (bottom)

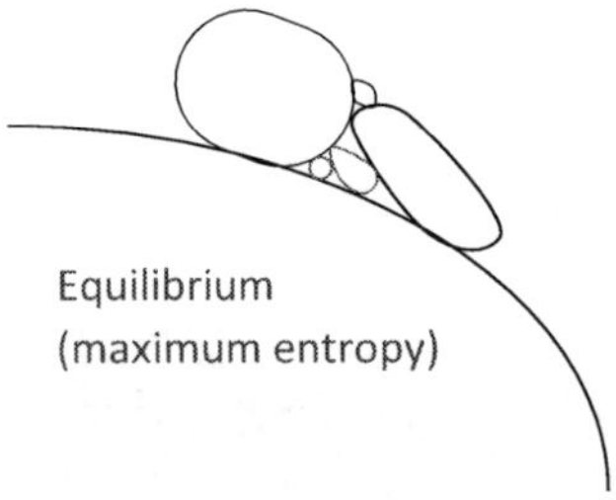

Fig. 3.12. Comparison of the sizes of different macrostate cells in phase space

The distinction between distributions and arrangements may be criticized by the supporters of the principle of the identity of indiscernibles, which we discussed when analyzing Leibniz's shift argument in section 2.5. If we assume that the particles that make up a given statistical complex do not differ physically from each other except for having different positions and possibly velocities, then it is easy to see that all the arrangements that support a given distribution will become completely indistinguishable. For a follower of Leibniz, there is no difference between the arrangement: particle number 1 in the first cell, particle number 2 in the second, and the "swapped" arrangement: particle number 2 in the first cell, particle 1 in the second. With this approach, all arrangements included in a given distribution should be identified – each distribution has exactly one qualitatively defined arrangement. But doesn't this mean that the argument that uniform distributions correspond to much larger areas in phase space collapses?

Fortunately, Boltzmann's argument is independent of how we count qualitatively indistinguishable arrangements. If we follow Leibniz, we will also consequently have to modify our interpretation of phase space. We must identify all points of phase space that differ only in the "exchange" (permutation) of particles. This corresponds to dividing each area by the factor $N! = 1 \cdot 2 \cdot 3 \cdot \ldots \cdot N$ (the number of permutations of the N-element set). It can be seen that the area in the "reduced" phase space, corresponding to a uniform distribution, will still be much larger than the area of an uneven distribution. I encourage the readers to examine this for themselves using the simplified example described in footnote 13. In this case, the phase space of the indistinguishable particles will be a half of the plane below (or above – it doesn't matter which) the diagonal $x = y$. The region of phase space corresponding to an uneven distribution-arrangement in which both particles are in one cell will be a triangle (half a square) with the length of the shorter side equal 1, while the region corresponding to the uniform distribution will be a single square, i.e. it will be twice as large.

To proceed to the justification of the second law of thermodynamics, we must statistically reinterpret the concept of entropy. This is not difficult – according to the statistical definition, the entropy of a given macrostate is simply the volume of the corresponding area in phase space. More specifically, Boltzmann characterized entropy as the logarithm of volume with an appropriate proportionality factor:

$$S = k \log V$$

Since, as we have already mentioned, the volume of the area corresponding to a given macrostate is proportional to the number of arrangements realizing this macrostate, entropy can also be defined as the logarithm of this number. Thus, we can see that the entropy of uniform distributions will be higher, and that of non-uniform distributions will be lower.[14]

We can now explain the popular view of entropy as a measure of disorder. A given distribution is called disordered when there is a very large number of realizations of this distribution created by swapping objects. For example, if we consider a gas container divided into small cells, and there are roughly the same number of particles in each cell, this distribution will have a high degree of disorder. On the other hand, the particle distribution, in which almost all particles are in a single cell (e.g., in the upper left corner of a container), is ordered. The number of realizations of such a distribution, and therefore its entropy, is minimal. The

[14] The use of a logarithmic function in the definition of entropy is a convention, but one that is adopted for good reasons. First, the logarithms of large numbers are relatively small quantities, which allows entropy to be operated more easily. Second, and more importantly, the logarithmic function secures an important property of entropy, namely its additivity. If we consider two independent distributions for which the numbers of the arrangements implementing them are N and M, respectively, then the sum of these distributions will have $N \cdot M$ arrangements. Since the logarithm of the product of numbers is the sum of their logarithms, the entropy of the entire system will be the sum of the entropy of its components.

same applies to velocity distributions (remember that the phase space in which we consider micro- and macrostates contains information about both the positions and velocities of the particles). The state of a gas in which most of the particles move in one direction at a similar velocity is ordered, whereas the state in which the particles move chaotically in different directions at different speeds has high entropy. There is a strong connection of entropy and disorder with probability. If we assume that every realization (arrangement) is equally probable, then high-entropy states will be very likely, because there are many ways to implement them, as opposed to low-entropy states, which are unlikely.

The concept of entropy has strong links with information theory. Let us imagine that we have a random event *A* with several possible outcomes $a_1, ..., a_n$ (e.g. a dice roll). The information related to the specific outcome of such an event is determined by its probability – very probable (close to certain) results carry little information, because we knew that they would most likely happen anyway. On the other hand, unlikely results give correspondingly more information. The information "contained" in one result of a random event with probability p_i can be defined as follows (this is so-called Shannon's definition of information):

$$H(a_i) = -\log_2 p_i$$

In the above formula, we use the base 2 logarithm, which is standard in computer science (that is, we measure information in bits). The minus sign comes from the fact that the logarithm of a number less than 1 (probability) is negative. Then we can calculate the information contained in the entire random event *A*. It will be an average of the information of individual results, weighted of course by their probabilities:

$$H(A) = -\sum_i p_i \log_2 p_i$$

The above formula behaves analogously to the entropy function and is in fact identical to entropy. It takes the highest value for symmetric distributions, for which the probabilities p_i are approximately equal, and the minimal value for non-symmetric distributions, where one probability dominates the others. For example, for two outcomes (coin toss), the information contained in the symmetric distribution $p_1 = p_2 = 1/2$ is 1 bit, while the information contained in the asymmetric distribution $p_1 = 1$, $p_2 = 0$ is zero.

We are now ready to present Boltzmann's statistical justification of the second law of thermodynamics. Consider a physical system in a low-entropy macrostate – for example, a gas contained in one part of a container right after the barrier is removed. From the area representing this macrostate in phase space, we can draw many trajectories. Since, as we have already noted, the overwhelming part of the phase space will be occupied by the cell corresponding to the high-entropy macrostate, the vast majority of the trajectories will have to "fall" into this megacell and remain there for a very long time (Figure 3.13).[15] The probability

[15] To justify this conclusion, the aforementioned Liouville theorem is useful, which ensures that the area occupied by the target points of the trajectory will have the same volume as the initial macrostate. If this area were to increase in volume significantly, it would not be obvious that the vast majority of endpoints would be in a macroscopic equilibrium state.

that a randomly chosen trajectory will lead us to even smaller areas than the initial macrostate is completely negligible. Such trajectories exist, but are in the minority.

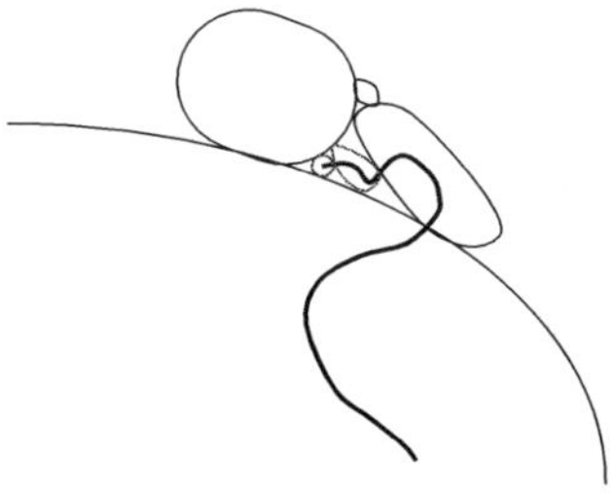

Fig. 3.13. A typical trajectory in phase space leading from the low entropy macrostate to the high entropy macrostate (equilibrium state)

This illustrative, qualitative argument can be supplemented with precise calculations that show how unlikely is the scenario in which, for example, the gas enclosed in a vessel spontaneously concentrates in one part of it, or one of its parts spontaneously heats up and the other cools. It follows that the average waiting time for such an event exceeds the known age of the entire universe, so it is not surprising that we do not observe this in practice. So it seems that our goal has been achieved. The second law of thermodynamics is not an unexceptional law of nature, such as, for example, Newton's laws of motion. However, it is extremely unlikely that we will ever encounter a process that violates this rule.

Unfortunately, Boltzmann's argument suffers from serious flaws, which call into question the correctness of his justification of the irreversibility of thermodynamic phenomena. First, one can have philosophical objections to the way in which Boltzmann uses the concept of probability. The above argument is based on the assumption of equal probability of all microstates included in a given initial macrostate. On what basis can we make this assumption? A typical answer here is to adopt an interpretation of probability based on ignorance. Since we do not know what exactly the microstate of our system is, it is safe to assume that all possibilities are equal. However, such an assumption says nothing about the world, but only characterizes the state of our knowledge. In a situation of complete ignorance, it will be rational to accept equal probabilities, but why should nature adapt to our rationality? It is conceivable that there is some hidden, unknown mechanism that prefers certain micro-realizations of a given macrostate. Allowing for this possibility ruins the whole argument, because we can no longer rely on simply counting trajectories to determine the probable course of the evolution of a system. Some of the trajectories may turn out to be completely improbable or even excluded in themselves (we will see in a moment that such a hypothesis is accepted by many scientists).

The second objection to Boltzmann's argument is more fundamental. As a matter of fact, it is easy to show that this argument in no way justifies the temporal asymmetry of thermodynamical processes. In fact, it is completely symmetrical! The illusion of asymmetry comes from the fact that we only considered the evolution of the system towards the future. However, let us try to think about how the system should have evolved in the past. Let us therefore consider all possible trajectories that lead to the current macrostate with low entropy. It is obvious that these will be essentially the same trajectories that we considered in the argument above. (Trajectories in phase space have no "arrows" to indicate their direction—they can represent evolution in one direction or the other way.[16]) Thus, exactly the same statistical argument convinces us that the overwhelming majority of trajectories will come from states with higher entropy. This means that it is extremely likely that the current state of the system evolved from a higher entropy state, and so the second law of thermodynamics does not apply to past processes. Moreover, this conclusion is completely inconsistent with experience. Take, for example, the instantaneous state of a gas in which it occupies one-third of the volume of the entire vessel. Which is more likely: that a minute earlier it took up a quarter of the volume, or that it took up the entire volume? It is obvious that the first option is much more likely – we have never seen the gas occupying the entire volume of the vessel contract to one third of the volume. However, Boltzmann's argument applied to the past strongly favors the second option. It shows that the most likely scenario for the evolution of a given system is that the entropy of the present state is the lowest, and the entropy of the past and future states is higher.[17]

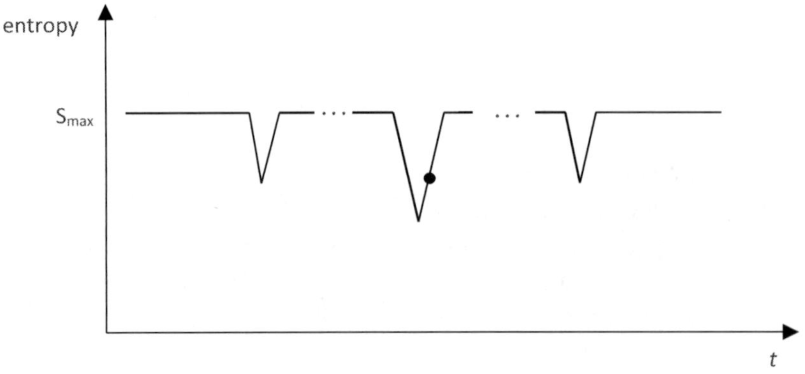

Fig. 3.14. Boltzmann's fluctuation hypothesis. The present state of the universe is marked by a point on the ascending curve

Boltzmann himself was aware of the importance of the problem outlined above. In order to deal with it, he resorted to considerations of a cosmological nature, i.e. concerning the entire universe. He assumed that in the history of the universe, the state of maximum entropy, i.e. the so-called heat death, should be dominant. It is a state in which the total energy of the universe is maximally dispersed – there are no clusters of matter and energy such as stars or galaxies. The universe is simply filled with uniform, dark dust with an energy density barely

[16] What about the arrows representing generalized velocities, calculated from Hamilton's equations (Fig. 2.15 from the previous chapter)? Their choice depends on the tacitly accepted direction of the passage of time, determined by the parameter t used. If we replace the time parameter t with its "mirror" image $-t$ and solve the corresponding Hamilton equations, the directions of the arrows will change to the opposite. Therefore, the trajectory itself does not in any way determine which way we will move along it.

[17] It can be seen that this conclusion is in fact absurd. On the one hand, the argument shows that the entropy of the state of the system at the moment t_1 five minutes after the present moment t_0 should be higher than the current entropy. On the other hand, applying the same argument to the moment t_1, we conclude that the entropy five minutes earlier at the moment t_0 should be higher. So we got a contradiction.

above the absolute zero. However, from time to time, small fluctuations appear in it, which result in a decrease in global entropy – matter begins to concentrate in the form of stars and planets. The universe reaches a certain minimum value of entropy, after which it begins to increase, according to the second law of thermodynamics, until a state of heat death is reached again. This cycle repeats at the next fluctuation, eons later (Figure 3.14). The current state of the universe is on an upward curve, so we observe processes around us that increase entropy. The question of why we are lucky to be in such a privileged state, whose duration is incredibly shorter than the duration of the maximum entropy period, is that the formation of complex organisms is possible only in a world with rich sources of low-entropy energy (for example, in the interior of hot stars, whose explosions in the form of supernovae create the building blocks of life). In the state of the heat death, no complex physico-chemical structure can be formed.

Boltzmann's fluctuation hypothesis gives a global explanation of the second law of thermodynamics and temporal directionality of processes. It is worth noting that it is irrelevant whether the current state of the universe is on an ascending or descending entropy curve (cf. the entry in the box on p. 94). If we relate the arrow of time to the global increase in entropy, the curve called descending will actually be upward for us. The only important thing is that the global entropy gradient is to be non-zero.

Despite its attractiveness, Boltzmann's hypothesis is open to serious objections. First of all, it should be noted that it is based on the assumption of the temporal infinity of the universe, which was widely accepted in Boltzmann's time. This assumption is needed to introduce fluctuations, the occurrence of which is extremely unlikely. However, according to the current state of knowledge, the universe is several billion years old, and this is far too short a time to combine periods of heat death with fluctuations in entropy. Moreover, Boltzmann's cosmological argument does not seem to solve the fundamental problem of the temporal symmetry of the statistical argument. According to our statistical-combinatorial considerations, the most probable state of the universe during the fluctuation period is the state with the lowest entropy, for which both the past and the future are characterized by a positive entropy gradient. However, this does not agree with our observations. So we still need an explanation of why we are on an upward curve and not just at the entropy minimum.

This explanation is provided by another hypothesis, known as the low initial entropy hypothesis. Note that Boltzmann's time-reversed argument is based on the assumption that all phase-space trajectories leading from the past to the present moment are physically possible. However, most of them can be eliminated by assuming that the system under consideration in the distant past occupied a macrostate with an even lower entropy than today. For example, if we consider a situation in which the gas occupies one-third of the entire volume of the vessel, we can enter the information that an hour ago the gas was even more compressed, say, to one-tenth of the volume. Then the statistical argument shows that the most likely evolution of the gas in the last hour is expansion combined with a constant increase in entropy. Most anomalous trajectories starting with high entropy were simply eliminated by an additional assumption. In the case of the universe, we need an assumption of a global nature. This is the hypothesis that the state of the universe at the initial moment (the Big Bang) was characterized by extremely low entropy. By accepting this thesis, we can easily explain why the entropy of past states is lower than the entropy of the present – because it was even smaller in the more distant past (Fig. 3.15).

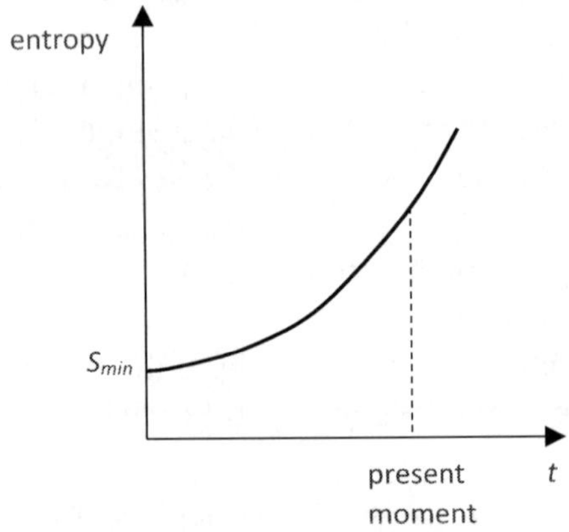

Fig. 3.15. The hypothesis of the low initial entropy of the universe. The curve represents the most probable evolution from the initial moment to the present, according to Boltzmann's statistical argument

Why was the entropy of the initial state of the universe so small? Science has no answer to this question – it is simply a fact that cannot be explained. Some philosophers believe that the low initial entropy thesis has the status of a law of nature, similar to Newton's law of universal gravitation or the law of motion. No one asks why planets attract each other with a force inversely proportional to the square of the distance – it's just the way it is. Others, on the other hand, believe that singular statements, such as "The entropy of the universe at a given moment is such and such", should not be called laws. Only general claims can achieve the status of laws. Regardless of who is right in this dispute, the fact remains that we can explain the time asymmetry of physical processes only by assuming the asymmetry of individual facts about entropy – entropy at the initial moment was lower than the current entropy.

Questions and problems

1. Explain the difference between heat and temperature using the heuristic concept of caloric.

2. Did Joule's experiments really show that heat and mechanical work are equivalent (they are the same quantities)? What result of his experiments would we consider as refuting such a hypothesis?

3. In the formulation of the first law of thermodynamics, the concept of internal energy appears. Can this concept be used to effectively protect this principle from the possibility of experimental refutation? In any situation in which the work done does not balance with the heat supplied, we interpret the difference between the two quantities as an increase or decrease in the corresponding portion of internal energy. Doesn't the first law then become an analytic proposition?

4. Formulate the three versions of the second law of thermodynamics. Show that each of these versions introduces a kind of irreversibility of phenomena in time. Discuss the equivalence of these formulations.

5. Prove that the thermodynamic entropy of the gas occupying the entire volume of the container is greater than the entropy of this gas occupying half of the container by considering the process of slow compression of the gas by means of a piston.

6. Is the statement "For every isolated system its entropy does not decrease over time" an *a priori* analytic or *a posteriori* synthetic proposition? Consider this problem assuming that the time order is given independently of the increase in entropy, as well as assuming that the time order is determined by the increase in entropy.

7. List the basic conditions that two theories T_1 and T_2 must meet in order to conclude that theory T_2 has been reduced to T_1. Show that these conditions are satisfied in the case of the reduction of thermodynamics to statistical mechanics.

8. How can we define in statistical mechanics the heat transferred to a given body, the temperature of the body, and the pressure of a gas sample?

9. How do we define macrostates and microstates of a given physical system? What is the relationship between them? Discuss the representation of macrostates and microstates in phase space.

10. Discuss Loschmidt's reversibility argument and the case of Maxwell's demon. What exactly do these arguments show with respect to the validity of the second law?

11. What is the difference between phase space and μ-space? Explain the concepts of distribution and arrangement in relation to cells in μ-space. Which element of phase space does a given distribution in μ-space correspond to?

12. Present a statistical argument for the claim that macrostates in phase space will differ significantly from each other in size. What is the characteristic feature of the distinguished macrostate that occupies the largest volume in phase space?

13. Critically analyze Boltzmann's argument that there is a huge probability that the entropy of a given system will increase over time. What are the basic objections to this argument?

14. Compare Boltzmann's fluctuation hypothesis with the hypothesis of low initial entropy of the universe. Which of these hypotheses better explains the fact that there is a positive entropy gradient in the current state of the universe?

Further reading

A fascinating history of the development of the concept of temperature and techniques for measuring it, combined with a deep philosophical and methodological analysis, is presented in the book: H. Chang, *Inventing temperature: measurement and scientific progress,* Oxford University Press, Oxford 2004.

Much of this chapter is based on an accessible but in-depth discussion of the relationship between thermodynamics and statistical mechanics in the book: D. Albert, *Time and Chance*, Harvard University Press, Cambridge, Mass. 2000.

A discussion of thermodynamic concepts, including entropy, in the context of cosmology can be found in the popular book by an eminent physicist: R. Penrose, *The Emperor's New Mind. Concerning Computers, Minds and the Laws of Physics*, Oxford University Press, Oxford 1989, chapter 7.

The following monograph by an Australian philosopher contains an incisive analysis of the problem of the arrow of time in statistical mechanics and other physical theories: H. Price, *The Time's Arrow and Archimedes' Point*, Oxford University Press, Oxford 1996.

A very insightful and extensive analysis of philosophical problems of statistical mechanics (including the problem of reduction of scientific theories) is done in the work: L. Sklar, *Physics and Chance*, Cambridge University Press, Cambridge 1993.

Chapter 4. Electricity and magnetism

In the previous chapter we considered thermal phenomena which, although apparently different from mechanical processes (motions, collisions, etc.), after a closer examination turned out to be reducible to the latter. We shall now deal with another category of physical phenomena for which such a reduction to purely mechanistic processes is impossible. We are talking about electrical and magnetic interactions. Moreover, science has made a 180-degree turn in this case, essentially reducing purely mechanical phenomena to electromagnetism. For example, the form of mechanical interaction through contact during the collision of two material bodies, which was fundamental e.g. for the Cartesians, actually turns out to be an electrostatic interaction at a distance between the electron shells of the atoms that make up these bodies. The impenetrability of material bodies, which is the basis of the world picture of classical mechanics, is a consequence of the existence of enormous repulsive forces between charged particles. If the electrons in the atomic shells did not interact with a force that increases to infinity as they approach one another, the material bodies could penetrate each other like ghosts, since most of the space between the electrons, protons, and neutrons contained in atoms is practically empty.

The theory of electromagnetism is important to the philosopher for many reasons. First, it introduces a new ontological concept to the description of the world – the concept of a physical field. The fundamental ontology of classical mechanics is the ontology of material bodies – impenetrable objects endowed with mass and possessing well-defined temporal and spatial locations. Although gravitational interactions between material bodies can be described using the concept of gravitational field, this description is just a useful tool, facilitating the formulation of appropriate laws. In fact, we do not need to assume the ontological reality of the gravitational field – it is entirely sufficient to admit that there are only bodies that attract each other with the appropriate force. In the case of electromagnetic phenomena, the situation is different, because certain laws of the theory of electromagnetism explicitly require the application of the concepts of electrostatic and magnetic fields and their mutual relations. This leads to further questions about the nature of fields – extended and often overlapping entities that fill up space, with blurred space-time boundaries.

The theory of electromagnetism is also important for the historian and methodologist of science. It turns out that electric and magnetic phenomena are related in such a way that it is possible to unify these seemingly different aspects into a single description of reality, using the concept of an electromagnetic field. This poses the question of what it means to unify two theories, two descriptions, or two types of phenomena. Is it enough to give a set of laws connecting these phenomena, as is the case with Maxwell's laws of electromagnetism, or is something more needed? Another methodological aspect of the theory of electricity and magnetism in Maxwell's version is its outstanding success in formulating the hypothesis of the existence of electromagnetic waves, which was subsequently confirmed in experiments and used in practice for communication purposes. Such episodes are rare in the history of science, so it is worth taking a closer look at how Maxwell derived such a far-reaching consequence from his equations.

This is not the end of the list of philosophically important aspects of electromagnetic phenomena. It is well known that certain theoretical problems within Maxwell's theory gave rise to the formulation of a radically novel conception of time, space, and mechanical processes in the form of Einstein's special theory of relativity. Maxwell's theory does not behave correctly when moving from one inertial frame of reference to another, as required by Galileo's principle of relativity, discussed in Chapter 1. The solution to this problem may be either to assume that the equations of electromagnetism apply only in a certain privileged reference frame, defining absolute rest, or to modify the rules of the transformation of coordinates from one reference frame to another. The latter option was chosen by Einstein, who accepted the principles of the Lorentz transformation, according to which time and space are no longer absolute, as in classical physics, but depend on the adopted frame of reference. Therefore, if it had not been for the analysis of electromagnetic phenomena, we would probably never have learned that time and space have a different structure from what common experience tells us. Finally, it cannot be overlooked that the theory of electromagnetism uses beautiful mathematics, whose cognitive and aesthetic values are worthy of the philosopher's attention.

4.1 Electrostatics and fields

The ancients observed already that by rubbing certain substances, e.g. amber against cloth, it is possible to cause the rubbed objects to attract or repel other small bodies. Today we know that these phenomena can be explained by the existence of two types of charges, such that charges of the same type repel each other, and opposite charges attract. It seems natural that the existence of two types of electric charges results directly from the fact that there are two types of electrostatic forces. In contrast, in the case of gravity, there is exactly one type of gravitational "charge", i.e. mass, because the force of gravity only attracts. However, the existence of attractive and repulsive electrostatic forces does not yet guarantee the existence of exactly two types of charge – additional experimental data are needed to draw this conclusion. Theoretically, it could turn out that there are five or even twenty types of charges in nature, each of which would only interact with itself. If you are interested in what empirical facts suggest that the number of distinct charges is two, take a look at the text in the box.

Experience teaches us that the repulsive interaction between charged bodies has the following property: if one body repels another, and the other in turn repels yet another, the interaction of the first body with the third will also give rise to the force of repulsion. In the language of logic, this property is called "transitivity". This means that the bodies involved in the phenomena of repulsion can be classified into groups, each of which will include all and only those objects that repel each other. (Strictly speaking, for such a classification we also need the assumption that the repulsion relation is symmetrical – this is quite obvious – and reflexive. There is a problem with the latter, because reflexivity means that the object repels itself, which is clearly false. However, this is a purely formal difficulty, because we can simply "add" the identity relation to the repulsion relation, and the resulting relation will be a full-fledged equivalence relation, which allows objects to be divided into subgroups.)

In principle, there can be any number of such groups, and therefore the number of different charges could also be arbitrary. However, we still have at our disposal facts about the force of attraction. First of all, we know that if body *a* attracts body *b* and *b* attracts body *c*, then bodies *a* and *c* will repel each other (this is the second empirical fact we will use in addition to the transitivity of the repulsion relation). Finally, we know the third important fact: for any two bodies in the group of "repulsors" (i.e., those that repel some other bodies – in other words, they are simply electrically charged bodies, as opposed to neutral ones), they will either attract or repel each other, while some of them attract each other.

From the three facts mentioned above we can already conclude that there are exactly two groups of "repulsors", such that within each of these groups each two objects repel each other, while objects from different groups attract each other. We know for sure that there are at least two classes closed in respect of the relationship of repulsion, since some of the "repulsors" attract each other, so they cannot belong to the same category. So let's assume that we could distinguish more than two such groups – let's call three of them A, B and C. By definition, the elements of group A attract the elements of group B, and the elements of group B attract the elements of group C. But from the second of the previously mentioned facts it follows that in such a situation the elements of group A will repel the elements of group C, and this is contrary to the assumption that groups A, B, and C are closed with respect to the repulsion relation. Therefore, there can be only two such groups. One of them can be conventionally called positively charged objects, and the other negatively charged.

A quantitative description of electrostatic attraction and repulsion is expressed in Coulomb's law, which is analogous to Newton's law of universal gravitation. Coulomb's law states that the strength of the electrostatic interaction between two point charges is proportional to their charges q_1 and q_2 and inversely proportional to the square of the distance r between them:

$$F = k\frac{q_1 q_2}{r^2} \tag{4.1}$$

The constant k is much greater than the gravitational constant G, and therefore the forces of the electrostatic interaction are incomparably greater than the forces of gravity. The gravitational interaction between two medium-sized electrically charged bodies can be practically ignored. As for the electric charges themselves, experience seems to suggest that one and the same physical body can have different values of its charge, from a value equal to zero (an electrically neutral body) to an arbitrarily large value (achieved by the process of electrostatic charging, for instance with the help of a van der Graaf generator). This would therefore be another difference between electrostatics and gravity, since the mass of a body is its unchangeable property. (Of course, a given body can lose some of its mass, but this happens at the cost of losing some of its matter, so we can assume that we are dealing with a different physical object.) However, the development of science has shown that our intuitive approach to electric charge as a property that can be changed for a given body is wrong. Today we know that the charging of a body can only take place by adding or subtracting from it a certain number of elementary components of matter endowed with a constant charge. Under normal circumstances, the number of positively charged particles (protons) balances with the number of negative particles (electrons). If a body has an excess number of electrons, it will become negatively charged, and if a certain number of electrons is taken away from it, the resultant charge will be positive. However, the charge of a single electron or proton cannot change – elementary particles cannot be "charged". In this respect, charge does not differ fundamentally from mass, which also cannot change for a given particle.[1]

Let us now make a small modification of formula (4.1) for the Coulomb force:

$$F = q_2\left(k\frac{q_1}{r^2}\right)$$

The expression $k\frac{q_1}{r^2}$ in brackets will be called the value (intensity) of the electric field E from the charge q_1 at the point where the charge q_2 is located. (It should be remembered that the electric field at a given point is given by a vector parallel to and having the same direction as the force that acts on the positive charge placed at that point.) If we were to remove the charge q_2 (i.e. if we lay $q_2 = 0$), the force F would obviously disappear. Nevertheless, the value of $E = k\frac{q_1}{r^2}$ would still remain non-zero. This suggests that even in the case of a "solitary" electric charge, there is some kind of physical reality at the point where the second charge could be. Because each point in space could be occupied by a second charge q_2, we can talk about an electric field from the charge q_1 in the entire space around it.

However, the question arises whether the mere fact of the presence of a certain non-zero factor in the formula for the Coulomb force is sufficient to recognize the existence of a new entity, which is the electrostatic field. The basic argument against postulating the existence of fields can be empirical: we are not able to observe the electric field itself until we put a certain charge in it (such a charge is called a test charge) and find that it begins to accelerate under the influence of the acting force. This, however, does not prove that there was some physical reality at a given point before we placed the test charge there. Experience shows the existence of the force acting between the two charges, while the status of the field is much more problematic.

In philosophy, we encounter similar situations when analyzing certain properties, called dispositional. A classic example of a dispositional property is the fragility of objects (e.g. glass), which is not observable until we apply a sufficient force to the object, for example by hitting it with a hammer. A manifestation of fragility is the shattering of the object under the influence of impact. On the other hand, does fragility also exist before the impact? Opinions are divided here. Some believe that the claim about the brittleness of a material does not make empirical sense until the material is properly tested. Most, however, admit that fragility exists as a kind of potentiality, even if it is never revealed. The same applies to the existence of a

[1] We will deal with the principle of conservation of charge (and mass) in more detail in the chapter devoted to special relativity.

field – it can be thought that its presence is the same as the possibility of obtaining an appropriate observable effect when placing a test charge.

Radical empiricists tend to treat physical fields purely instrumentally as a convenient tool, which, however, does not have a counterpart in reality. We have already come across the concept of instrumentalism when discussing the interpretation of astronomical theories. The instrumentalist approach to the Ptolemaic or Copernican model is based on the assumption that we should treat the theoretical elements of these models (e.g. epicycles and deferents for Ptolemy or circular orbits for Copernicus) as merely useful tools for calculating the positions of the planets in the sky. Similarly, in the case of fields, their usefulness lies in predicting the behavior of objects placed near other bodies when interacting with them. However, in the end, objective reality characterizes only physical bodies and their interactions.

One of many possible arguments in the debate on the reality of physical fields takes into account the concept of non-locality, or action at a distance. If we adopt the instrumentalist position on the existence of fields, then we must consequently assume that the interaction between charged bodies occurs at a distance without any physical mediation, in violation of the intuitive principle of locality. (In the box below you can read more about possible interpretations of the concept of "non-locality".) A test charge placed at a distance from another charge "feels" its presence despite their spatial separation. The assumption of the existence of a physical intermediary in the form of a field allows us to maintain the intuition of locality. In this approach, the direct cause of the force acting on the test charge (and consequently the acceleration of the charge) is the physical field present at the location of this charge. In turn, the value of the field strength at a given location is related to the value of the intensity in the areas closer to the field source. There is a physical continuity between the source of the field and the location of the test charge, which is the ontological basis of the cause-and-effect chain leading from the source to the charge feeling the action of the force. The causal relationship does not occur "in leaps and bounds": there is an intermediate cause between every two elements of the cause-and-effect chain.

The concept of locality and its negation – non-locality – can be interpreted in many ways. In the sense that can be called temporal, the causal relationship between spatially distant events *A* and *B* is non-local when *A* and *B* occur at the same time. In other words, the non-local causal interaction between *A* and *B* propagates at an infinite speed, without any delay. On the other hand, an interpretation based on the concept of spatial discontinuity characterizes the causal relationship between *A* and *B* as non-local, when there are no intermediate causes in the spatial area between *A* and *B* – i.e. there is no event *C* between *A* and *B* that would be an effect of *A* and at the same time a cause of *B*. In addition, if *B* occurs later than *A*, we can also speak of nonlocality based on temporal discontinuity: in the interval between the occurrence of *A* and *B*, there is no *C* such that *A* is the cause of *C* and *C* is the cause of *B*. Causal relationships that are non-local in the latter sense seem to violate the principle of determinism, since the state of the entire universe at the interval between *A* and *B* does not determine that *B* will occur (it is possible to imagine two worlds that are identical at any point in time between *A* and *B* and such that in one of them events *A* and *B* happen, while in the second they do not).

Non-locality based on the lack of spatial continuity is a consequence of the rejection of the reality of physical fields. If we also assume that the electrostatic interactions are not non-local in the temporal sense (i.e. there is a time delay between the cause – the source of the field – and the effect in the form of a force acting on the test charge), then as a consequence we must assume that there is also non-locality in the sense of temporal discontinuity, and therefore also some kind of indeterminism. This indeterminism can be avoided if we assume that electrostatic interactions propagate at an infinite speed. The problem of non-locality takes on particular importance in the context of the phenomenon of quantum entanglement, which we will discuss in the later parts of the book.

4.2 Potential and lines of force

When considering the ontological problem of the reality of physical fields, it is worth recalling examples of situations in which physicists themselves question the objective existence of the correlates for certain concepts that can be introduced in the formalism of a given theory, despite the fact that these concepts render considerable service. An example of such a concept in electrostatics is electric potential. Potential is a scalar quantity (represented by a number) that can be used to calculate the vector of field strength at each point. The relationship between potential and field strength is as follows: the field at a given point x is determined by how quickly the potential changes with a small change in position x, and in which direction this change is the greatest. This intuition can be illustrated using the example of a gravitational field. Imagine a hill with different slopes on which we can place an object, e.g. a ball (Fig. 4.1). Depending on how steep the slope is at a given point, the ball will feel a different strength of gravity – from a zero value at the foot of the mountain or at its very top, to a maximum value at the steepest part of the slope. The effective strength of the gravitational field therefore depends on the slope of the hill at a given point. The height of the hill at a point is the equivalent of the potential.[2]

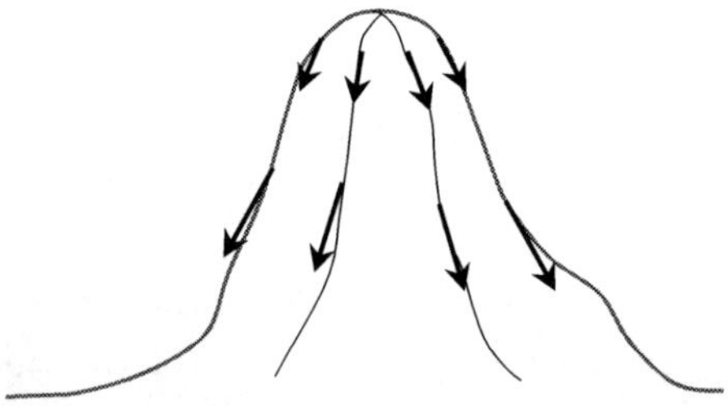

Fig. 4.1 Illustration of potential using an example of a hill in a gravitational field.

[2] An illustration of the concept of potential using the example of a mountain in a constant gravitational field appears in many textbooks and popularizing books. However, this example should not be taken too literally. It is not so that the hill directly represents the potential function with respect to the spatial coordinates on the surface of the Earth. The component of gravitational force tangent to a given surface cannot exceed the value of mg, while the derivative of the height function at a given point relative to the coordinate representing the horizontal distance (i.e. the gradient) can take an arbitrarily large value. For example, when the slope of the hill approaches vertical, the gradient increases to infinity, so the force should also be infinite. Therefore, the height of the hill at a given point is not strictly speaking identical with the potential function, but only illustrates this concept visually.

The mathematical formula illustrating the relationship between the potential and the strength of the field uses the concept of a derivative, which, as we already know, characterizes the rate of change of a given quantity with respect to the corresponding variable. We can also use the well-known geometric interpretation of the derivative as defining the angle that the tangent to a given function at a point makes with the horizontal line (more precisely, the tangent of this angle). In the case where the potential is a function of only one position variable (one-dimensional case), the corresponding relation between the potential V and the strength of the field E is as follows:

$$E = -\frac{dV}{dx}$$

The minus sign is related to the fact that the force acting on the object is directed in the opposite direction to the direction of growth of the function V. In the general case, potential V is a function of three spatial coordinates x, y and z. In such a situation, the derivatives with respect to each of the variables should be calculated separately, treating the other two as constant (these are partial derivatives, which we have already seen in Chapter 2). This gives us three numbers that represent the three components of the intensity vector E_x, E_y, and E_z (the rate of the change of potential in directions x, y, and z):

$$E_x = -\frac{\partial V}{\partial x}$$

$$E_y = -\frac{\partial V}{\partial y}$$

$$E_z = -\frac{\partial V}{\partial z}$$

These equations can be simplified by introducing the so-called gradient operation on a scalar quantity, which is based on the following vector operator, called nabla (the three quantities in parentheses represent its three components):

$$\boldsymbol{\nabla} = \left(\frac{\partial}{\partial x}, \frac{\partial}{\partial y}, \frac{\partial}{\partial z}\right)$$

$$\boldsymbol{E} = -\boldsymbol{\nabla} V \qquad (4.2)$$

The bold print symbolizes the fact that the field strength $\boldsymbol{E}$ is a vector (it has three components, determining its direction in space). Nabla operator $\boldsymbol{\nabla}$ applied to any scalar quantity yields a vector.

The concept of electric potential plays a very important practical role when considering problems in electrostatics. The basic problem of electrostatics is the question of how to calculate the value of a field at a given point, knowing the distribution of electric charges in the entire space. Theoretically, such a problem can be solved by applying the differential form of Coulomb's law, adding up the contributions from small (infinitesimal) electrically charged areas. In practice, however, such a task is often difficult to perform mathematically. It turns out that the introduction of the concept of potential often allows for a significant simplification of the situations under consideration, as long as we are dealing with certain symmetries. An example of one such problem with a solution based on the concept of potential is given in the box (the approach used is called the method of images).

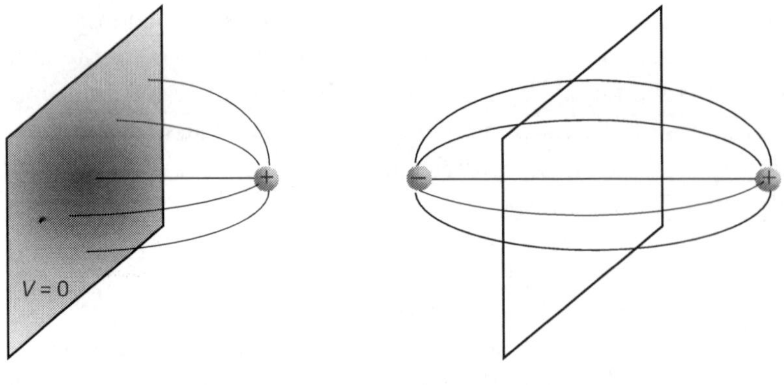

Fig. 4.2 Application of the method of images to solve an electrostatic problem.

Imagine the following system: a positive point charge placed at a certain distance from an infinite, grounded metal plate (Fig. 4.2). What will the distribution of the electric field look like in such a set-up? As you perhaps know, the earth can be considered an unlimited source of free electrons. Thus, grounding the plate will cause negative charges to flow into it, due to the attraction from the positive charge. The plate will become electrically charged, but the distribution of charges on the plate will not be homogeneous – the point closest to the positive charge will have the highest charge density, and the further areas will have a correspondingly lower density. In such a situation, it seems rather difficult to calculate the resultant electric field from all charges. Note, however, that the electric potential of the plate is zero, due to its grounding. This means that there are no forces acting on the charges on the surface of the plate in the plane of the plate – they are in an equilibrium.

Now we can use an important theorem that can be proven in electrostatics. It states that the distribution of charges in any area and the potential at the boundaries of this area unambiguously determine the electric field in the entire volume. What happens outside the selected area has no effect on the value of the field inside (this can be seen as yet another instance of the principle of locality). Therefore, we can replace the plate-and-charge set-up with another condition, as long as the boundary potential and charge distribution match. The equivalent set-up is a simple electric dipole, consisting of our positive charge and an equal but opposite negative charge placed on the other side of the plate at the same distance. Note that at each point of the plate, the electric potential from both charges is zero, because the distances from the charges are the same, and the charges themselves are equal but of the opposite sign (the potential from the point charge q is $-k\frac{q}{r}$). Thus, the entire half-space on the right is identical as in the previous case (the same charge distribution and the same boundary potential). But of course, calculating the electric field from two point charges using Coulomb's law is a trivial matter, so the problem is solved.

Despite its undoubtful usefulness, electric potential is not treated as an objectively existing physical reality. The reason for this is that the value of the potential at a given point is not uniquely determined by the value of the field strength. For a particular function V satisfying equation (4.2) with a field $\boldsymbol{E}$ fixed, there are an infinite number of alternative functions that differ only by the addition of a constant: $V' = V + c$, which also satisfy (4.2) (this is due to the fact that the derivative of a constant function is equal to zero). Using our illustrative example of gravity, we can say that the height of a hill can be determined from any zero-level – sea level, the foot of a mountain, or any other place. A change in the level of zero potential has no effect on the calculation of the field strength and, consequently, on the forces acting, which are the only tangible manifestation of the electric field. The transformation of the electric potential consisting in adding a constant value to it belongs to the class of the so-called gauge transformation. Gauge transformations and gauge symmetries are an important element in the description of many physical theories.

The fact that the absolute value of the potential is "non-physical" has concrete empirical consequences. For example, electrostatic charge added to a closed metal box (called a Faraday cage) has no physical effect inside the box. The potential of the entire system increases in relation to the potential of the earth, but until we come into contact with two bodies with different potentials, we will not observe any effect of the increased potential of the cage. It can also be assumed that the potential of the cage is still zero, and the potential of the earth decreases to an appropriate negative value. Only the difference in potential has a physical meaning, not its absolute value.

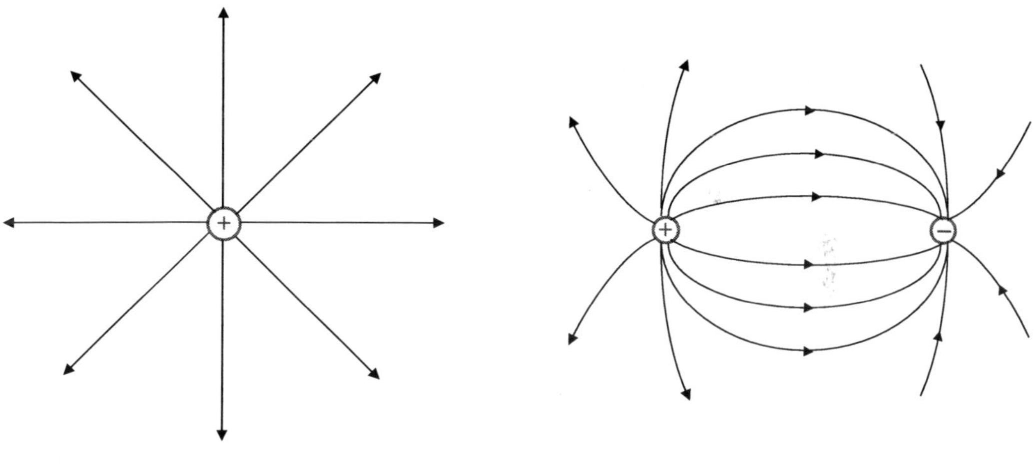

Fig. 4.3 Lines of force of the electrostatic field.

The question of the existence of electrostatic fields represented by the vector $\boldsymbol{E}$ is still open. We will return to it a little later, when discussing the laws that connect electrical and magnetic phenomena. For the time being, however, let us mention one more helpful concept in electrostatics, which, however, has only the status of a useful fiction. I have in mind the lines of force of the field. Textbooks on the theory of electromagnetism often depict electric or magnetic fields in the form of lines spreading over a given area. For example, a field from a point charge has the form of straight lines originating from that charge and extending to infinity. Similarly, the field from the dipole (two electric charges with opposite signs) is depicted as consisting of curved lines connecting the two charges (Fig. 4.3). Lines of force represent the paths of test particles of negligible mass placed in a given field. In addition, we can adopt the interpretative principle according to which the density of lines in a given area represents the value of the field strength in this area. For example, in the case of a field coming from a point charge, it can be seen that the lines of force "diverge" as the distance increases, reflecting the fact that the field strength weakens the farther away we are from its source.

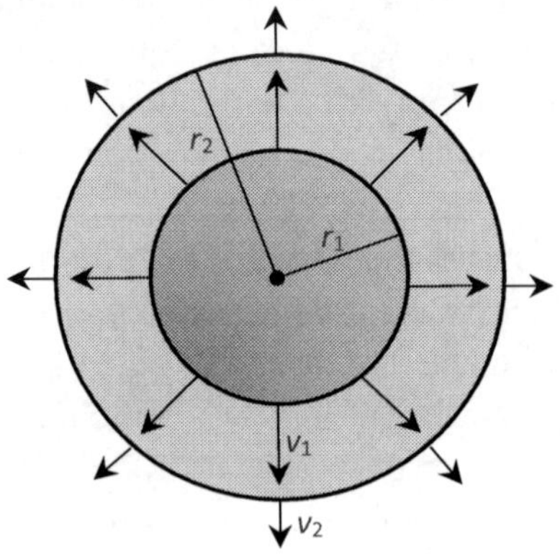

Fig. 4.4 Model of liquid outflow through concentric spheres.

The latter qualitative observation can be used to derive a strict mathematical formula describing how the field strength decreases with increasing distance, i.e. Coulomb's law. Imagine that the lines of force of the field represent the outflow of some incompressible liquid, and the velocity of the flow of this liquid at a given point corresponds to the strength of the field. Consider two concentric spheres with radii r_1 and r_2 with a charge placed in the center (Fig. 4.4). Due to the symmetry of the system, the velocity of the outflow of the liquid should be the same at every point of a given sphere. The amount of liquid flowing through a sphere per unit time is equal to the area of the sphere multiplied by the velocity. This can be justified by the fact that the volume of a layer of liquid flowing through a sphere in a short time Δt is equal to the thickness of that layer $v\Delta t$ times the area S, and therefore the amount of liquid flowing per unit time equals Sv. However, due to the incompressibility of the liquid, which is also not replenished or removed, the amount of it flowing through each enclosed surface must be constant, and thus we have $S_1v_1 = S_2v_2$. Since the area of a sphere is proportional to the square of its radius, we have the equality $\frac{v_1}{v_2} = \frac{r_2^2}{r_1^2}$, i.e. the velocity (and therefore the intensity) decreases with the square of the distance, exactly as in Coulomb's law.

However, the fact that Coulomb's law can be derived from the assumption of the existence of an "electrostatic fluid" does not prove that such a fluid really exists. To substantiate the thesis about the existence of a physical object, independent evidence is needed to confirm this claim. The putative fluid does not produce any observable effects that could independently corroborate its existence. The same is true of the concept of the lines of force – it is difficult to imagine that there are infinitely thin "filaments" in space running out of one charge and converging on another. Despite this, many nineteenth-century physicists (including the English physicist Michael Faraday) believed in the objective existence of the lines of force. Faraday used the concept of the lines of force to explain the phenomenon of electromagnetic induction and the phenomenon of the interaction of moving electric charges with a magnetic field (we will talk about these phenomena in the following sections). He introduced the idea that when a conductor cuts through the lines of force of a magnetic field, this physically causes the event of current flow in that conductor, and this proves the objective reality of the lines. However, it turned out that Faraday's theory of the lines of force was inconsistent. It required the assumption that the lines of force of the field created by a bar magnet move with the magnet when it shifts in a given direction, while when the magnet rotates, the lines of force remain stationary.[3] Since it is difficult to reconcile these two assumptions, Faraday's idea was abandoned.

4.3 Magnetism

Magnetic phenomena have been known to mankind for at least as long as electrostatic phenomena. We know that certain types of minerals have the property of attracting metal objects. We also know that both attractive and repulsive forces can be present between magnets. On the surface, it would seem that there is a full analogy with the interactions of electrically charged objects, so we should also introduce two types of magnetic charges. However, the matter is not so simple. Naturally occurring magnets are always bipolar: the attachment of two magnets on one side causes them to be attracted, and the flipping of one magnet in relation to the other results in a repulsive force. Dividing a single magnet into two parts creates two new magnets, which again have two poles. We are not able to "extract" individual magnetic charges, which would always be subject to either attractive or repulsive forces, regardless of their position in space. Today we know how to explain such behavior of magnets – the field from natural magnets is produced not by single charges, but by small current flows in the entire volume of a ferromagnetic. However, it took quite a long time for scientists to arrive at the correct interpretation of magnetic phenomena. In this brief description, we will skip the historical aspects and go straight to presenting the basic laws of magnetism as they are accepted today.

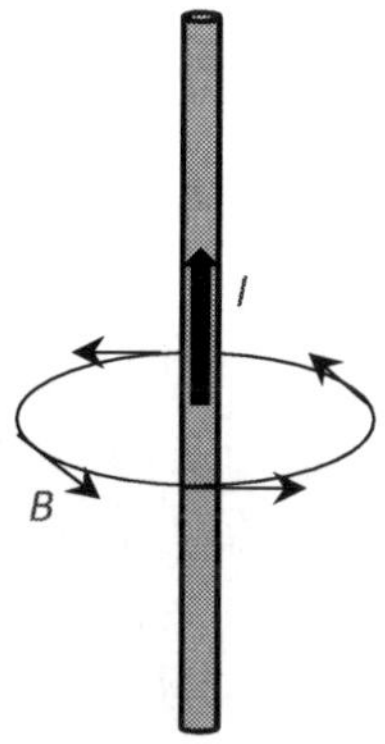

Fig. 4.5 Magnetic field around a conductor with current

[3]Details of Faraday's conception along with its critique can be found in M. Lange's book, *An Introduction to the Philosophy of Physics. Locality, Fields, Energy and Mass*, Blackwell, Oxford 2002, pp. 51-59.

Let us start by presenting the interactions between magnetic fields and moving electric charges. Experience teaches us that a magnetic field can be caused by a flow of charges, i.e. an electric current. It is a well-known fact that when we bring a compass needle close to a conducting wire, we will observe the rotation of the needle. The magnets near a wire conducting electricity align themselves concentrically, forming lines in the shape of a circle. This suggests that the magnetic field vector at a given point is tangent to the circle passing through that point with the center at the conductor site (Fig. 4.5). A quantitative description of this phenomenon is contained in the Biot-Savart law, which gives a formula determining the value and direction of the magnetic field vector $\boldsymbol{B}$ coming from a small section of a conductor through which a current of a given intensity flows. The magnetic field vector is perpendicular to the plane containing the conductor and the radius connecting the conductor to a given point, while its value is proportional to the current intensity and inversely proportional to the square of the distance (if the conductor under consideration is assumed to be infinite, the value of the field is inversely proportional to the distance itself).

We also know from experience that magnetic field affects electric charges. However, this interaction looks different than in the case of an electric field. Firstly, a necessary condition for the interaction of a charge with a magnetic field is that this charge remains in motion – stationary electric charges do not "feel" the magnetic field. The force from the magnetic field always acts in the direction perpendicular to the direction of the particle's motion, so the magnetic field cannot change the value of the body's velocity, only the direction of its motion (such a motion will take place in a circle – Fig. 4.6). Secondly, the direction of the magnetic force is also perpendicular to the magnetic field vector $\boldsymbol{B}$. Finally, the value of the magnetic force depends on the orientation of the velocity vector of the charged body in relation to the direction of the magnetic field – it is greatest when the velocity is perpendicular to the field lines of force, and decreases to zero in the case of a parallel position. All these facts can be summarized synthetically in the form of the equation for the magnetic Lorentz force:

$$\boldsymbol{F} = \frac{q}{c}(\boldsymbol{v} \times \boldsymbol{B})$$

The symbol $\times$ denotes the cross product that produces a vector perpendicular to both the multiplied vectors, while its length is given by the formula $vB\sin\alpha$, where α is the angle between the vectors $\boldsymbol{v}$ and $\boldsymbol{B}$. The letter c stands for a certain constant, which we will talk about in more detail later.

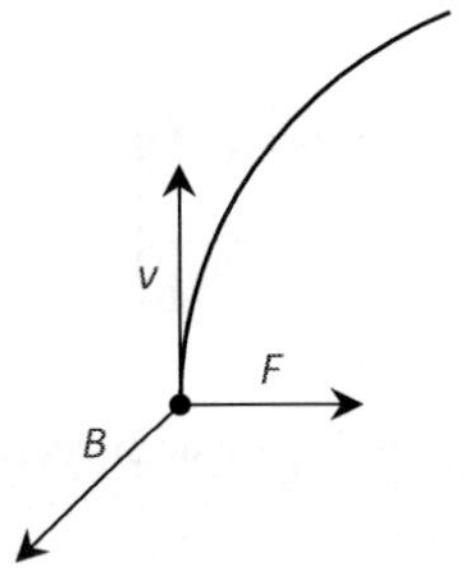

Fig. 4.6 The Lorentz force acting on a charge in a magnetic field.

As can be seen from the above discussion, magnetic field is closely related to the phenomenon of motion – both the creation of the magnetic field and its interaction with charges require the movement of bodies. This indicates that magnetic phenomena (more broadly electromagnetic ones) can affect the interpretation of Galileo's principle of relativity, which in a nutshell can be expressed in the statement that motion is relative. From a cursory glance, it can be concluded that we are dealing with a certain problem here. If any inertial frame of reference is equally good for describing physical phenomena, then it may seem strange that by going from a frame of reference in which an electric charge moves to a frame in which this charge rests, we can cause the magnetic force to be "switched off". Such a consequence is disturbing, because it either implies that the occurrence of force is not an objective, real fact, or it suggests that there are certain distinguished frames of reference in which physical phenomena happen differently than in other frames. We will have to return to this matter after discussing the nomic relationships between magnetic and electric fields.

Let us also mention the issue of magnetic potential at the end of this short introduction to magnetic phenomena. As we remember from the previous section, a stationary electric field $\boldsymbol{E}$ can alternatively be represented by a scalar quantity V, whose rate of change in space and the direction of maximum change determine the field $\boldsymbol{E}$. It turns out that there is no analogous scalar potential field for the magnetic field $\boldsymbol{B}$. This fact has a deeper physical justification. The presence of a scalar potential means that the field characterized in this way has the ability to do some work. A ball rolling down from the top of a mountain acquires more and more speed, and therefore also kinetic energy, at the expense of potential energy in the gravitational field. On the other hand, we have already seen that magnetic field is incapable of doing work on an electric charge. A charge moving in a magnetic field changes its direction of motion, but its velocity value – that is, also the kinetic energy – remains unchanged.

However, there is an equivalent of the electrostatic potential, in the form of the so-called magnetic vector potential $\boldsymbol{A}$. As the name suggests, it is a vector, not a scalar. Its relation to the magnetic field vector is a little more complicated than in the case of an electrostatic field. To represent it accurately, you need to have a new mathematical concept at your disposal – the concept of vector rotation – which we will discuss later. It is worth adding, however, that magnetic vector potential suffers from a similar ambiguity as the electrostatic potential. Any vector field in the form of the gradient of some scalar can be added to a given potential $\boldsymbol{A}$, and the result will be "as good" a potential as the original one. Admittedly, it is difficult to illustrate this case with an intuitive example similar to the one with the conventional choice of the zero-level in a gravitational field, but in fact it is exactly the same kind of freedom (this is another example of the so-called gauge freedom).[4]

4.4 Flux and field circulation

We have already stated that magnetic phenomena are linked with electricity by the relations of mutual influence. It turns out that these connections are even stronger, and moreover, they

---

[4] However, it should be remembered that there is an effect in quantum mechanics that seems to suggest that magnetic potential itself can have observable effects. This is the so-called Aharonov-Bohm effect, in which a particle running near a solenoid (coil) changes the phase of its wave function with current, even though there is no magnetic field outside the solenoid.

do not require direct participation of electric charges. The electric and magnetic fields themselves are in mutual causal relations, the special nature and elegant mathematical description of which will be discussed in this section. The existence of such close relations firstly provides us with a powerful new argument for the reality of physical fields, and secondly suggests that electric and magnetic fields are not separate entities, but rather certain aspects or manifestations of a single entity: the electromagnetic field.

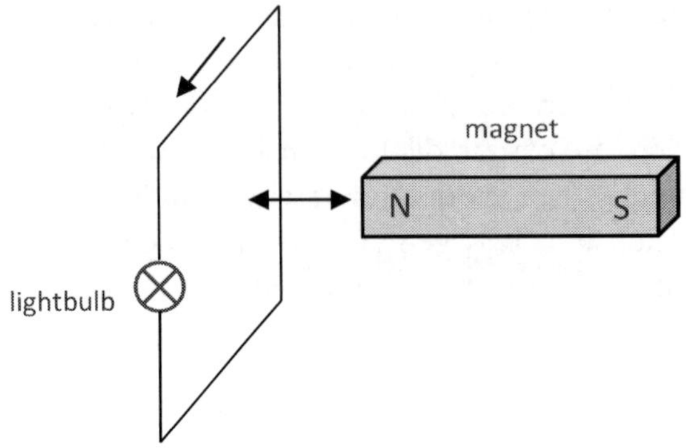

Fig. 4.7 The phenomenon of electromagnetic induction.

Let us start with a description of a phenomenon that is extremely important not only from a theoretical but also from a practical point of view. It is the phenomenon of electromagnetic induction. Its occurrence can be observed by anyone who has a piece of wire, a light bulb and a magnet. Let's consider a closed circuit with a light bulb connected to it. Bringing the magnet close to this circuit, we will observe a momentary flash of the bulb (Fig. 4.7). By continuing the movement of the magnet relative to the circuit, we can maintain the glow of the bulb, the intensity of which depends on both the speed of the magnet's movement and the power of the magnet itself. Evidently, the movement of the magnet induced the flow of an electric current in the circuit. This simple experiment is the basis for the operation of all devices that convert mechanical energy into electric current, such as dynamos or generators. The electricity supplied to our homes from power plants is created in this way, regardless of whether it is a coal, wind, hydroelectric or nuclear power plant (solar panels are an exception). Irrespective of its practical aspect, the phenomenon of electromagnetic induction gave rise to the formulation of an important law of electromagnetism, known as Faraday's law.

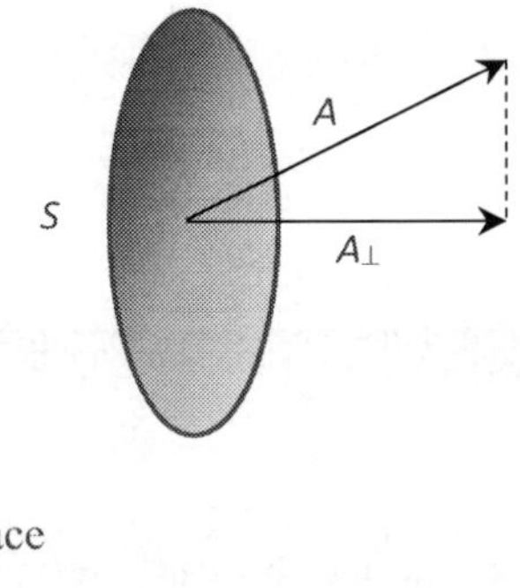

Fig. 4.8 Field flux through the surface

To present Faraday's law in its full form, we need to introduce some elegant new mathematical concepts. These are the concepts of the field flux through a given surface and the circulation of the field along a closed curve. Let us start with the former. Consider any flat surface enclosed by a curve – it can be the surface of a circle, a square, or any other figure. Suppose that at each point of this surface a certain vector field $\boldsymbol{A}$ is defined (it can be an electric field, a magnetic field, but also, for example, the field of a liquid flow given by the velocity vector). For the sake of simplicity, let's assume that vector A is constant at every point of the selected surface (its length and direction do not change when moving from one point to another). The flux of the field A through a given surface S is the product of the component of the vector $\boldsymbol{A}$ perpendicular to S (the so-called normal component) and the surface area S (Fig. 4.8):

$$\Phi_A(S) = A_{\perp}S$$

The physical interpretation of the flux is simple in the case of the velocity field. In this situation, the flux represents the amount of liquid flowing through a given surface per unit time. In the case of other fields, such as electric, magnetic, or gravitational, this interpretation cannot be taken too literally, because, as we have already mentioned, we have no additional arguments for the fact that these fields are really based on the flow of some invisible substance (fluid). However, the flow imagery is helpful and can be used to visualize relevant physical processes.

In general, we cannot assume that the field strength remains the same over the entire surface. In such a situation, we need to use the well-known mathematical method of integration. The idea behind this method is simple, although its application in specific situations can be complicated. A given surface S is divided into small fragments ΔS, such that within each fragment the field is practically constant. Then we calculate the flux flowing through each such small fragment and add up all the numbers obtained in this way. In the limit, when ΔS converges to zero, the sum of the individual terms turns into an integral over the surface, which we write generally as follows:

$$\int_S \boldsymbol{A} \cdot d\boldsymbol{S}$$

In the above notation, we used a simplifying notation using the dot product of vectors. The symbol $d\boldsymbol{S}$ denotes a vector perpendicular to the (infinitesimal) surface whose length is equal to the area of that surface. It is important to choose one direction for all sections of the total surface S ("outwards" or "inwards") and to use it consistently – this is called "orienting" the surface. The dot product $\boldsymbol{A} \cdot d\boldsymbol{S}$ is a number that is equal to the product of the length of both vectors multiplied by the cosine of the angle between them – that is, this product will yield the perpendicular component of the field to the surface times the area of the selected surface.

The concept of field circulation is in a sense formally analogous to the concept of flux, except that it concerns a one-dimensional situation, not a two-dimensional one. Consider any closed curve O (it can be a circle, a square, or any other curve – Fig. 4.9). First, let's determine a certain direction in which we will follow the curve – clockwise or counterclockwise. Next, let's divide the curve into small fragments of length Δl and for each such fragment calculate the product of its length and the component of field $\boldsymbol{A}$ parallel to Δl and directed towards the assumed orientation of the curve (so the result will be positive if the component has the same direction as the direction of orientation, and negative if it is opposite). Finally, let's sum up the results obtained in this way. Again, in the limit when Δl approaches zero, the sum of all terms will take the form of an integral along a closed curve (this is indicated by a small circle on the integral symbol):

$$\oint_O \boldsymbol{A} \cdot d\boldsymbol{l}$$

As before, the dot represents the dot product, and *dl* is the vector directed towards the orientation of the curve. Field circulation is a quantity that determines the tendency of a field to rotate. For example, in the case of fluid flow, non-zero circulation means that a vortex has formed in a certain area. On the other hand, if the field is stationary (but non-zero), its circulation around any closed curve is zero (make sure that you understand why).

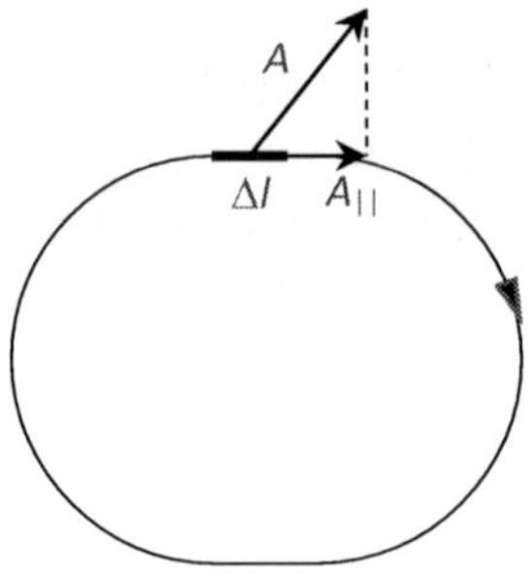

Fig. 4.9 Field circulation along an oriented closed curve

We can now formulate Faraday's law, which is the theoretical basis for the phenomenon of electromagnetic induction. Let us start with a verbal formulation:

> A change in the flux of the magnetic field through a given closed surface *S* results in the formation of an electric field, the circulation of which around the curve closing the surface *S* is proportional to the rate of change of the magnetic flux through the surface *S*.

In the above version, this law tells us that by changing the total flux passing through a given surface (for example, by moving a magnet closer or further away), we will create a new quality: an electric field that will "circulate" along the boundary of this surface. What is more, the value of the circulation of the electric field (which is of course related to the value of the field itself) is unambiguously correlated with the speed of flux change – the faster we change the flow of the magnetic field, the greater the circulation and, consequently, the greater the electric field. Faraday's law explains the creation of current in a circuit as a result of the phenomenon of induction – the free electrons contained in the conductor feel the existence of an electric field that has been induced, which "pushes" them in one direction, thus creating an electric current. However, this electric field is produced even when there is no conductor. The electric field is treated as a reality that exists regardless of whether it causes observable effects for us or not.

The mathematical expression of Faraday's law is as follows:

$$\frac{1}{c}\frac{d}{dt}\int_S \boldsymbol{B} \cdot d\boldsymbol{S} = -\oint_O \boldsymbol{E} \cdot d\boldsymbol{l} \tag{4.3}$$

The expression on the left denotes the rate of change of the magnetic flux (the derivative over time always represents the instantaneous rate of change of the differentiated function). The right side of the equation is recognized as the circulation of the electric field along the curve

O, which closes the surface S (the minus sign is a kind of convention). Symbol c denotes a particular constant, which we will discuss later. Note that the mathematical formula does not specify which process is the cause of which. This is typical for equations in mathematical physics, which require to be supplemented by the description in terms of causes and effects. The mathematical relationship between quantities itself does not yet determine in which direction the causal process runs, or even whether we are dealing with a causal process or with another type of correlation (e.g., co-occurrence). We will return to this issue later in the chapter.

4.5 Ampère-Maxwell's law and Maxwell's equations

Faraday's law is not the only law that determines the mutual causal correlations between electric and magnetic fields. There is an inverse relationship that shows us how to create a circulating magnetic field. From section 4.3 we remember that a magnetic field can be created by passing an electric current through a conductor. This observation can be generalized in the form of so-called Ampère's law, which has a form similar to Faraday's law. Namely, it states that the flow of electric charges through any surface causes the formation of a magnetic field, the circulation of which along the curve closing this surface is proportional to the intensity of the flowing electric current. The mathematical formula expressing this relationship is as follows (I is the electric current):

$$\frac{4\pi}{c} I = \oint_{O} \boldsymbol{B} \cdot d\boldsymbol{l}$$

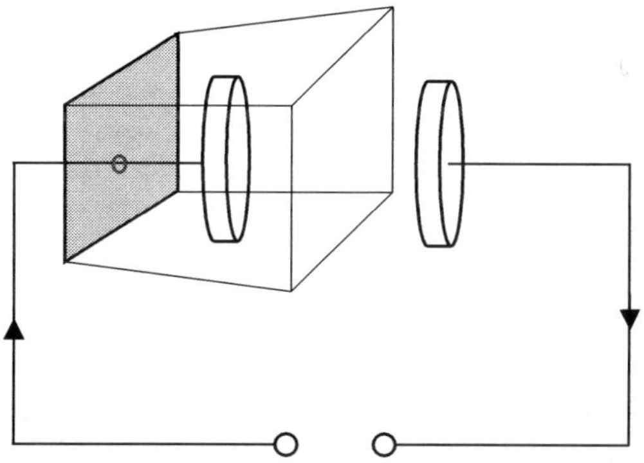

Fig. 4.10 An example illustrating a contradiction in Ampère's law.

However, it turns out that Ampère's law has a fundamental flaw. Namely, in certain situations, it leads to inconsistent predictions, depending on the arbitrary choice of the closed area. Consider a situation in which a power source has been connected to the plates of a capacitor, as in Fig. 4.10. For a period of time, there will be a current in the circuit – the so-called charge current – until the capacitor reaches its maximum capacity. Now consider a certain surface cutting through the conductor (marked in gray in the diagram). As the charging current flows through this surface, according to Ampère's law, a non-zero magnetic field will begin to circulate along the boundary curve (the rectangle in the figure). However, we can choose another surface bounded by the same curve (in the figure this surface consists of the walls of the polyhedron except for the side marked in grey), which passes between the capacitor plates and therefore does not intersect the conductor with current. In such a situation, Ampère's law predicts that the circulation of a magnetic field along the same curve will be equal to zero! So we have a contradiction.

James Clerk Maxwell has proposed to solve this problem by introducing an additional term in Ampère's equation, called the displacement current. This is in fact an exact equivalent of the term in Faraday's law, determining the rate of change of the electric field flux. Thus, the full form of Ampère-Maxwell's law will be as follows:

$$\frac{4\pi}{c}I + \frac{1}{c}\frac{d}{dt}\int_S \boldsymbol{E}\cdot d\boldsymbol{S} = = \oint_O \boldsymbol{B}\cdot d\boldsymbol{l} \tag{4.4}$$

The left side of the equation defines two ways of creating a circulating magnetic field: either by passing a current through a given surface or by changing the flux of the electric field. The charging current problem is solved by the fact that the electric field of the capacitor being charged increases, and therefore the flux through the second surface will increase, giving us exactly the same circulation of vector $\boldsymbol{B}$ as determined by the current flow through the first surface.

In words, we can express Ampère-Maxwell's law as follows:

> A change in the electric flux through a closed surface *S* or a current flow through *S* causes the formation of a magnetic field, the circulation of which around the surface *S* depends on the rate of change of the flux and the magnitude of the current flowing.[5]

Comparing Faraday's laws (4.3) and Ampère-Maxwell's laws (4.4), we notice that they are not exactly symmetrical. This is because there are no magnetic charges in nature, and therefore there are no magnetic currents either. The only way for magnetism to produce electrical phenomena is as described by Faraday's law: by changing the flux of the magnetic field. On the other hand, electrical phenomena can cause the circulation of the magnetic field in two ways: by means of changes in the flux of the electric field ("field-like") or by the flow of charges ("charge-like").

We can now summarize all the basic laws of electromagnetism, which encompass the totality of our knowledge of this field. To Faraday's and Ampère-Maxwell's laws, which characterize the mutual relations between electricity and magnetism, we will add two more laws that describe electricity and magnetism separately. The first is a generalization of Coulomb's law (sometimes referred to as Gauss' law). It states that the total flux of the electric field through a closed surface (e.g. a sphere) is proportional to the total electric charge Q contained inside this surface. Mathematically, we will express this relationship as follows:

---

[5] As an aside, we can note that the use of causal terminology in the formulation of the Ampère-Maxwell law leads to certain problems. Considering the example in Fig. 4.10, note that here we have two mutually incompatible hypotheses about what is the cause of the circulating magnetic field along the marked curve. Depending on which surface we consider, the answer will be: either the flow of current or a change in the flux of the electric field. To avoid this problem, we can abandon the causal formulation in favor of correlation: the circulation of the magnetic field is numerically correlated with both the flow of current and the change in flux. Such an acausal interpretation of Maxwell's equations will be accepted when we write them in differential form.

$$\int_S \boldsymbol{E} \cdot d\boldsymbol{S} = 4\pi Q \tag{4.5}$$

Although the expression on the left side of the equality is identical to the expression for flux in Ampère-Maxwell's law, we should remember that in Gauss' law the surface S must be closed, and therefore it is not bounded by any curve. Gauss' law implies that the total value of the flux through a closed surface is not affected in any way by the charges outside the surface.

An analogous law in the case of a magnetic field has the following form:

$$\int_S \boldsymbol{B} \cdot d\boldsymbol{S} = 0 \tag{4.6}$$

This is a direct result of the assumption that there are no magnetic charges. A magnetic field is a sourceless field. For any enclosed surface, exactly the same amount of magnetic flux flows in as it flows out.

For the sake of completeness, it is useful to present all four laws of electromagnetism (4.3) – (4.6) (also known as Maxwell's laws or equations) in the so-called differential form. The differential version of the laws of electromagnetism formulates all relationships between electric and magnetic fields, charge and electric current, in a "point-by-point" fashion. They tell us precisely what mathematical relations connect the electric field strength at a given point with the magnetic field strength at the same point, and with the charge and current density also at the same point. Let us write down the appropriate equations, which we will then describe for an illustrative purpose. A more detailed mathematical analysis of these formulas will be given in the section with an asterisk.

$$\begin{gathered} \boldsymbol{\nabla} \cdot \boldsymbol{E} = 4\pi\rho \\ \boldsymbol{\nabla} \cdot \boldsymbol{B} = 0 \\ \boldsymbol{\nabla} \times \boldsymbol{E} = -\frac{1}{c}\frac{\partial \boldsymbol{B}}{\partial t} \\ \boldsymbol{\nabla} \times \boldsymbol{B} = \frac{4\pi}{c}\boldsymbol{j} + \frac{1}{c}\frac{\partial \boldsymbol{E}}{\partial t} \end{gathered} \tag{4.7}$$

The symbols $\boldsymbol{\nabla} \cdot \boldsymbol{E}$ and $\boldsymbol{\nabla} \cdot \boldsymbol{B}$ denote the so-called divergences of the respective fields. The divergence of a given field at a point is a number that determines the "outflow" of a given field from that point (this concept can be interpreted by analogy with the flow of a liquid – a non-zero divergence in a given place means that the source of the liquid outflow is located at this point). Thus, Maxwell's first equation says that the point source of the outflow of an electric field is the density of the electric charge at that point. Similarly, the second law states that the outflow of a magnetic field is always zero (because there are no magnetic charges). The other two laws are, of course, Faraday's and Ampère-Maxwell's law. On the left side of both equations there is the operation of the so-called field rotation (also called "curl") denoted by $\boldsymbol{\nabla} \times \boldsymbol{E}$ and $\boldsymbol{\nabla} \times \boldsymbol{B}$. The operator $\boldsymbol{\nabla} \times \boldsymbol{E}$ represents the "propensity" of the $\boldsymbol{E}$ vector to rotate in a small area surrounding a given point. (Like any differential operator, rotation also tells us something about the behavior of a given object not only at the point itself, but above all in the "infinitesimal" region around the point. Of course, at the point itself, the vector $\boldsymbol{E}$ does not rotate – it has one fixed value and direction.)

The right side of the third equation (Faraday's law) represents the rate at which the value of field ***B*** changes at a point. Thus, the whole equation connects the tendency of the electric field vector to rotate at a point with the rate of change of the magnetic field at the same point. Similarly, the fourth equation (Ampère-Maxwell's) tells us that the rotation of the magnetic field at a point is equal to the rate of change of the electric field plus the density of the electric current ***j***. Maxwell's differential equations are formed from integral versions by contracting extended regions of integration to a single point. Let's take Faraday's integral law as an example. By choosing smaller and smaller areas, we will reach a situation in which the circulation of the electric field around an infinitesimally small closed curve (per unit area) will change into rotation at a point. In turn the flux of the magnetic field through the surface when moving to a point of zero size changes simply into the product of the value of the field itself times this surface. A more precise mathematical explanation of the relationship between the differential and integral versions can be found in the section marked with an asterisk.[6]

Maxwell's set of equations also illustrates a kind of transition to a higher level of abstraction compared to original Faraday's and Ampère-Maxwell's laws. As we mentioned earlier, these laws originally described certain asymmetric causal relationships between changes in relevant fields on the one hand, and the production of other fields on the other. Currently, however, we treat the laws of electromagnetism as mutual mathematical interdependencies between vectors ***E*** and ***B***, without additional assumptions of a cause-and-effect nature. We simply state that whatever electric and magnetic fields exist in a given area, they must satisfy all four equations. Whether one field causes the other, or vice versa, is no longer the subject of our analysis. Of course, in concrete applications we can return to thinking in terms of cause and effect (for example, by giving an intuitive description of the propagation of electromagnetic waves, which we will discuss later), but causal assumptions are not part of Maxwell's equations understood in an abstract way.

4.6 Unification of electromagnetism

Now we will take up the important topic of unification. It is widely believed that by formulating his equations, Maxwell unified independent electrical and magnetic phenomena into a single type of electromagnetic phenomenon. To put it more ontologically: instead of introducing two types of physical fields, Maxwell assumed the existence of a single entity in the form of an electromagnetic field. However, the question arises whether the very form of Maxwell's equations entitles us to such a far-reaching ontological conclusion. It is undoubtedly a fact that Maxwell's equations unify our knowledge of electrical and magnetic phenomena in the sense of performing a synthesis, i.e., subsuming many different phenomena under a small number of fundamental laws. But is such a synthesis sufficient to claim that the different concepts occurring in these laws represent one and the same entity? This would probably be too far-fetched a thesis. In science, there are many examples of synthesizing laws

[6] The ontological advantage of differential versions of Maxwell's laws over laws in the integral version is that the former are explicitly "local". For example, Gauss' law in the integral version seems to suggest that the flux through a certain closed surface depends directly on the charge, which can be located at any distance from that surface. On the other hand, the differential version makes a certain localized property of the electric field (its divergence at a point) dependent only on the charge density at the same location.

which combine together various quantities, but no one draws a conclusion about the identification of the objects described by these quantities. Although the ideal gas law expresses close correlations between pressure, temperature, and volume, this does not lead to the conclusion that these quantities represent one and the same aspect of reality: pressure-volume-temperature. So what additional arguments can be presented in favor of the thesis that in the case of electrical and magnetic phenomena we should make an appropriate ontological unification?

Let us note at the outset that it is always possible to trivially unify phenomena by means of the operation of a logical sum (alternative). We can, for example, consider the electromagnetic phenomenon to be an electrical or magnetic phenomenon. Or we can also call an electromagnetic field any electric field and any magnetic field. However, this would be a completely uninteresting form of unification, without any deeper theoretical foundations. In this way, gravity and electricity could just as well be unified, creating an artificial concept of a "gravitational-electric" field, which is either a gravitational or an electric field. It is clear that the sources of the unification of electromagnetism should be sought in the details of the theory, and not in arbitrary terminological stipulations. The two laws linking electricity and magnetism, Faraday's and Ampère-Maxwell's law, interpreted causally may give us a clue. The relation of "bringing about" that occurs between the change of one field and the occurrence of another may suggest that these two types of fields are closely related, perhaps even ontologically. However, the matter is still not completely clear. A loud explosion can cause panic in people nearby, but it is not believed that explosions and panic are "two sides of the same coin".

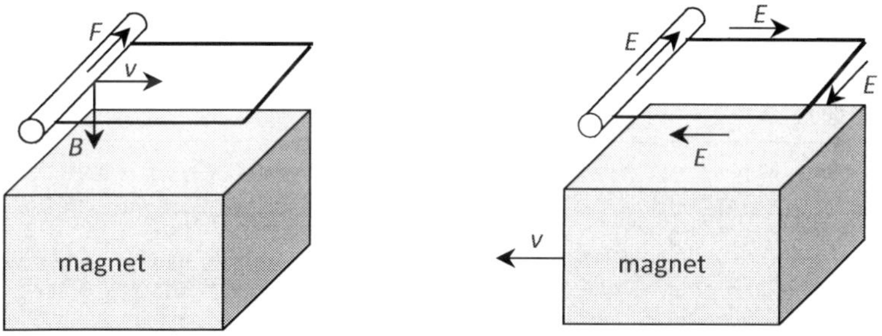

Fig. 4.11 Two descriptions of current induction in the case of relative motion of a conductor and a magnet

The justification for electromagnetic unification should be sought elsewhere – in a particular way of describing electrical and magnetic phenomena in various reference frames. Consider the following example: we move a conductor over a magnet (Fig. 4.11). Let us describe what happens in the conductor, assuming that our frame of reference is attached to the magnet (diagram on the left). In such a situation, the conductor containing free electrons moves in a constant magnetic field perpendicularly to the force lines, and therefore the electrons will begin to feel the Lorentz magnetic force acting on them, which will cause them to move, and this in turn will create a positive charge of one end and a negative charge of the opposite end (thus there will be a potential difference between the ends of the conductor). On the other hand, let us try to look at the situation from the perspective of the wire (diagram on the right). In this approach, the electrons in the conductor are stationary, while the magnetic field shifts. If we draw a surface with our conductor on one side, then we can conclude that the flux of the magnetic field passing through that surface is altered, and thus, according to Faraday's law, a circulating electric field is induced here. And it is the electric field that causes the uniform movement of electrons in the conductor.

Let us compare both descriptions with respect to the existence of the respective fields. In the first frame of reference, there is only a magnetic field and it causes the observed effect. In the second frame, however, there is both a magnetic and an electric field. Thus, by changing the frame of reference ("perspective"), we can cause one of two fields, namely the electric one, to "disappear". Using the principle that we formulated in the analysis of classical mechanics, we can say that true reality belongs only to objects or phenomena that are invariant with respect to the transformations from one frame of reference to another. As a result, we should assume that the electric field itself is not an objective component of reality, since its role can be taken over by the magnetic field.

In fact, there is full symmetry here – magnetic field can also be "converted" into an electric field by choosing an appropriate frame of reference. Consider, for example, two identical electric charges at rest relative to each other (Fig. 4.12). In the reference frame associated with these charges, there is only an electric field that repels these charges. However, when we switch to a frame moving relative to charges, we must take into account the magnetic field, which arises according to Ampère-Maxwell's law due to the motion of each charge, and which causes the corresponding force acting on the other charge. So this time, the existence of the magnetic field depends on the frame of reference adopted. What remains invariant under the change of reference frame is, firstly, the presence of an effective force acting on objects (regardless of whether its source is electricity or magnetism),[7] and secondly, the presence of at least one of the two fields: electric or magnetic. We are not able to completely eliminate both the electric and magnetic fields by choosing the right frame of reference.

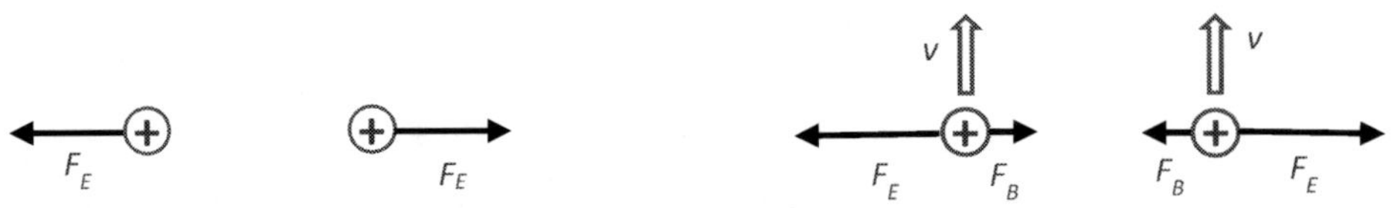

Fig. 4.12 Interaction of two charges in different reference frames.

The above facts strongly suggest that electric and magnetic fields treated separately do not have a full physical reality. On the other hand, there must be a physical entity responsible for all observable electromagnetic phenomena. This entity is a kind of combination of electricity and magnetism, which we will call an electromagnetic field. The electromagnetic field manifests itself in different frames of reference in different ways: sometimes as the electric field itself, sometimes as the magnetic field itself, and sometimes in the form of both fields.

[7] We are omitting an important problem here: in different frames of reference, the value of the force acting on charges calculated on the basis of the laws of electromagnetism seems to be different. A full solution to this problem is possible only in the framework of special relativity, in which both the formula for Lorentz force and Newton's law linking the acting force to the acceleration of a body must be modified (see sections 5.7 and 5.8).

However, the question arises as to how this new electromagnetic field can be mathematically represented. Shouldn't we introduce a new vector for this purpose, marking it e.g. as ***EB?*** Such a vector should be expressible by means of separate vectors ***E*** and ***B***, and should also have the feature of invariance when moving from one frame of reference to another.

It turns out that matters are not so simple. A correct description of the electromagnetic field requires the introduction of a new mathematical concept – a tensor. Tensors play an extremely important role in many areas of physics. They are very useful in formulating the relativistic laws of mechanics resulting from the special theory of relativity, and the general theory of relativity could not do without them in any way. But what exactly is a tensor? Let us start by building some intuitions. The concept of a tensor can be understood as a kind of generalization of the concept of a vector. A vector in a given space is characterized by its components – three in the case of three-dimensional space, and four in space-time, which has the fourth time dimension. Therefore, vectors can be abstractly represented as n-tuples of numbers, where n is the dimension of a given space. (Of course, we must remember that these numbers depend on the choice of coordinate system – one and the same vector will have different components in different coordinate systems, e.g. those whose axes are rotated or shifted relative to each other). In short, a given vector can be symbolized in the form of a letter with one index, e.g. V_p, where p takes on values from 1 to n.

Tensors are objects that also have components in different coordinate systems, but have more of them than vectors. More precisely, the number of their components is an appropriate power of the dimension n of the space. The simplest example is a "two-dimensional" tensor, which has two indices: T_{pq}. Since any index p and q can take any value from 1 to n, in general such a tensor will have $n \times n$ components. It is convenient to represent two-dimensional tensors in the form of a matrix, i.e. an array of numbers with n columns and n rows. However, it should be remembered that – just like vectors – tensors in different coordinate systems look different. Therefore, we should not identify a tensor with a concrete matrix of numbers, but rather with some abstract object whose representations in different systems are the corresponding matrices. Moving from one coordinate system to another, we need to perform an appropriate transformation of the tensor components.[8]

The tensor describing the electromagnetic field is of "size" 4×4. [9] This is because the correct formulation of the theory of electromagnetism requires the application of the theory of relativity, which is based on the concept of four-dimensional space-time. This tensor is symbolized as $F_{\mu\nu}$. (In physics, there is a convention according to which indices in the form of Greek letters range over four values 0, 1, 2, 3, where index 0 is the time coordinate and 1,

---

[8]Let us add that there are two general ways of transforming tensors with a given change of coordinates, which lead to the distinction of two types of tensors (as well as vectors, which are a special case of the tensor). These types are called covariant and contravariant tensors – they are distinguished by superscripts (contravariant tensors) and subscripts (covariant tensors). Thus, T_{ab} symbolizes a covariant tensor, and T^{ab} symbolizes a contravariant tensor. We will talk more about this distinction in the chapters on special and general relativity.

[9] Strictly speaking, it is possible to completely represent an electromagnetic field with a vector – let's add, a vector in four-dimensional space-time, and therefore having four components. It is the so-called four-vector of the electromagnetic potential, which is a generalization and synthesis of the electrostatic potential V and the 3-dimensional vector of the magnetic potential ***A***. However, like the electrostatic and magnetic potentials, the electromagnetic potential is given only up to a certain transformation (gauge transformation), and therefore, according to the arguments presented earlier, it is not entitled to full physical reality.

2 and 3 are three spatial coordinates. On the other hand, the letters of the Latin alphabet always denote spatial coordinates: 1, 2 and 3.) The components of the electromagnetic field tensor are the components of the electric and magnetic fields E_x, E_y, E_z, B_x, B_y, B_z with an appropriate sign + or –. It turns out that if we perform a transformation of the tensor $F_{\mu\nu}$ by moving from one frame of reference to another, a combination of electric and magnetic field components may appear in a given place of the matrix where previously there was only one component of the electric field or magnetic field. This corresponds to the aforementioned fact that by changing the frame of reference we can "turn on" or "off" the appropriate fields – electric or magnetic. However, tensor $F_{\mu\nu}$ remains the same object, representing one and the same physical entity, although its components are different in different frames of reference. It is worth noting that tensor $F_{\mu\nu}$ transforms according to the principles of special relativity, which we will discuss later in Chapter 5.

To summarize the topic of unification in the theory of electromagnetism, let us add that all four laws of electromagnetism can be expressed in a synthetic form, using only the electromagnetic tensor $F_{\mu\nu}$ (and the 4-component electric current density, which together represents the electric charge and current). Without going into the mathematical details of the notation, I will give Maxwell's equations in the tensor version, so that the reader can appreciate their brevity and formal elegance:

$$\partial_\sigma F_{\nu\tau} + \partial_\nu F_{\tau\sigma} + \partial_\tau F_{\sigma\nu} = 0$$

$$\partial_\nu F^{\mu\nu} = J^\mu$$

The symbol ∂_σ is an abbreviated notation of the partial derivative with respect to the x^σ coordinate: $\frac{\partial}{\partial x^\sigma}$. The first of the above equations is equivalent to the conjunction of Maxwell's second and third laws – these are the so-called homogeneous equations, in which there is neither charge nor electric current. The second represents the other two Maxwell equations (so-called inhomogeneous). The symbol J^μdenotes a four-vector current density, the first component of which is the electric charge density, and the other three represent the current density in the spatial directions *x*, *y*, and *z*. Only the above two equations give the essence of the unification that Maxwell initiated with his equations. (For more on that, see the asterisked sections in chapter 5.)

4.7 Electromagnetic waves

Maxwell's equations – both in their "traditional" and tensorial versions – beautifully unify our knowledge of electromagnetic phenomena. However, is this their only scientific value? Shouldn't they bring some added benefit that would expand our knowledge in previously unforeseen ways? It turns out that this is the case. A surprising consequence of Maxwell's equations is that they suggest the existence of a new and previously completely unknown phenomenon: the propagation of electromagnetic waves. This is one of the most spectacular examples from the history of science of how the development of a theory can lead to the formulation of new empirical predictions, which are then fully confirmed experimentally.

Maxwell's reasoning culminating in the hypothesis of the existence of electromagnetic waves was strictly mathematical in nature – we will present it in general outlines in a moment, and in more detail in the section with an asterisk. Before we get to that, however, let us note that the idea of the propagation of "alternating" electric and magnetic fields can easily be derived from causally interpreted Faraday's and Ampère-Maxwell's laws. Imagine a single electric charge suspended from a spring that "jumps" up and down at a variable speed (Fig. 4.13). According to Ampère-Maxwell's law, a circulating magnetic field will appear around this charge, as the charge will cut through the plane perpendicular to the spring, thus creating an electric current passing through this plane. However, let us note that the circulating magnetic field will be changing its intensity over time, because it depends on the magnitude of the electric current, i.e. the speed of the charge, and this varies its value and even direction. Thus, the circulating magnetic field will alternately increase and decrease, responding to changes in the velocity of the charge. By selecting any surface perpendicular to the magnetic field vector at a given point, we will notice that the field flux through that surface will vary over time, resulting in a circulating electric field, according to Faraday's law. But again, the value of this field will change depending on the rate of change of the magnetic field flux, so another application of Ampère-Maxwell's law implies the formation of a "secondary" magnetic field appropriately oriented in space. This process will continue in space, creating a kind of an infinite chain, the "eyes" of which will be successive circulating electric and magnetic fields. So we have a qualitative description of what is called an electromagnetic wave. It is also worth noting that this process will continue even when the charge producing the original magnetic field is removed. This gives us an additional argument for the reality of fields that have the ability to "push" each other through empty space.

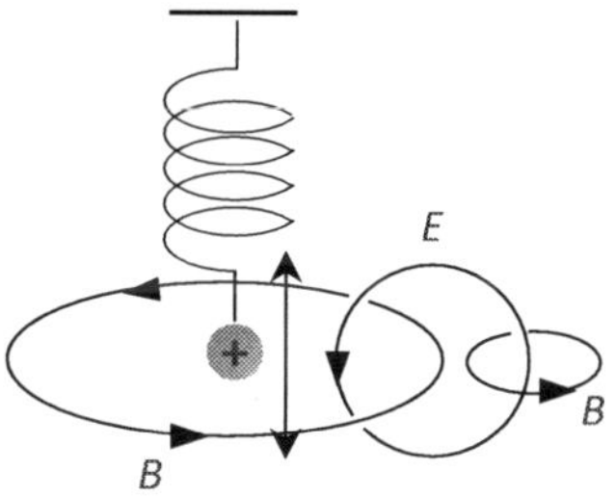

Fig. 4.13 Explanation of the formation of electromagnetic waves

Let us now try to describe this phenomenon a little more precisely. Let's start with a mathematical approach to wave phenomena. A wave is any disturbance of a certain medium that changes in an appropriate way over time. Such a disturbance may be, for example, a displacement of a vibrating string at a given point or raising or lowering the surface of the water in relation to the equilibrium position. Let a function F represent this disturbance. Suppose that F is the following function of position x and time t: $F(x - vt)$. It follows that, for example, at the moment $t = 0$, the disturbance at any point x will be equal to $F(x)$. Since we have an obvious equality $F(x) = F[(x + vt) - vt]$, after time t the disturbance at point $x + vt$ will be exactly the same as it was at point x at time zero. Thus, the function $F(x - vt)$ describes the propagation of a certain disturbance with velocity v in the direction of increasing values of x.

Let us now write down a mathematical equation that connects the second derivatives of the function F with respect to the space variable and the time variable. Those of you who know the basic principles of differentiation of functions will easily check that any periodic function $F(x - vt)$ will satisfy this equation:

$$\frac{\partial^2 F}{\partial x^2} = \frac{1}{v^2}\frac{\partial^2 F}{\partial t^2}.$$

This is what is known as the wave equation in the one-dimensional case. For the three spatial dimensions, the wave equation takes the following form:

$$\frac{\partial^2 F}{\partial x^2} + \frac{\partial^2 F}{\partial y^2} + \frac{\partial^2 F}{\partial z^2} = \frac{1}{v^2}\frac{\partial^2 F}{\partial t^2}. \tag{4.8}$$

There are many ways to derive this equation for various special cases of waves: sound waves in air, stress waves in metals, and even physical field waves described by an appropriate Lagrangian. To get a "feel" of the above differential equation of the second order, we can think of it as a variant of Newton's second law. Notice that the right side of the equation contains an expression that is very similar to acceleration (if we consider the function F as determining the position of a certain object, e.g. a single weight attached to a long rope with similar weights). Then the left side of the equation can be interpreted as representing the total force (per unit mass) acting on the object and coming from the neighboring objects. However, we will be interested in the relationship of this equation with Maxwell's equations. If we introduce the assumption that there are no electric charges and electric currents ($\rho = 0, \boldsymbol{j} = 0$), Maxwell's equations simplify to a form "symmetrical" with respect to electric and magnetic fields:

$$\nabla \cdot \boldsymbol{E} = 0$$

$$\nabla \cdot \boldsymbol{B} = 0$$

$$\nabla \times \boldsymbol{E} = -\frac{1}{c}\frac{\partial \boldsymbol{B}}{\partial t}$$

$$\nabla \times \boldsymbol{B} = \frac{1}{c}\frac{\partial \boldsymbol{E}}{\partial t}$$

By performing appropriate mathematical operations on the above equations and using the laws combining the rotation and divergence operations, we can show that both the vector $\boldsymbol{E}$ and $\boldsymbol{B}$ satisfy the wave equation (see section 4.9* for details), with the constant c appearing in place of the velocity v:

$$\frac{\partial^2 \boldsymbol{E}}{\partial x^2} + \frac{\partial^2 \boldsymbol{E}}{\partial y^2} + \frac{\partial^2 \boldsymbol{E}}{\partial z^2} = \frac{1}{c^2}\frac{\partial^2 \boldsymbol{E}}{\partial t^2}$$

$$\frac{\partial^2 \boldsymbol{B}}{\partial x^2} + \frac{\partial^2 \boldsymbol{B}}{\partial y^2} + \frac{\partial^2 \boldsymbol{B}}{\partial z^2} = \frac{1}{c^2}\frac{\partial^2 \boldsymbol{B}}{\partial t^2}$$

These equations seem to describe waves propagating at the velocity c. We will return to the analysis of the physical sense of the constant c in a moment, but for now let us mention that Maxwell's equations provide us with additional information about the physical properties of these waves (if they occur in reality). Suppose our wave propagates in the direction of the z-axis. By decomposing the vectors of electric and magnetic field strength into three components along the x, y and z axes, it is possible to prove, using the assumption of zero divergence, that the components E_z and B_z will be 0. This means that both the electric and magnetic field vectors are perpendicular to the direction of wave propagation. Waves of this kind are called transverse (similar to waves on the surface of water or on a vibrating string).

Finally, we can see that the magnetic field vector $\boldsymbol{B}$ of the propagating wave will always be perpendicular to the vector $\boldsymbol{E}$. This follows directly from Maxwell's fourth equation – rotation $\nabla \times \boldsymbol{B}$ is a vector (cross) product, and therefore it is perpendicular to $\boldsymbol{B}$, and the equation shows that it must be equal to the direction of change of the field $\boldsymbol{E}$, and therefore also to the direction of $\boldsymbol{E}$ itself. Thus, the propagation of electromagnetic waves can be represented as two perpendicular waves, whose direction of displacement from equilibrium is perpendicular to the direction of propagation, as in Figure 4.14.

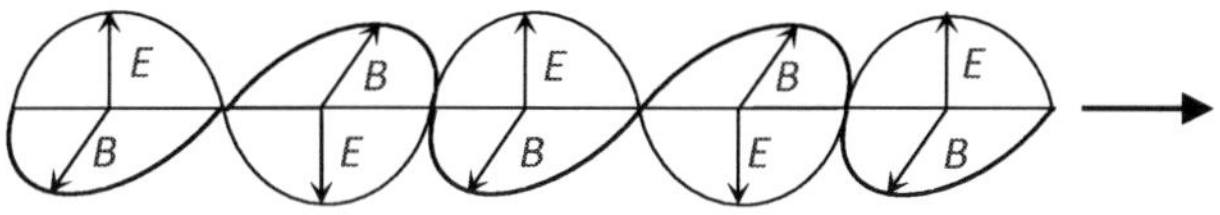

Fig. 4.14 Electromagnetic wave propagating in space

4.8 Light as an electromagnetic wave

Let us now return to the issue of the interpretation of the constant c, which appears in Maxwell's equations and also in the wave equation. In textbooks, this constant is usually defined as the speed of light in a vacuum. However, this is an ahistorical approach, because the relationship between the laws of electromagnetism and the speed of light was detected relatively late. Originally, the constant c had to be interpreted simply as a certain physical constant, the value of which could be determined experimentally on the basis of appropriate measurements of electrostatic and magnetic interactions. The fact that the speed of light as such does not occur directly in the equations of electromagnetism is best seen when these equations are expressed in other units – instead of the so-called Gaussian units (also called CGS – centimeter, gram, second), as in the above approach, in SI units (or MKSA: meter, kilogram, second, ampere). In Maxwell's equations expressed in metric units, two constants appear: the dielectric permittivity of vacuum ε_0 and the magnetic permeability of vacuum μ_0. The dielectric permittivity of vacuum can be determined using Coulomb's law, as it enters into the proportionality factor $k = \frac{1}{4\pi\varepsilon_0}$. However, for practical reasons, its value is precisely determined from the equation describing the electrical capacitance of a parallel-plate capacitor with a given area and distance between the plates. On the other hand, constant μ_0 is assigned a conventional value in MKSA units, due to the fact that in this system we define the unit of current (ampere) by the forces exerted by the magnetic interaction between conductors and current. Notwithstanding these details, which may be too convoluted for impractical philosophers, it is important to emphasize that the constants in the equations of electromagnetism are in no way defined by the speed of light.

Moving from the MKSA system to the CGS, we get the following relationship between the constants used in them:

$$c = \frac{1}{\sqrt{\varepsilon_0 \mu_0}}$$

Thus, the constant *c* characterizes the electrical and magnetic properties of the vacuum. However, if we calculate its value in units m/s, it turns out that it is surprisingly close to the known value of the speed of light in a vacuum, determined in various experiments (e.g. by Ole Rømer as early as the 17th century, and later with much greater accuracy in the experiments of Armand Fizeau and Jean Foucault). Added to this is the fact that constant *c* appears in Maxwell's derived wave equation in the place of velocity. These two facts cannot be a mere coincidence – the probability that they happened by chance is unimaginably low. Imagine, for example, that while studying the writings of an ancient author, we came across a text that is deceptively similar to Shakespeare's *Hamlet.* Would it be justified if we shrugged our shoulders "well, it's just a coincidence that is not worth bothering about"? It is obvious that we would immediately look for an explanation – most likely the examined writings are a modern forgery, or perhaps Shakespeare copied his drama from a work by an earlier author. Excluding these hypotheses, we can finally, following the guidance of the great detective Sherlock Holmes[10], assume that someone from the future built a time machine and for some unknown reason gave this author a copy of *Hamlet*, who made such an impression on him that he recreated it in his work. Whatever solution to the mystery would eventually be accepted, there is no doubt that the detected coincidence would not be ignored.

The same is true of the coincidence between the value of a certain electromagnetic constant and the speed of light. The most natural explanation for this fact is the assumption that electromagnetic waves propagate in a vacuum at speed *c*, and secondly, that light is an electromagnetic wave. However, the matter does not end there. To propose an explanation of a given fact is one thing, and to justify the truth of this explanation is another. Ptolemy beautifully explained the retrograde motion of the planets with his epicycles, but this explanation turned out to be false. How do we know that a similar situation cannot be repeated in the case under consideration? The principles of scientific methodology oblige us to subject new hypotheses to comprehensive empirical tests. This is what happened in the case of the electromagnetic wave hypothesis. Its first experimental test was conducted by a German physicist, Heinrich Hertz. Hertz used a primary conductor in his experiment, in which with the help of a coil he produced a high voltage to cause electromagnetic oscillations, and then noticed that in a secondary circuit placed at a certain distance, spark discharges appear with a frequency correlated with the frequency of the primary oscillations. Moreover, Hertz confirmed the wave nature of the propagating electromagnetic force by creating a standing wave by means of a metal plate reflecting the wave coming from the original circuit. Guglielmo Marconi's later application of electromagnetic waves to wireless radio communication sealed the success of Maxwell's theory.

The claim that visible light is an electromagnetic wave with a frequency in a certain range is today a truism comparable perhaps to the thesis about the sphericity of the Earth. However, it is difficult to point to one particular experimental fact confirming this thesis. Rather, a whole range of phenomena revealing the wave nature of light (interference, diffraction) are involved here, which are consistent with Maxwell's description of electromagnetic waves. We know from experiments that light is a transverse wave, which agrees with the predictions

---

10 As Holmes explained to his friend Watson, "When you have eliminated all which is impossible, then whatever remains, however improbable, must be the truth".

of Maxwell's theory. Another fact confirming the electromagnetic nature of light is the phenomenon of polarization, commonly used, for example, in sunglasses. Finally, Maxwell's theory allows us to derive many observable optical regularities: the law of refraction, reflection, or facts about the scattering of light by media such as air. Therefore, it can be assumed that the confirmation of the hypothesis about the electromagnetic nature of light radiation, although indirect, is strong due to its wide range and diversity.

What impact should the episode with the discovery of electromagnetic waves have on our assessment of the theory of electromagnetism? It is widely believed that the theoretical prediction of the existence and properties of electromagnetic waves was a huge success and the crowning achievement of Maxwell's theory. Thanks to this success, the theory gained additional strong empirical support, strengthening our confidence in it. Of course, Maxwell's theory explains and accurately describes many well-known electrical and magnetic phenomena: interactions between charged bodies and between conductors with electricity, the phenomenon of electromagnetic induction, current flows in circuits, and many others. However, the significance of the discovery of electromagnetic waves on the basis of Maxwell's theoretical considerations is incomparably greater. Is the only reason for this fact the spectacular nature of this discovery and its wide practical applications? After all, we cannot imagine modern technical civilization without the widespread use of radio waves. However, there are other reasons why theories that predict the occurrence of new, previously unknown phenomena should gain a much stronger degree of confidence than theories that only reproduce previously known facts. It is quite obvious that at the formative stage of a given theory, all previously known data and facts should be taken into account and included in the theoretical description. It was not surprising, for example, that Ptolemy's epicycles explained the phenomenon of retrograde planetary motion well, because that was exactly what they were introduced for. The same is true of the wealth of facts about electromagnetic phenomena collected by generations of researchers – all of them were used to construct the appropriate theory, and thus were included in it at the stage of creation. This is not to say that the gathering of a variety of facts and phenomena into a single synthetic mathematical formulation is not in itself a significant scientific achievement. However, the theory obtained in this way is still confirmed because it was constructed in this way. To reinforce our belief that the theory is correct, we need something completely new. Such a new, unexpected confirmation for Maxwell's theory was precisely the discovery of electromagnetic waves.

In the box below you will find a slightly different formal argument that the influence of previously unknown evidence on the assessment of the probability of a hypothesis should be greater than the impact of known facts. This argument is formulated within the so-called Bayesian approach to confirming hypotheses. This approach is based on simple laws of probability calculus, to which is added an interpretation of the concept of probability in terms of the subjective degree of belief. For example, when we say that the probability of a hypothesis is equal to 1/2, we mean that we have as strong a conviction in its truth as in its falsity.

In probability theory, a simple theorem, known as Bayes' theorem, can be proven. For any two events H and E, the probability of event H conditional on E is equal to the probability of E assuming H times the probability of H divided by the probability of E occurring:

$$P(H|E) = \frac{P(E|H)P(H)}{P(E)}.$$

This formula can be directly derived from the definition of conditional probability: $P(A|B) = \frac{P(A \cap B)}{P(B)}$. In application to the confirmation of hypotheses by evidence, it is assumed that *H* is a general hypothesis (e.g. in our case it is Maxwell's theory) and *E* is a certain experimentally confirmed fact (the existence of electromagnetic waves). Conditional probability $P(H|E)$ expresses numerically the degree of belief that hypothesis *H* is true, assuming that we have verified the occurrence of fact *E* (this is the so-called posterior probability of the hypothesis). On the other hand, probability $P(E|H)$ can be considered equal to unity in a situation in which the hypothesis *H* logically implies the occurrence of the event *E* (as is arguably the case with Maxwell's theory and the existence of electromagnetic waves). The probability *P*(*H*) is the so-called prior probability of the hypothesis (how reliable we consider our hypothesis before the evidence *E* appeared).

The most important piece of the puzzle is the probability *P*(*E*), which determines our estimate of the chance that the evidence *E* is true, regardless of the truth of the hypothesis *H*. This parameter sharply distinguishes previously known facts from new, unexpected facts. For example, the phenomenon of electromagnetic induction was known before Maxwell formulated the whole set of equations, so it can be assumed that the degree of certainty with which we accept the occurrence of this phenomenon is one. In such a situation, the posterior probability of the theory, assuming the occurrence of the phenomenon of electromagnetic induction, is equal to the prior probability, i.e. the occurrence of the phenomenon of induction does not increase the credibility of our hypothesis. On the other hand, the existence of electromagnetic waves was in no way known (or even suspected) before the formulation of Maxwell's hypothesis, so we can assume that the probability of *P*(*E*) had to be relatively low (but not lower than the probability *P*(*H*) of the hypothesis itself, since by assumption *H* implies it). Therefore, the ratio $\frac{P(H)}{P(E)}$ can be a relatively large number, approaching unity within the limit. This shows that the occurrence of a previously unknown but theoretically predicted event *E* significantly strengthens our degree of confidence in the hypothesis.

Deriving a mathematical description of electromagnetic waves from Maxwell's equations was an unquestionable success, but paradoxically it contained an element of destruction, which eventually led to the blowing up of the foundation of classical physics – Galileo's and Newton's concepts of time and space. The ignition spark here was the interpretation of the velocity *c* in the wave equation describing the propagation of electromagnetic waves. As we know from Galileo, the concept of velocity always requires relativization, i.e. determining what we measure a given speed against. In the case of equations describing mechanical waves, this is not a problem, because the speed of the wave is always measured in relation to the medium in which the wave propagates – air, water or metal. The problem arises when describing electromagnetic waves, for which there is no clearly defined medium of their propagation. The disturbance for an electromagnetic wave is the value of the electric field and magnetic field at a given point in space, and not the corresponding physical state of a substance (e.g. pressure or density of a small portion of gas). Electromagnetic waves travel without any problems in a vacuum, in which there is no physical medium.

But is it really the case? Many nineteenth-century physicists, including Maxwell, believed that electrical and magnetic interactions were transmitted through an invisible substance called the ether. The ether belongs to the category of invisible fluids, one example of which that we have already learned about is caloric – an alleged carrier of thermal phenomena. The concept of the ether appeared already in ancient astronomical systems (in particular in Aristotle) as representing a substance filling the supralunar world (in the Middle Ages also called *quinta essentia*). It was used, among others, by Newton to describe optical phenomena. In the modern version, the ether was supposed to be a substantial basis for electric and magnetic fields, which in this view could be treated as a "disturbance" of a certain physical medium. As a curiosity, it can be added that Maxwell himself created a very ingenious model of the electromagnetic ether based on the idea of transferring interactions by means of something like interlocking wheels or vortices. This was therefore an attempt to reduce electromagnetic phenomena to mechanical interaction by contact. Let us note that accepting Maxwell's model of the ether as an adequate description of reality solves the problem of action at a distance (non-locality), which we discussed earlier. For example, the action of a current in a conductor on a distant magnetic needle can be explained by the fact that the moving electrons in the conductor excite the adjacent wheels (vortices) of the ether, which again transmit this motion accordingly until it reaches the magnetic needle.

However, the concept of a mechanical ether that carries electromagnetic interactions and also allows electromagnetic waves to propagate has encountered considerable theoretical difficulties. As we remember, electromagnetic waves are transverse waves. We know from experience that transverse waves propagate in elastic mediums such as metals. Inelastic media, such as air, mainly transmit longitudinal waves (expanding and contracting the medium in the direction of wave propagation). However, it is difficult to imagine a mechanical medium that would be elastic enough to allow the propagation of transverse waves, and at the same time would offer no noticeable resistance to the physical bodies moving within it. Skepticism about the ether hypothesis gradually grew, but the final blow to this hypothesis was dealt by experiments determining the speed of light in different reference frames. We will discuss these experiments in detail in the next chapter, devoted to an introduction to special relativity.

4.9* Mathematical foundations of Maxwell's theory

As you have probably noticed, the theory of electromagnetism uses a fairly advanced mathematical apparatus, the basic elements of which are vectors and operations on vectors, including operations referring to differential and integral calculus. Let us try to take a closer look at these concepts, starting with two fundamental operations: dot and cross product. We will consider vectors in three-dimensional space, i.e. having three components along arbitrarily selected perpendicular axes x, y and z. Let $\boldsymbol{A}$ and $\boldsymbol{B}$ be vectors with the components A_x, A_y, A_z and B_x, B_y, B_z (remember that the components of a vector are ordinary numbers). The dot (scalar) product of vectors $\boldsymbol{A}$ and $\boldsymbol{B}$ is the following number:

$$\boldsymbol{A} \cdot \boldsymbol{B} = A_x B_x + A_y B_y + A_z B_z. \tag{4.9}$$

It is possible to prove, although we will not do it, that this formula is equivalent to the definition known from the school course: "the product of the length of both vectors times the cosine of the angle between them". It is important that the dot product allows us to define the concept of perpendicularity. Two vectors are perpendicular when their dot product is zero. This can be proven in general using somewhat complicated trigonometric functions, or by applying the following trick: since the choice of the coordinate system is a matter of convention, it is always allowed to choose three axes so that only one coordinate of the vector $\boldsymbol{A}$ is non-zero, e.g. A_x. Then the dot product of $\boldsymbol{A}$ with any vector $\boldsymbol{B}$ is reduced to the product of $A_x B_x$, and thus its zeroing means that $B_x = 0$. If so, then the only possible non-zero components of the vector $\boldsymbol{B}$ are B_y and B_z, i.e. it is a vector perpendicular to $\boldsymbol{A}$ (it lies in the yz plane, perpendicular to the x-axis).

The cross product of two vectors is a bit more complicated. First of all, the result of it is a vector, not a number (scalar). The components of this vector are as follows:

$$
\begin{aligned}
(\boldsymbol{A} \times \boldsymbol{B})_x &= A_y B_z - B_y A_z \\
(\boldsymbol{A} \times \boldsymbol{B})_y &= A_z B_x - B_z A_x \\
(\boldsymbol{A} \times \boldsymbol{B})_z &= A_x B_y - B_x A_y
\end{aligned}
\tag{4.10}
$$

As you can see, each component of the cross product is given in the form of some combination of the other two components of the multiplied vectors. It is not difficult to show that the vector $\boldsymbol{A} \times \boldsymbol{B}$ will be perpendicular to both $\boldsymbol{A}$ and $\boldsymbol{B}$. Please check for yourself, using formulas (4.9) and (4.10), that the following equalities hold:

$$(\boldsymbol{A} \times \boldsymbol{B}) \cdot \boldsymbol{A} = 0$$

$$(\boldsymbol{A} \times \boldsymbol{B}) \cdot \boldsymbol{B} = 0$$

Let us now introduce the important differential operations of divergence and rotation. The key to these concepts is to treat the differentiation operation with respect to the three coordinates x, y, and z as a vector. This is, of course, a certain "abuse" of notation, but it turns out to be very convenient. So we will define the following differential "vector" $\boldsymbol{\nabla}$, by giving its components

$$\nabla_x = \frac{\partial}{\partial x}$$

$$\nabla_y = \frac{\partial}{\partial y}$$

$$\nabla_z = \frac{\partial}{\partial z}$$

or "collectively":

$$\boldsymbol{\nabla} = \left(\frac{\partial}{\partial x}, \frac{\partial}{\partial y}, \frac{\partial}{\partial z}\right)$$

As you can see, the components of a vector $\boldsymbol{\nabla}$ (called "nabla") are not numbers, but operators. However, it can be used for ordinary scalar and vector product operations based on multiplication of numbers, if we replace multiplication with the action of the operator on the corresponding vector. To do this correctly, however, we need to have not one vector, but a whole vector field. A vector field is nothing more than a function that assigns a vector to each point in space. Mathematically, it is best to represent such a function as consisting of three functions with three real arguments, such that each function assigns to the coordinates x, y, a component of the vector in a given direction. Thus, in general, the vector field $\boldsymbol{A}$ is given as follows:

$$A(x,y,z) = \left(A_x(x,y,z), A_y(x,y,z), A_z(x,y,z)\right).$$

In practice, the arguments x, y, and z are often omitted, and a given vector field is represented by a single symbol A. However, it should be remembered that it is not one vector, but a whole infinite group of them, each of which is assigned to one point.

We can now define divergence and rotation operations on the vector field A using the formulas (4.9) and (4.10):

$$\nabla \cdot A = \frac{\partial A_x}{\partial x} + \frac{\partial A_y}{\partial y} + \frac{\partial A_z}{\partial z} \tag{4.11}$$

$$\nabla \times A = \left(\frac{\partial A_z}{\partial y} - \frac{\partial A_y}{\partial z}; \frac{\partial A_x}{\partial z} - \frac{\partial A_z}{\partial x}; \frac{\partial A_y}{\partial x} - \frac{\partial A_x}{\partial y}\right) \tag{4.12}$$

Remember that the expressions A_x, A_y, A_z denote functions of three arguments, so they can be differentiated with respect to each argument separately. In this way, we will obtain two new functions defined on points (triples of real numbers): the divergence function, which assigns a number to each point (so it is a scalar field), and the rotation function, which assigns an appropriate vector to the points.

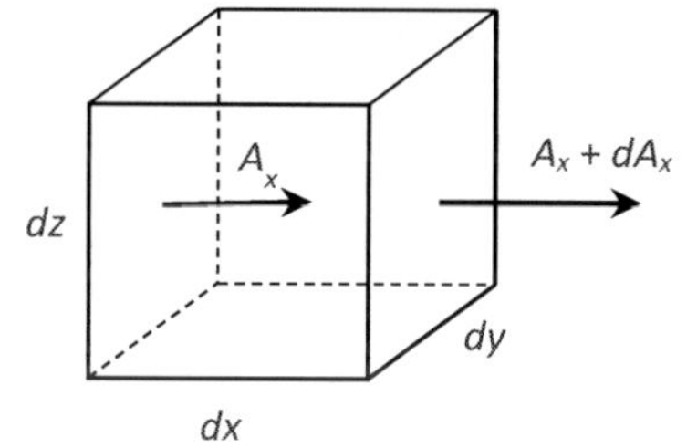

Fig. 4.15 Explanation of the concept of divergence

What is the physical meaning of both operations? Let us start with divergence. Consider an "infinitesimal" cube with sides dx, dy, and dz (Fig. 4.15). Imagine that the vector field A represents the flow of a certain substance (for example, the velocity field of water in a stream). Let dA_x denote the change in the field component A_x with a shift along the x-axis by the value dx. Multiplying this quantity by the area of the side surface of the cube perpendicular to x we get $dA_x dydz$. It is, as you can easily guess, a measure of the amount of outflow of a given substance from a cube through the walls perpendicular to x per unit of time (e.g. if the component A_x does not change at all, i.e. $dA_x = 0$, the total outflow will be zero: the same amount flows in as it flows out). If we want to count this amount relative to the unit of volume, we have to divide the quantity $dA_x dydz$ by the volume element $dxdydz$. But in this way, we will get the rate of change of the component A_x along x: $\frac{dA_x}{dx}$, or rather $\frac{\partial A_x}{\partial x}$, given that A_x depends on three variables. Analogous reasoning shows that the partial derivatives $\frac{\partial A_y}{\partial y}$ and $\frac{\partial A_z}{\partial z}$ represent the outflows of a given substance along the y- and z-axes per units of time and volume. Adding up these three numbers we get the total field outflow per unit volume from the infinitesimal cube.

Rotation, as the name suggests, represents the tendency of a vector field to rotate with small shifts. We will not justify this in general, so as not to overload our argument with various trigonometric functions. However, we can see that if field $\boldsymbol{A}$ has a constant direction and its magnitude does not change in perpendicular directions (cf. Fig. 4.16), its rotation will be zero. Suppose $\boldsymbol{A}$ has only one non-zero component A_x, which of course can change its value, but only in the direction of the x-axis. In this situation, all partial derivatives appearing in the definition (4.12) of rotation $\nabla \times \boldsymbol{A}$ will be zero. The derivatives of zero components A_y, A_z are zero for obvious reasons, while the only derivatives of the component A_x included in the equation are derivatives in the y and z directions, in which, as we assumed, this component does not change, i.e. the result is also zero. In order for the rotation of a vector field to be non-zero, the field must change its direction with small shifts, or change its value with shifts in the direction perpendicular to the vector (this creates a shear force which will cause objects to rotate, if the field is a real flow of some substance), as in the figure.

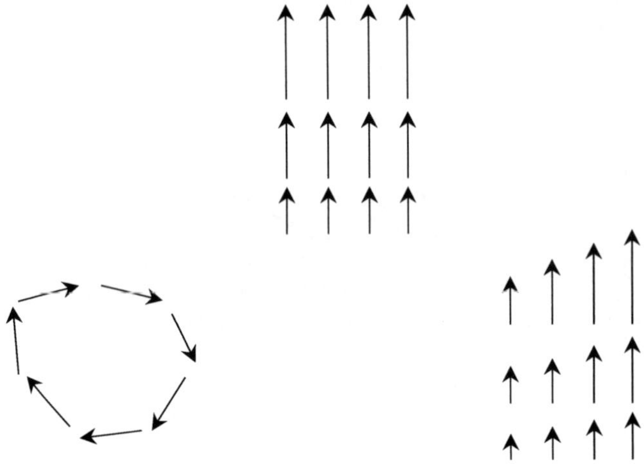

Fig. 4.16 Example of a vector field with zero rotation (top) and vector fields with non-zero rotation (bottom).

As we remember, the divergence and rotation operators are found in the differential version of Maxwell's equations. However, more intuitive from a physical point of view are the integral versions, which describe well-known phenomena – electromagnetic induction or the production of a magnetic field by conductors with electricity. How to obtain their "abstract" differential forms from the integral laws of electromagnetism? Two important mathematical theorems are helpful here: Gauss' theorem and Stokes' theorem. In general, these theorems equate an integral over an n-dimensional manifold with an integral over the manifold of one more dimension (n+1). For example, Gauss' theorem connects integrals over a two-dimensional closed surface with integrals over a three-dimensional area enclosed by that surface. Stokes' theorem, on the other hand, compares an integral over a one-dimensional closed curve with an integral over a two-dimensional area enclosed by this curve.

It is interesting that the same class of theorems (some believe that it is actually one general theorem) includes the so-called fundamental theorem of differential and integral calculus. Probably many of you have come across the slogan that differentiation is the "inverse" of integration. This means that if we take the derivative of a given function and then calculate the integral from this derivative, we get back the same function, which can be expressed as follows:[11]

$$\int \frac{df(x)}{dx}dx = f(x).$$

The integral on the left side of the equation is one-dimensional, as can be seen from the infinitesimal element of length *dx*. The function on the right, on the other hand, can be interpreted as an integral on a zero-dimensional surface, i.e. a point. Since the integral over a given area of a function is the sum of products of the value of this function times a small element of the area, then in the case of a point such a "zero integral" is simply the value of this function itself.

Let us note an interesting regularity – in order to express a "lower-dimension" integral by means of an integral one degree higher, a differential operator should be used in the latter. So the scheme of the general theorem is as follows: the n-dimensional integral of any function is equal to the $n+1$-dimensional integral of some differential operator of this function. It only remains to determine what kind of operator it will be. Let us take the case of the transition from a one-dimensional integral to a two-dimensional integral. Let O be a closed curve and S be the surface inside this curve. Consider the integral along the curve O from some vector field $\boldsymbol{A}$. Then the appropriate differential operator will be the rotation of the field $\boldsymbol{A}$, and Stokes' theorem, expressing the connection between the integrals after O and S, will have the form:

$$\oint_O \boldsymbol{A} \cdot d\boldsymbol{l} = \int_S (\boldsymbol{\nabla} \times \boldsymbol{A}) \cdot d\boldsymbol{S}. \tag{4.13}$$

Note that in the expressions under the integrals there is the dot product of vectors that we already know, which means that we will be integrating functions of real numbers. In the case of a higher-dimensional case, the operator that allows us to "transition" from a two-dimensional surface to a three-dimensional volume is, of course, divergence. Hence we have Gauss' theorem (also called the Green-Ostrogradski-Gauss theorem):

$$\int_S \boldsymbol{A} \cdot d\boldsymbol{S} = \int_V \boldsymbol{\nabla} \cdot \boldsymbol{A} dV \tag{4.14}$$

As usual, the regions of integration are related to each other in such a way that the volume V is contained in the closed surface S.

We can now apply the above substitutes to Maxwell's equations in the integral version. Let us start with the first law of electrostatics, also called, albeit confusingly, Gauss' law.[12]

$$\int_S \boldsymbol{E} \cdot d\boldsymbol{S} = 4\pi Q$$

[11] Strictly speaking, any constant C can be added to the function $f(x)$ on the right side of the equation. This constant can be eliminated by calculating the definite integral within the ranges from a to b, which will give us the equation .$\int_a^b \frac{df(x)}{dx}dx = f(b) - f(a)$

[12] Let us remember that Gauss' theorem is a thesis of mathematics, i.e. an *a priori* truth, while Gauss' law is an empirical thesis that could be false in the physical world.

Recall: the expression on the left is the total electric field flux flowing through the closed surface S, while on the right side we have the total electric charge Q contained in the surface S. We can now replace the surface integral with the volume integral of the divergence of field $\boldsymbol{E}$ using Gauss' theorem (4.14). On the other hand, the charge Q can be calculated by integrating the charge density ρ over the entire volume V. We will then get:

$$\int_V \nabla \cdot \boldsymbol{E} dV = 4\pi \int_V \rho dV$$

By equating the functions under the integrals (which corresponds to "shrinking to a point"), we get Maxwell's first equation in the differential version (cf. 4.7):

$$\nabla \cdot \boldsymbol{E} = 4\pi\rho$$

Let us show what the analogous transition looks like in the case of Faraday's law (4.3):

$$\frac{1}{c}\frac{d}{dt}\int_S \boldsymbol{B} \cdot d\boldsymbol{S} = -\oint_O \boldsymbol{E} \cdot d\boldsymbol{l}$$

Using Stokes' theorem (4.13), we replace the integral on the right:

$$\frac{1}{c}\frac{d}{dt}\int_S \boldsymbol{B} \cdot d\boldsymbol{S} = -\int_S (\nabla \times \boldsymbol{E}) \cdot d\boldsymbol{S}$$

By moving the operation of time differentiation "under" the integral and at the same time converting it into a partial derivative (because function $\boldsymbol{B}$ also depends on spatial coordinates), we get, after dropping the integrals:

$$\nabla \times \boldsymbol{E} = -\frac{1}{c}\frac{\partial \boldsymbol{B}}{\partial t}$$

that is, another Maxwell's equation in the differential version. In the same way, we will obtain the remaining equations.

Let us now proceed to derive the wave equation from Maxwell's laws assuming that the charges and currents are zero. Here are these laws for the record:

$$\begin{gathered} \nabla \cdot \boldsymbol{E} = 0 \\ \nabla \cdot \boldsymbol{B} = 0 \\ \nabla \times \boldsymbol{E} = -\frac{1}{c}\frac{\partial \boldsymbol{B}}{\partial t} \\ \nabla \times \boldsymbol{B} = \frac{1}{c}\frac{\partial \boldsymbol{E}}{\partial t} \end{gathered} \tag{4.15}$$

Let us consider the fourth of the above equations and apply the time differentiation operation to both sides of it. Taking advantage of the fact that time differentiation commutes with rotation, we get:

$$\nabla \times \frac{\partial \boldsymbol{B}}{\partial t} = \frac{1}{c}\frac{\partial^2 \boldsymbol{E}}{\partial t^2}.$$

Substituting the expression for $\frac{\partial \boldsymbol{B}}{\partial t}$ from the third law, we get

$$-\nabla \times (\nabla \times \boldsymbol{E}) = \frac{1}{c^2}\frac{\partial^2 \boldsymbol{E}}{\partial t^2}. \tag{4.16}$$

We now need to apply a mathematical relationship between rotation, divergence operations, and one more operation, called the Laplace operation (abbreviated as the Laplacian). It is defined as follows:

$$\nabla^2 = \frac{\partial^2}{\partial x^2} + \frac{\partial^2}{\partial y^2} + \frac{\partial^2}{\partial z^2}$$

The mathematical relationship is as follows:

$$\nabla \times (\nabla \times \boldsymbol{E}) = \nabla(\nabla \cdot \boldsymbol{E}) - \nabla^2 \boldsymbol{E} \tag{4.17}$$

By laboriously writing both sides of the equation in terms of their three components, using the above definitions of divergence and rotation (4.11) and (4.12), we can confirm that such an equality does indeed hold. We will leave this issue to interested readers to verify on their own, but we should perhaps note one small fact: the operation $\nabla(\nabla \cdot \boldsymbol{E})$ on the right side of the equation is not the same as Laplacian $\nabla^2 \boldsymbol{E}$, despite the superficial similarity (thus nabla does not behave like a number). Divergence $\nabla \cdot \boldsymbol{E}$, as we know, gives us a scalar of the form $\frac{\partial E_x}{\partial x} + \frac{\partial E_y}{\partial y} + \frac{\partial E_z}{\partial z}$, while applying to it the nabla operator ∇ (i.e. gradient) produces a vector whose x component looks like this:

$$\frac{\partial^2 E_x}{\partial x^2} + \frac{\partial^2 E_y}{\partial x \partial y} + \frac{\partial^2 E_z}{\partial x \partial z}$$

This is of course a different result from the x component of the vector $\nabla^2 \boldsymbol{E}$, which has the form:

$$\frac{\partial^2 E_x}{\partial x^2} + \frac{\partial^2 E_x}{\partial y^2} + \frac{\partial^2 E_x}{\partial z^2}$$

Returning to the mathematical relationship (4.17), let us note that from Maxwell's first equation of zero-charge and zero-current equations (4.15) it follows that $\nabla(\nabla \cdot \boldsymbol{E}) = 0$. So the double rotation of the electric field will simply be equal to the Laplacian with a minus. We can therefore insert this Laplacian into formula (4.16), resulting in:

$$\nabla^2 \boldsymbol{E} = \frac{1}{c^2}\frac{\partial^2 \boldsymbol{E}}{\partial t^2}$$

It is none other than the wave equation for the $\boldsymbol{E}$ vector. In the same way, we obtain the equation for the magnetic field vector. Thus, in fact, Maxwell's equations contain a description of a new phenomenon: the electromagnetic wave.

Therefore, we can write down the components of the electric (and magnetic) field at a given point as a certain disturbance F propagating in the direction of the z axis and dependent on time t, as we did in the previous section:

$$E_x = \mathcal{E}_x F(z - vt)$$

$$E_y = \mathcal{E}_y F(z - vt)$$

$$E_z = \mathcal{E}_z F(z - vt)$$

The function F can be, for example, the well-known periodic function sine or cosine, and in this case the numbers $\mathcal{E}_x$, $\mathcal{E}_y$, $\mathcal{E}_z$ will be amplitudes in the x, y and z directions, respectively. Note that since the components of the field in the x and y directions depend only on the z direction, their derivatives with respect to x and y are zero:

$$\frac{\partial E_x}{\partial x} = 0$$

$$\frac{\partial E_y}{\partial y} = 0$$

Remember, however, that the divergence of the ***E*** field is also equal to zero (the first equation in 4.15):

$$\frac{\partial E_x}{\partial x} + \frac{\partial E_y}{\partial y} + \frac{\partial E_z}{\partial z} = 0$$

From this we conclude that

$$\frac{\partial E_z}{\partial z} = 0.$$

But this is only possible if $\mathcal{E}_z = 0$, and therefore the entire component E_z vanishes. From this we conclude that the vector of the electric field is always perpendicular to the direction of wave propagation. The same is true with respect to the magnetic field vector, and in addition, we know that the vector ***E*** must be perpendicular to ***B***. Thus, the electromagnetic wave looks like the one shown in Fig. 4.14.

Questions and problems

1. Does the fact that there are two types of electrostatic forces (attractive and repulsive) imply the existence of exactly two types of electric charge? Consider the theoretical possibility of three types of charges, such that like charges repel each other while distinct charges attract. What known experimental facts show the falsity of such a hypothesis?

2. Explain the concept of an electrostatic field. What is the relationship between the assumption of the reality of electrostatic fields and the problem of non-locality of electrostatic interactions?

3. Compare the problem of the reality of the electrostatic field with the philosophical problem of the reality of dispositional properties. What kind of dispositional property could a field be?

4. Discuss the basic arguments for the unreality of electrostatic potential and electric lines of forces.

5. How can we derive Coulomb's law (the dependence of the field strength on the inverse of the square of the distance) from the false assumption that an electric field arises as a result of the outflow of an incompressible fluid from charges?

6. Is there a law analogous to Coulomb's law in electrostatics describing magnetic interactions? What is the reason for the different treatment of electric and magnetic fields?

7. How does the magnetic field affect electric charges? What is the necessary condition for an electric charge to "feel" a magnetic force acting on it?

8. Discuss the concepts of field circulation along a closed curve and field flux through a surface. Does zero circulation around a given curve and zero flux through a closed surface imply that the field is also zero?

9. Discuss the phenomenon of electromagnetic induction and Faraday's law describing it. Does the mathematical form of Faraday's law contain information about the causal relationships between magnetic and electric fields?

10. Why does Ampère's law describing the formation of a magnetic field caused by the flow of charges need to be modified? What are the two possible sources of the magnetic field according to Ampère-Maxwell's law?

11. Present the content of all four laws of Maxwell's electromagnetism. What is the difference between the integral and differential version of Maxwell's equations?

12. Does the existence of laws linking electrical and magnetic phenomena mean that electricity and magnetism should be unified? What does the unification of certain types of phenomena mean?

13. How do electric and magnetic fields behave during the transition from one inertial frame to another? Give examples of this behavior. What does this have to do with the problem of the unification of these fields?

14. How can the electromagnetic field be described mathematically? How does such a description solve the problem of the variability of separate electric and magnetic fields due to a change of the reference system?

15. Outline the reasoning leading to the electromagnetic wave hypothesis. How was this hypothesis verified? What impact did the discovery of electromagnetic waves have on the status of the theory of electromagnetism?

16. Does the existence of electromagnetic waves provide an argument in the dispute between realistic and instrumentalist interpretations of fields?

17. What is the basis for the claim that light is an electromagnetic wave? What is the main problem with the interpretation of the constant c as the speed of light?

Further reading

A very good discussion of the theory of electromagnetism and its philosophical implications can be found in the book: M. Lange, *An Introduction to the Philosophy of Physics. Locality, Fields, Energy and Mass, Blackwell*, Oxford 2002.

Part V of the book J.T. Cushing, *Philosophical Concepts in Physics*, Cambridge University Press, Cambridge 1998, contains interesting historical details regarding the development of the classical theory of electromagnetism, including a sketch of Maxwell's original theory of the mechanical ether.

The presentation of the basic laws of electromagnetism along with a very clear explanation of their mathematical apparatus in Richard Feynman's famous textbook is well-worth reading: R.P. Feynman, R.B. Leighton, M. Sands, *Feynman's Lectures in Physics*, volume II, chapter 1 (numerous editions, also available on-line for free: https://www.feynmanlectures.caltech.edu/II_01.html).

An advanced textbook in classical electromagnetism, commonly used in university courses for physicists, is J.D. Jackson, *Classical Electrodynamics*, John Wiley & Sons 1962 and later editions.

CHAPTER 5. SPECIAL RELATIVITY

5.1 Experimental tests of the ether hypothesis

As we remember from the previous chapter, the most natural interpretation of the velocity of electromagnetic waves appearing in Maxwell's equations is to assume it to be the velocity relative to the medium of their propagation, i.e. relative to the ether (sometimes also called the luminiferous, i.e. light-bearing ether). However, independent confirmation of the ether hypothesis proved problematic. The expected effect of the existence of the medium in which light propagates is the dependence of the measured speed of light on the state of the observer's motion relative to it. Consider the example of sound waves propagating through the air. For an observer resting relative to the air, the speed of sound is about 340 m/s. On the other hand, an observer moving in the air will register a different speed – higher or lower, depending on the direction of movement. In particular, when a supersonic jet is breaking the sound barrier, the sound wave propagating in the direction of flight will have zero velocity relative to the aircraft. All this is of course due to the well-known law of velocity composition: in order to calculate the velocity of a given object relative to a moving reference frame, one must subtract the velocity of this system from the velocity of the object with respect to the resting frame.

In the case of the ether, however, the problem is that we cannot determine our speed relative to this substance, because it is invisible and does not directly affect our measuring instruments. Therefore, we must refer to indirect methods, using light rays, which will be our "detectors". The first question we should answer is whether physical bodies in motion are able to "drag" the invisible ether with them. This is an important problem, because if the dragging effect occurs, it may be difficult to observe the variability in the speed of light depending on the state of the observer's motion, since each time a layer of the ether of some thickness will travel with the observer. The dragging hypothesis was subjected to an experimental test in the so-called Fizeau experiment The question posed by Fizeau was whether the movement of water has an effect on the ether. If the water drags the ether with it, then the speed of light transmitted "upstream" should be different than the speed of light "downstream". Since light must travel at the constant speed c relative to the ether, and the ether by hypothesis moves with water at velocity v, the speed of light in the laboratory should be $c - v$ for the "upstream" case and $c + v$ in the "downstream" motion.

The experimental setup used in Fizeau's experiment is shown in Fig. 5.1. Two light rays were passed through tubes of flowing water – once in the direction of the water flow, and once in the opposite direction. Because of the huge difference between the speed of water v and the speed of light c, the differences in transit times for the two rays were negligible, but Fizeau used a very sensitive interference phenomenon to determine this difference. The result of the experiment was essentially negative – the theoretically predicted effect of the ether dragging was not confirmed. Therefore, it can be assumed that the ether does not show a tendency to interact with material bodies, unlike denser mediums such as water or air.

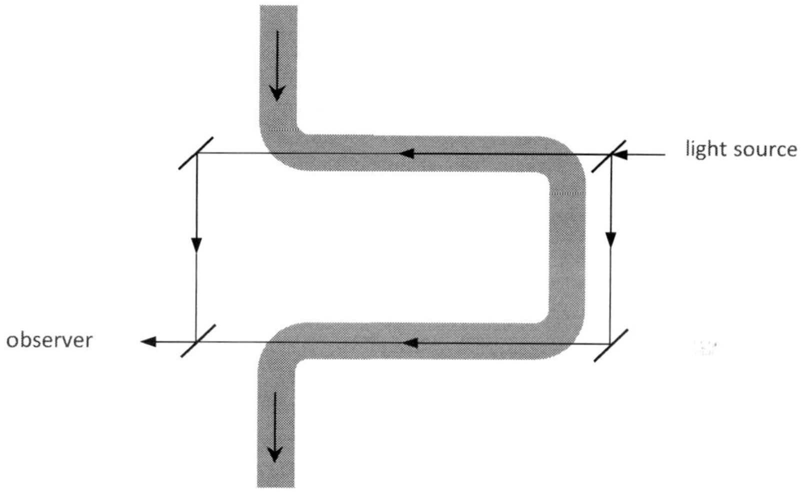

Fig. 5.1 Diagram of the Fizeau experiment

The second possibility regarding the behavior of the ether is expressed by the so-called hypothesis of the sea of ether. It assumes that the ether is a single whole, the fragments of which do not move relative to each other, and material objects simply "penetrate" through it, without changing its state of motion. In this view, the ether would be a kind of Newtonian absolute space, as it would determine the state of absolute rest. If this hypothesis is true – and in the absence of the dragging effect it seems to be the only logical possibility – then we should be able to observe the variation in the speed of light by changing the velocity of the measuring device (e.g. changing its direction of motion). This observation became the basis for the famous Michelson-Morley experiment. Let us take a closer look at the theoretical foundations of this experiment, as they will prove important later in our discussions. Consider a situation in which a light ray emitted from a certain source reaches a mirror placed at a certain distance l, and then after getting reflected returns to the point of origin. We will try to calculate the time in which the light ray will make its entire journey, depending on the relative motion of the system with respect to the ether.

The first case under analysis assumes that the entire device moves relative to the ether parallel to the course of the ray with a velocity v (Fig. 5.2). In such a situation, the velocity of the ray traveling in the direction towards the mirror will be $c - v$ (remember that by definition c is the speed of light relative to the ether, not relative to the laboratory), and in the opposite direction $c + v$. It is easy to calculate that the total time of the round trip will be as follows:

$$T_{\parallel} = \frac{l}{c-v} + \frac{l}{c+v} = \frac{2l}{c}\frac{1}{1-\frac{v^2}{c^2}}. \tag{5.1}$$

The second case is when the light ray moves perpendicularly to the direction of motion relative to the ether (Fig. 5.3). Then the path of the ray from the perspective of the ether will look like in the diagram (the ray will run at an angle in relation to the vertical line). Here we get a right triangle, whose sides are l and vt, and the hypotenuse equals ct (because it is the path of the ray relative to the ether). From the Pythagorean theorem we get

$$l^2 + (vt)^2 = (ct)^2.$$

From this we can calculate t, which is half the time of the total journey there and back. Thus, the total time $T_{\perp} = 2t$ is:

$$T_{\perp} = \frac{2l}{c}\frac{1}{\sqrt{1-\frac{v^2}{c^2}}}. \tag{5.2}$$

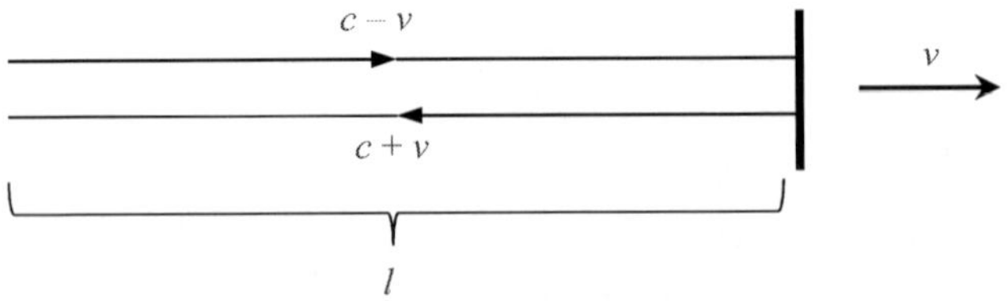

Fig. 5.2 A light ray running parallel to the motion of the system relative to the ether

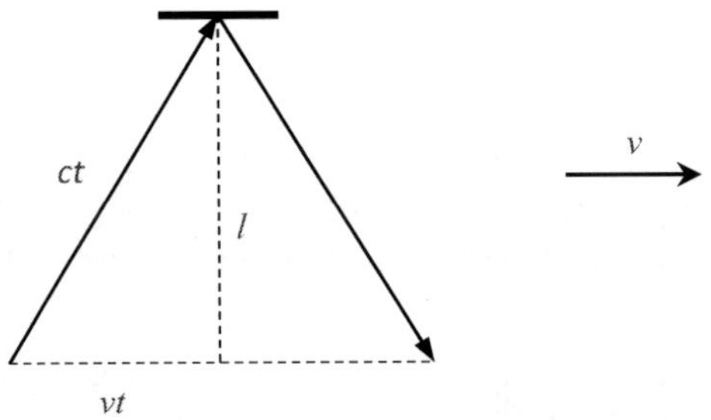

Fig. 5.3 A light ray traveling perpendicularly to the direction of motion relative to the ether

As you can see, times $T_{\parallel}$ and $T_{\perp}$ are different from each other. The Michelson-Morley experiment was designed to capture this difference. The outline of the experimental system – the so-called Michelson-Morley interferometer – is schematically presented in Fig. 5.4. The light ray coming out of the source passes through a half-silvered mirror, which splits it into two rays running perpendicular to each other. After reflection from the mirrors, the two independent light rays merge again, and the effect of their overlap can be observed through a scope. Since light is a wave, as a result of superimposing two wave sequences, we will observe the phenomenon of interference – amplification or weakening of intensity depending on the difference in optical paths. It is important that the interference effect depends very strongly on the time of arrival of both rays. Even a minute difference in the time of arrival between the arms of the interferometer will cause a visible shift in the interference fringes.

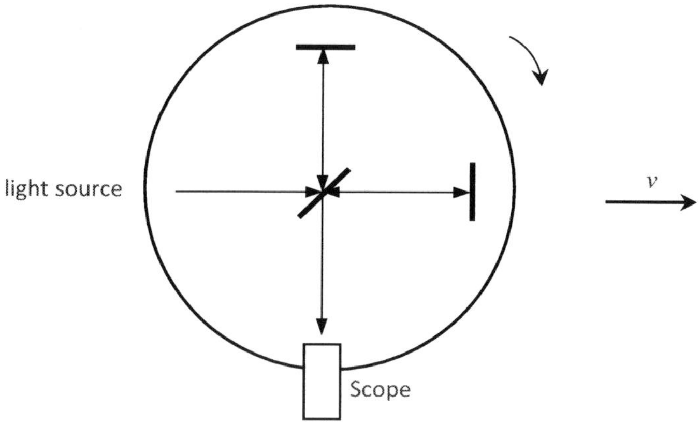

Fig. 5.4 Michelson-Morley's interferometer.

Michelson and Morley placed their device on a rotating base that allowed the position of both arms of the interferometer to smoothly change in relation to the environment. Their reasoning was as follows: if the laboratory moves relative to the ether, then by rotating the whole system we should change the transit times of the rays in both perpendicular directions depending on their position in relation to the direction of motion of the laboratory. These changes, in turn, should be observed in the form of a shift in the interference fringes. However, despite repeated observations, it was not possible to detect any change in the position of the interference fringes, and therefore no difference between the time of arrival of the light ray running along the direction of motion of the system, and the time of arrival of the ray perpendicular to the direction of motion.

The logical structure of the Michelson-Morley experiment is as follows. We assume that the ether as a whole is at rest relative to a certain frame of reference (the sea of ether hypothesis), and the Earth, together with the entire laboratory and the interferometer, is moving relative to it. From this premise we conclude that times $T_{\parallel}$ and $T_{\perp}$ differ from each other. Despite this, we do not detect any difference. So how do we explain the negative result of the experiment? There are several possible strategies, which we will briefly discuss. The simplest explanation is to reject the premise of the motion of the Earth relative to the sea of ether. This, however, means a return to the thesis about the privileged status of the Earth in relation to other celestial bodies, because the planets, the Sun and the stars are in motion with respect to the Earth, and therefore also with respect to the hypothetical ether. Besides, such a solution could be subjected to a definitive experimental test. For this purpose, the Michelson-Morley experiment would have to be carried out in a laboratory moving relative to the Earth – e.g. in an airplane or on a space station orbiting our planet. As far as I know, no one has ever come up with the idea of such an experiment, probably because of a slim chance that such an experiment would detect motion relative to the ether.

The second possibility is to return to the hypothesis of the ether being dragged by material bodies, including the Earth's surface. Fizeau's experiment showed that this hypothesis is highly implausible, but it is logically possible that the Earth as a whole acts on the ether in a different way than the water flowing in a tube. On the other hand, since most of the Earth's surface is occupied by oceans, it would be strange if the ether "stuck" to them during the daily and annual movement of our planet, but did not show such a tendency in the case of water flow in the laboratory. Let us add that again such a solution can be subjected to another test. For this purpose, it would be necessary to construct an interferometer, the mirrors of which would be in a state of motion relative to the light source. Then the ether would be dragged at a different speed near the mirrors and near the light source, and therefore the light ray would have to pass through the layers of the ether with different relative speeds, and this would have to give an observable result in the form of variable arrival times. However, I do not think that any physicist today would be persuaded to conduct such an experiment, since there is a well-confirmed answer to the negative result of the Michelson-Morley experiment in the form of special relativity.

But before we get to that, let us mention the third possible solution, which has been seriously considered by physicists. It is known as the Lorentz-FitzGerald contraction hypothesis. It concerns the influence of the ether on the bodies passing through it. Perhaps the ether creates the effect of "flattening" the bodies moving in it in such a way that the path of the light ray in the direction of motion relative to the ether is shortened by an appropriate factor. On the other hand, the arm of the interferometer, which is perpendicular to the direction of movement, does not change its length. Let us assume that as a result of the contraction effect, a body of length l, moving with velocity v with respect to the ether, will have a length given by the following formula:

$$l' = l\sqrt{1 - \frac{v^2}{c^2}}.$$

Inserting this expression instead of l in formula (5.1) for the "parallel" time $T_{\parallel}$, we get exactly the same formula as in the case of the "perpendicular" time $T_{\perp}$ (5.2). So we achieve an agreement with the experiment.

However, is this a satisfactory solution to the problem? It is not difficult to see that the shortening hypothesis suffers from serious flaws. Firstly, it is clearly an *ad hoc* hypothesis, invented for just one purpose of explaining the negative outcome of the experiment. We have no independent evidence for the existence of a physical contraction effect, other than the negative result of the Michelson-Morley experiment. Secondly, it is extremely unlikely that the effect described by the above equation would affect all physical objects equally, regardless of their structure, chemical composition, etc. The very idea of the ether influencing objects moving in it is not completely baseless. If the ether allows the transmission of electromagnetic interactions between electrically charged bodies, and material objects are mostly made of positively charged protons and negatively charged electrons, then it should not be surprising that the ether will affect the charges moving within it, and thus also macroscopic material bodies. However, the cumulative effect of such an interaction in the form of observable deformations of the shape of moving objects should strongly depend on their structural details: density, the way charges are distributed (e.g. whether there are free electrons in it or whether they remain on atomic shells), the type of chemical bonds, and many others. If the Lorentz-FitzGerald contraction effect really existed, it would belong to the type of so-called universal effects, which require specific theoretical explanations (we will encounter a similar example when analyzing gravitational interactions in general relativity).

At this point, it is worth clearing up some possible misunderstandings. Probably many readers have heard about the effect of length contraction, resulting from the special theory of relativity (this contraction is also referred to in the literature as Lorentz or Lorentz-FitzGerald contraction, which unfortunately leads to confusion). Mathematically, the relativistic contraction effect is described by the above equation, but its physical sense is quite different. First, velocity v in the relativistic equation is the velocity of the body relative to any system, not to the ether. Second, the contraction of an object is observable only from the perspective of the system relative to which the object moves. In a system moving with an object, there are no physical contraction phenomena that could be observed in any way. We will talk about this in more detail in the next sections of the chapter.

After eliminating the contraction hypothesis, we are basically left with only one solution: we should assume that the ether does not exist, since its predicted effect on the speed of propagation of electromagnetic waves does not happen. The reasoning here is simple: if the phenomenon A necessarily produces the effect B, and there are no additional factors preventing B from occurring, then from the fact that B does not occur, we draw the logical conclusion that A cannot occur. In our case, A is the fact of the existence of the ether, and B is the occurrence of the shift of interference fringes in the Michelson-Morley experiment. In this situation, however, we are back at the starting point: relative to what should we measure the velocity of electromagnetic waves derived from Maxwell's equations? At this point, Einstein's bold hypothesis comes into play, that c is the speed relative to *any* frame of reference. Einstein suggests that the speed of light is the same in every frame of reference.

Let us think about the consequences of such a surprising assumption. It implies that the transition from one frame of reference to another, moving at a certain speed, does not change the observable speed of light. This is clearly inconsistent with the intuitive rule of composing velocities, resulting directly from the Galilean transformation, which we derived in section 2.4. Therefore, in order to maintain Einstein's assumption of the constancy of the speed of light in all reference frames, the rules for the transformation of coordinates between inertial reference frames must be modified. We will address this issue in the next section.

5.2 Relativity of simultaneity and Lorentz transformation

The fundamental assumption of classical physics, visible in the Galilean transformation ($t' = t$), is the thesis that the time coordinate does not depend on the chosen frame of reference. To put it loosely, time flows the same regardless of the state of motion of the observer. Moreover, it is tacitly accepted that determining the time coordinate of events which are spatially distant from each other does not present any fundamental difficulty. This assumption, however, came under scrutiny in Einstein's famous reasoning. To assign the time coordinate to an event some light years away from us, we need to have a universal method for ascertaining simultaneity (to synchronize distant clocks that allow for a consistent time specification of events at any point in space). However, we know that due to the finite speed of light propagation, an observed event is not simultaneous with the act of observation, but is prior to it. For example, when we see a supernova explosion in a distant galaxy, we know that it took place hundreds of thousands or even millions of years ago, because this is how long it takes for the light ray carrying the information about the explosion to reach us. How, then, can we determine which events are truly simultaneous with the act of observation?

At first glance, it may seem that it should be enough to take into account the travel time of the light signal. After all, we know the value of the speed of light – both from experience and theoretically, based on the analysis of electromagnetic phenomena. In addition, by measuring the distance between a given event (e.g. the aforementioned supernova explosion) and the act of observation, we can calculate the delay and subtract it from our clock reading, thus obtaining the "true" time of the observed phenomenon. On the basis of similar reasoning, we concluded in the previous paragraph that the supernova explosion must have occurred sufficiently earlier than the appearance of the explosion in our sky. Unfortunately, the proposed method has one major drawback. As we remember from the considerations devoted to the concepts of time and space in Newtonian physics (in Galilean space-time), the spatial distance between two events is generally not invariant with respect to the Galilean transformation, as long as these events are not simultaneous. In one system, the spatial distance between non-simultaneous events can be millions of light years, and in the other, equally good, zero! To apply the appropriate correction, we would have to know already that the events are simultaneous, but this is what we want to determine. So we fall into a vicious circle.[1]

We have reached an impasse. On the one hand, we intuitively believe that for every spatially separated events it is an objective fact whether they are simultaneous or not. On the other hand, we do not have a method to determine whether such simultaneity occurs. Einstein believed that admitting a concept of which we do not know how to apply it in practice does not agree with the principles of scientific methodology. For such a fundamental notion as the concept of simultaneity, we must find an appropriate criterion to compare the time of events regardless of the distances between them. The signaling criterion proposed by Einstein is very simple and at first glance intuitive, but it turns out to have a very unintuitive consequence. The criterion is as follows: two events will be considered simultaneous when the light rays emitted from these events meet exactly halfway between these events. It seems very common sense. However, it is not common sense to accept that we can use the signaling criterion regardless of the frame of reference we are in and at what speed we are moving.

Einstein's definition of simultaneity is one of the crowning examples of the so-called operational definitions in science. An operational definition of a given concept consists in indicating a specific method of its application in the form of a certain operation, most often a measurement. There is a trend in the methodology of science called operationalism – initiated by the American physicist P.W. Bridgman – according to which only operationally defined terms have empirical meaning and can be accepted in science. Concepts for which no appropriate method of application has been given should not be admitted. In the case of the concept of simultaneity of events in Newtonian physics, there is no well-defined method of verifying whether two events are simultaneous, and therefore this concept does not meet the requirements of science. Einstein's definition, on the

[1] If the two compared events occur "on" objects persisting in time, which do not move relative to each other, then we can assume that the spatial distance between these events is determined with respect to the system in which both of these objects are at rest. But what about a situation in which objects are in relative motion, or in which events are not related to any temporarily extended bodies at all? For example, events can be flashes of light in a vacuum, without any material "basis". In such a situation, any frame of reference seems equally good.

other hand, gives such a method in the form of an appropriate operation, the use of which allows for the determination of simultaneous events. However, the thesis that scientific concepts are unambiguously defined by appropriate measurement operations seems too radical. It implies that two different measurement operations define two different concepts, even if the results of such measurements are the same. For example, measuring length with a ruler and with a laser device designate two different concepts of length, which seems unintuitive.

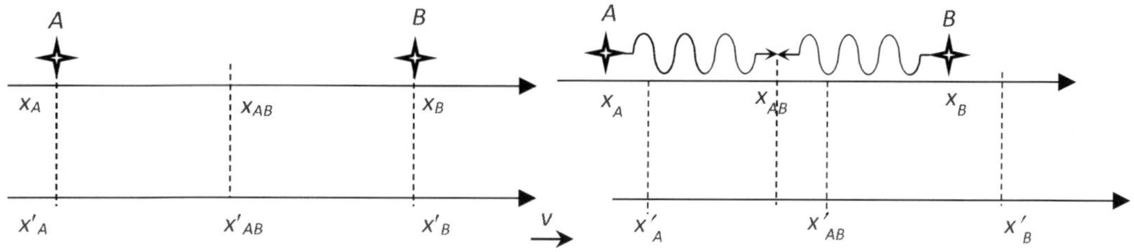

Fig. 5.5 Einstein's relativity of simultaneity

As we can easily see, determining the point dividing the distance between two events in half depends on the adopted frame of reference. Consider two systems moving relative to each other and two events A and B (cf. Fig. 5.5). In one system, half of the distance between A and B falls at the point with the coordinate $x_{AB} = \frac{x_A + x_B}{2}$. In the frame moving relative to the first, the coordinate of the mid-point is $x'_{AB} = \frac{x'_A + x'_B}{2}$. However, the point x'_{AB} on the axis of the second coordinate system will not coincide with the point x_{AB} all the time, because the second system moves to the right. As a result, if the light rays propagating from A and B come into contact at a spacetime point with the coordinate x_{AB}, this point will not have the coordinate x'_{AB} in the moving system, but will be "shifted" in the direction of the position of the event A (x'_A). Therefore, using Einstein's criterion, we must admit that events A and B are simultaneous in the first system, but not in the second. From the point of view of the moving system, event A occurred earlier because the light ray emitted from A traveled a longer distance than the ray from B before the two rays met.

It is worth emphasizing, in response to possible objections to the above reasoning, that Einstein's signaling criterion is based on the assumption of the constancy of the speed of light in all reference frames. It is thanks to this assumption that we have the right to claim that this criterion can be applied to any frame of reference. If the ordinary law of velocity composition were applied to light (as in the case of sound waves, for example), the observation in a moving system would have to be rejected, since the velocities of the light signals from A and B would differ from each other (the signal coming from A would be faster and from B slower). In such a situation, the signal-based criterion could only be used in a special frame in which the speed of light in all directions is c. However, as we already know, e.g. from the Michelson-Morley experiment, we are not able to detect such a privileged frame of reference. Thus either we stick to the classical criterion of simultaneity, which we will not be able to use in practice, or we settle for an applicable criterion, which nevertheless implies the relativity of simultaneity.

Accepting the latter solution, we see that the simple identity transformation of time $t' = t$ from classical physics is not tenable. Since two events are simultaneous in one frame of reference (i.e. $t_A = t_B$), and they are not such in the other ($t_A' \neq t_B'$), the time coordinate in the moving frame cannot be the same as in the "resting" frame. We need new rules for the transformation from one system to another. The relevant rules are called the Lorentz transformations – they form the basis of a new theory, i.e. special relativity. There are many interesting methods of deriving these transformations from certain general principles, including the principle of constancy of the speed of light. However, we will not do this – instead, we will present them in a ready-made form, and then we will see that they lead to the desired results. Here is the mathematical form of the Lorentz transformation:

$$x' = \frac{x - vt}{\sqrt{1 - \frac{v^2}{c^2}}}$$
$$t' = \frac{t - \frac{vx}{c^2}}{\sqrt{1 - \frac{v^2}{c^2}}}. \tag{5.3}$$

As usual, we assume that the reference frame (x', t') is moving relative to the frame (x, t) with velocity v in the direction of increasing values of x. Let us note that in both equations there is a factor $\sqrt{1 - \frac{v^2}{c^2}}$ already known to us from the analysis of the Michelson-Morley experiment. In a sense, it is a trademark of the theory of relativity – its presence always indicates that we are leaving the well-known territory of classical physics and venturing into the world of bizarre relativistic effects. The numerical value of this factor depends on how high the velocity v is in relation to the speed of light. If the velocity v is low – such as the speed of objects moving around us (cars, trains, and even jet planes or space rockets) – then this coefficient is very close to unity. In such a situation, the first equation turns into the well-known Galilean transformation rule: $x' = x - vt$.

And what about the second equation, characterizing the time coordinate? There is also a new factor $\frac{vx}{c^2}$ in the numerator. We can say that the ratio $\frac{v}{c^2}$ for small velocities v will be an extremely small number (the speed of light squared is a huge quantity), and thus we can omit the whole expression. However, this is not entirely correct reasoning. By choosing a sufficiently large x, we can "compensate" for a huge number in the denominator, which will cause the time t' in the moving system to be significantly different from the time t. Therefore, even when we shift between relatively slow-moving systems, the time coordinates of events can differ significantly from each other, if we consider events that are very distant from one another.

Summing up the above observations, we can conclude that if we limit ourselves to slow-moving systems and nearby events, the formulas of the Lorentz transformation turn into the Galilean transformations. This explains why for centuries people have not been able to observe any deviation from the principles of classical physics. We also see that special relativity had no chance of originating as a simple generalization of many observations. Applying the principle of induction (deriving future predictions from the past instances), we would have to conclude that since classical mechanics has worked well in an overwhelming number of past cases, it must be universally true. It was only a deep conceptual crisis in the foundations of the theory of electromagnetism which showed that a revision of the fundamental laws of time and space was needed.

Let us now establish that the new rules of transformation confirm the principle of the constancy of the speed of light. For this purpose, we will derive a new law of velocity composition, which allows us to calculate the velocity of a given object (including light) in a system moving at a certain speed. Let x/t be the velocity of the body equal to V determined in one system. Let us now calculate the ratio x'/t', representing the velocity of this body in another frame of reference, moving relative to the first with velocity v:

$$\frac{x'}{t'} = \frac{x - vt}{t - \frac{vx}{c^2}} = \frac{\frac{x}{t} - v}{1 - \frac{vx}{c^2 t}}.$$

Thus, the relativistic rule of the velocity composition will be as follows:

$$V' = \frac{V - v}{1 - \frac{vV}{c^2}}.$$

Suppose that the velocity V with respect to a given system is equal to c. We then get:

$$V' = \frac{c - v}{1 - \frac{v}{c}} = c\frac{c - v}{c - v} = c\,.$$

Thus, the speed of light in a system moving at a velocity v will still be c. Light propagates at a constant speed, regardless of the frame of reference.

Let us note another mathematical consequence of the Lorentz transformation, which is of great physical importance. What happens if we assume that one frame is moving at the speed of light? Substituting the velocity c for v in formulas (5.3) for x' and t' gives zero in the denominator, i.e. an impossible mathematical operation (division by zero). This fact is usually interpreted as showing that it is impossible for a frame of reference to move at the speed of light. Material bodies can approach the speed of light at will, but they can never reach it.

5.3 Two relativistic effects

I am sure every reader must have heard of two astounding effects predicted by relativity: time dilation and length contraction. These terms have infiltrated popular culture and science-fiction literature. Unfortunately, as a result, there have often been a lot of misunderstandings and misinterpretations as to the nature of these effects. In this section, we will show how the existence of both effects can be derived directly from the Lorentz transformation. We will also see more clearly what the aforementioned increase in time and shortening of length amount to.

Let us start with the phenomenon of time dilation. Imagine a process that lasts for a certain interval from 0 to t – for example, pouring sand from one hourglass chamber to another. Since we know that measurements of time and space must always be explicitly relativized to some frame of reference, let us choose the system in which the hourglass rests. The time for a given object measured in the frame in which this object rests is called the *proper time* of that object. Now consider the same process from the perspective of a frame moving relative to the hourglass. This process will last from the moment 0 to t', where the time t' is related to the coordinates in the first frame by the second equation of the Lorentz transformation (5.3). Thanks to the assumption of the hourglass resting in the first frame, we can assume that the position of both the beginning and the end of the process is $x = 0$. Thus, the transformation of time takes a simple form:

$$t' = \frac{t}{\sqrt{1 - \frac{v^2}{c^2}}}$$

Time measured in a system moving relative to a given body will always be longer than the proper time of that body, because we divide the number t by a factor less than one. The time dilation effect has been confirmed experimentally, e.g. in experiments with unstable elementary particles. On the basis of measurements in the laboratory, the lifetime for a particle of a given type resting in relation to the measuring system was determined. The same particle was then identified in cosmic rays traveling at enormous speeds relative to Earth. The time elapsed until decay for a particle in the rushing stream turned out to be longer than the previously measured life time at rest, as predicted by the above formula.

Deriving the formula for length contraction requires slight caution. First of all, let us define precisely what the length measurement of e.g. a rod is. In the arrangement in which the rod rests, the matter is obvious – we use an ordinary ruler to determine its length in centimeters, inches or any units. However, the problem is how to measure the length of a rod moving relative to us. We cannot simply stop it (or, for that matter, "jump" onto it), because in this way we will return to the case of the frame in which the rod rests. We need to come up with some method to measure the length of a rod without disrupting its motion. The solution may be to determine the instantaneous position of the beginning and end of the rod in our system, using for instance fast-frame photography triggered by a motion detector. However, it is crucial here that both measurements are taken simultaneously. Nothing will come of it if we mark the position of the end of the rod in one moment, and the position of the beginning in a few seconds – the result will not reflect the length, because the rod keeps moving. Let us thus assume that the time coordinate of both measurements in our system is $t = 0$. Conventionally, we can also assume that the recorded spatial coordinate x_0 of one end of the rod is equal to 0, and then the coordinate of the other end x_1 will simply be the measured length of the rod in motion (let's call it l_m). On the other hand, in a system in which the rod rests (and thus it is a frame that moves together with the rod relative to us), the spatial coordinates of the beginning and end of the rod can be fixed at $x_0' = 0$ and $x_1' = l_r$, where l_r is the length of the resting rod (Fig. 5.6). From the first Lorentz transformation (5.3) we have the following formula connecting the coordinate x_1' with the coordinates in our system (according to the assumption we put $t = 0$):

$$x_1' = \frac{x_1}{\sqrt{1 - \frac{v^2}{c^2}}}.$$

Hence we get the formula for the length of the rod in motion:

$$l_m = l_r \sqrt{1 - \frac{v^2}{c^2}}.$$

Therefore, it can be seen that the result of the measurement in motion will be smaller than the length of the rod at rest. The rod is "shortened".

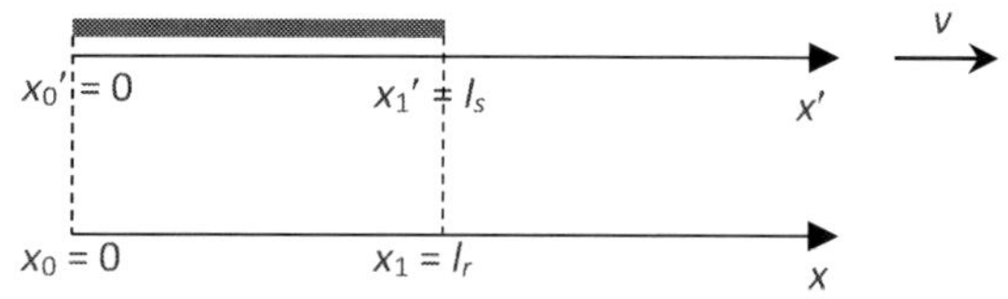

Fig. 5.6 Measuring the length of a rod in motion.

Are the phenomena of time dilation and length contraction real? But what does it mean that a phenomenon is real? The effects of shortening the length of objects and increasing the duration of processes occur when accurate and objective measurements are carried out with the use of appropriate devices, and therefore they are not "illusions" similar to well-known sensory illusions such as the Müller-Lyer effect. On the other hand, however, one should beware of a naïve and fundamentally incorrect interpretation of these relativistic effects as resulting from the causal influence of motion on the physical state of objects. It is sometimes suggested that objects moving at a speed close to the speed of light will be subjected to a special physical influence resulting from this motion. Physical, chemical or biological processes will slow down, and the objects themselves will shrink in the direction of movement. Sometimes it is added that these effects cannot be observed in a moving system, because the measuring devices (clocks, measuring rods) will undergo analogous deformations on their own (if, for example, if we measured a slowed-down process with a slow-running clock, the measurement result would remain the same as in the stationary case). Relativistic effects can only be observed "from the outside", because then the measuring devices are not disturbed.

This is a misinterpretation for a simple reason – motion is not an absolute property of objects, but only a relative one. It makes no sense to say that the rocket is "really" moving and the Earth is at rest – we might as well say that the rocket is standing and the Earth is rushing in the opposite direction. If this is the case, then we cannot claim that movement will objectively affect the processes taking place. This is best illustrated by the example of the Lorentz-FitzGerald contraction effect. As we remember, the contraction hypothesis appeared on the occasion of the negative result of the Michelson-Morley experiment, as an attempt to save the assumption of the existence of the ether. Central to this hypothesis was the claim that objects "really" and "objectively" move. The real motion is motion relative to the ether, and it is the ether that is responsible for the physical "flattening" of bodies. On the other hand, the relativistic length contraction discussed in this section applies to every two systems in mutual motion. From the Earth's point of view, the objects in the rocket are shortened, but from the rocket's point of view, the Earth's objects become shorter. Neither of these effects is "more real" than the other. Perhaps a more adequate description of what happens in the situations we describe is the statement that it is not physical objects and processes that are deformed, but time and space, which depend on the adopted frame of reference.

The most spectacular example of relativistic deformations of time and space is the so-called twin paradox. Here there should be no doubt that the observed effect is very real. We have a pair of twins who were born on the same day, and now one of them is a young man and the other an old man. It is difficult to find a more objective effect of relativistic transformations. We will examine this case in more detail in the next section, when we introduce the concept of space-time interval.

5.4 Minkowski spacetime geometry

Special relativity is, in a sense, a new theory of space and time, or space-time combined. It is therefore not surprising that the novelty of this theory should manifest itself in the modification of the geometric properties of space-time. As we have already mentioned in the analysis of the properties of Galileo's space-time, the geometry of a given space is characterized by a complete set of its "invariants", i.e. metric properties that do not change when the coordinate system changes. In the case of Euclidean geometry, such an invariant is the ordinary spatial distance between two points, which can be written for three dimensions in the "quadratic" version as follows:

$$\Delta s^2 = \Delta x^2 + \Delta y^2 + \Delta z^2.$$

Galileo's space-time geometry is slightly more complicated. There are two invariants in it: the spatial distance between points having the same time coordinate (simultaneous) and the time distance between any points. The theory of relativity is characterized by yet another geometry, called Minkowski geometry. It operates with a single invariant, the form of which is to some extent analogous to the Euclidean distance. This invariant is called *the space-time interval*:

$$\Delta I^2 = (c\Delta t)^2 - \Delta x^2 - \Delta y^2 - \Delta z^2. \tag{5.4}$$

It can be verified that the above expression will remain unchanged under the action of the Lorentz transformation. It is enough to rewrite the expression using the Lorentz transformations for x' and t' to see that the result will be equal to $(ct')^2 - x'^2$, i.e. it will be the same as the interval $(ct)^2 - x^2$ in the system u. Note: these calculations can be greatly simplified by adopting the commonly used convention $c = 1$. In other words, we use units in which the speed of light is 1 (i.e., we measure spatial distance in light years). Then the Lorentz transformations take a much simpler and more "symmetrical" form:

$$x' = \frac{x - vt}{\sqrt{1 - v^2}}$$

$$t' = \frac{t - vx}{\sqrt{1 - v^2}}.$$

It is now easy to calculate that the interval in the "primed" system will be exactly equal to the interval in the "non-primed" system:

$$t'^2 - x'^2 = \frac{1}{1 - v^2}[(t - vx)^2 - (x - vt)^2] = \frac{1}{1 - v^2}(t^2 - x^2)(1 - v^2) = t^2 - x^2 .$$

As can be seen from formula (5.4), the space-time interval "treats" time and space coordinates differently: the square of the spatial interval enters the equation with a minus sign, and the square of the spatial interval with a positive sign. Now we can consider three cases. In the first case, we assume that the temporal distance between the two events "dominates" over the spatial distance: $\Delta t^2 > \Delta s^2$ (remember the convention $c = 1$). In other words, the square of the space-time interval $\Delta I^2$is a positive number. The physical interpretation of this situation is simple: two events (or space-time points) are separated from each other by a smaller spatial distance than the distance traveled by a light ray in the time between the two events. In other words, both events can be connected by a signal moving at a subluminous speed.[2] This also means that there will be a frame of reference, moving at a speed less than c, in which the two events under consideration occur in the same place (their spatial distance will be zero). A space-time interval that satisfies the above conditions is called a *time-like interval*.

The second case occurs when $\Delta I^2 < 0$. We are then dealing with a *space-like interval*. Events separated by such an interval are too distant spatially from each other to be connected by a light signal. However, it can be shown that in such a case there is a frame of reference in which these events will be simultaneous, i.e. their time distance Δt will be equal to zero. Sometimes such a situation is also called the occurrence of the relation of *quasi-simultaneity*. Although the simultaneity relation is not invariant with respect to the Lorentz transformation, the quasi-simultaneity relation is, as it is defined by the concept of interval.

The third, borderline case is when $\Delta I^2 = 0$. Two events separated by a null interval (also sometimes called *light-like*) can be connected by a signal traveling exactly at the speed of light, because their spatial distance in each system is equal to the path that the light ray travels in the time between them. Time-like, space-like, and null intervals determine an important division of space-time in relation to a given event. However, to discuss this in more detail, we need to learn a simple but useful method of drawing space-time diagrams. Thanks to space-time diagrams, we will be able to easily and clearly illustrate many facts and relationships between events, the description of which in the language of algebra itself may be difficult to read.

Let us start with a graphical representation of the axes for a given reference frame (Fig. 5.7). For obvious reasons, we will be able to take into account at most two spatial dimensions, because we are not able to depict four dimensions (four axes perpendicular to each other) on a piece of paper. For now, let us simplify things even further by considering only one spatial dimension x in addition to the time dimension t. The temporal axis will be determined by the set of all space-time points (events) whose spatial coordinate is zero ($x = 0$). Similarly, the spatial axis is given by the equation $t = 0$. Each point in our two-dimensional "space-time" will receive its coordinates (t, x) in the usual way in the Cartesian coordinate system.

Now let us ask how to represent on our diagram a new frame of reference, which moves at a constant speed towards positive values of x ("to the right"). Drawing the t'-axis poses no problem – it is a line determined by the uniform movement of a point with a zero coordinate in the new system, so it will be a straight line inclined at an angle to the t-axis. For now, the

---

[2] This fact is independent of the choice of reference frame. Although the sizes of spatial and temporal intervals will generally be different in different frames of reference, the space-time interval as an invariant will remain unchanged, and therefore the condition $\Delta t > \Delta s$ will have to be met in each frame.

situation is the same as in classical physics. The difference will appear when we will want to draw the x'-axis, i.e. the set of all those events that have a t'-coordinate equal to zero. In Galilean spacetime, the x'-axis coincides with the x-axis, because time does not change when moving to another frame of reference. In special relativity, however, we must take into account Einstein's simultaneity criterion. The events on the x'-axis must be simultaneous with event 0 on the t'-axis. To determine them, let us use an alternative variant of the Einstein criterion, which is equivalent to the version presented in the previous section. The procedure is as follows: we send a ray of light at time t_A towards the mirror, the ray is reflected from the mirror and returns to the starting point at time t_B. We assume that the moment the ray reaches the mirror is exactly half the time between t_A and t_B: $\frac{t_A+t_B}{2}$.

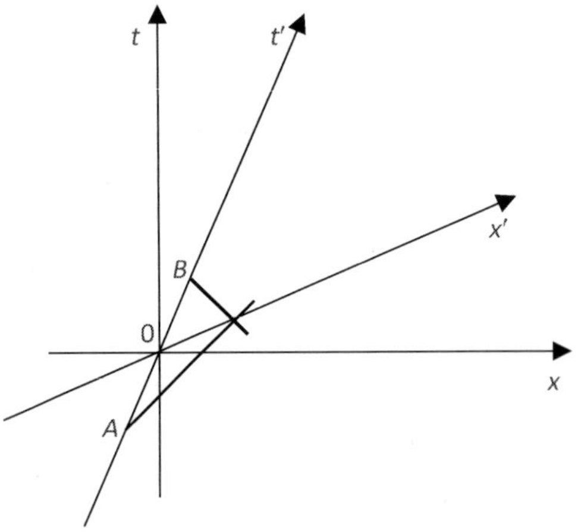

Fig. 5.7 The axes of two coordinate systems moving relative to each other.

We can now select any point A on the temporal axis t' prior to the zero point. How do you draw a light ray path in a diagram? Simple: it is a line inclined to the x-axis at an angle of 45 degrees (in the adopted units, the speed of light equals 1). So let's draw such a line from the selected point A. In turn, from the point B such that 0 lies half-way between A and B, let us draw a line ("the ray of light") downwards. The intersection of both lines represents the reflection event, which according to Einstein's criterion is simultaneous in a moving system with the zero point. So the x'-axis must pass through this point. As we can see from the diagram, this axis will be inclined in relation to the x-axis. It can be proved, on the basis of simple geometrical considerations, that the angle of inclination of the x'-axis in relation to the x-axis will be equal to the angle between the axes t and t'. Thus, the Lorentz transformation changes the inclination of both axes, temporal and spatial.

We could have guessed this result from the fact that the speed of light is constant (and therefore equal to 1) in any frame of reference. Thus, the line of light must divide the angle between the axes t' and x' in half, as we have shown before. The x'-axis, as well as all lines parallel to it, represent the "areas" of simultaneity in the new layout. Extending this result to the case of two spatial dimensions, we can see that the planes of simultaneity in moving systems will be inclined in relation to the initial plane, and their inclination will increase with the increase of the speed of the system. In the three-dimensional case, we are dealing not with a two-dimensional, but with a three-dimensional surface of simultaneity (which is why it is sometimes called a "hypersurface"). It is extremely difficult to imagine two three-dimensional hypersurfaces inclined at an angle to each other in the fourth dimension, but this is the relationship between the "simultaneity" in two systems resulting from the theory of relativity.

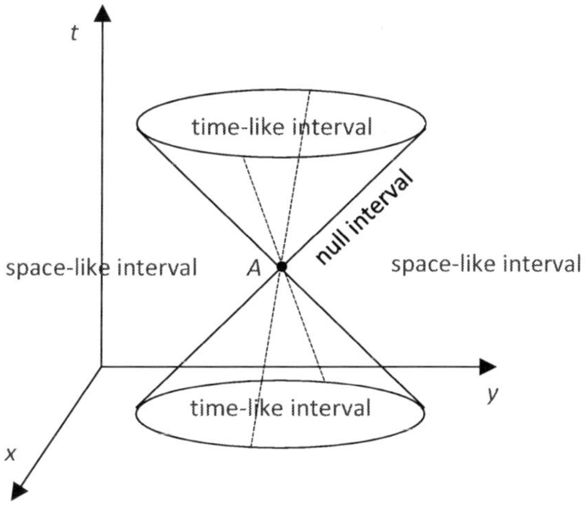

Fig. 5.8 Light cones in Minkowski spacetime.

Now let us go back to space-time intervals. Let us choose a certain point A in space-time and draw paths of light rays spreading out of and converging on it (Fig. 5.8). In the case of two spatial dimensions, these lines will form two cones, called light cones. Now consider the areas inside both cones. It is easy to see that the events contained in both of these areas are separated by a time-like interval from our chosen event A. This is because a straight line connecting event A to any event in these areas will have a smaller slope relative to the timeline than the path of light (the edge of the cone), and therefore such a line represents an object moving slower than light. It can also be seen that for any event inside the cone, there is a frame of reference in which this event occurs in exactly the same place as event A. It is a system whose t'-axis connects A with the given event. Note that in this arrangement the spatiotemporal interval is reduced to proper time, since $\Delta s = 0$. This is a general fact: the spatiotemporal interval for events occurring "on" a given physical body is simply equal to the proper time of this body.

Events that are located on the surface of both cones are separated from event A by a null interval. The remaining part of spacetime, located outside both cones, includes all and only those events that are separated from A by a space-like interval. It is impossible to link these events to event A with a signal traveling at most at the speed of light. At the same time, we can see that for any event from the area outside the cones, there is a frame of reference in which this event occurs simultaneously with A. It is, for example, a system whose x'-axis passes exactly through A and the selected event, and therefore, according to the previously discussed geometric representation of the Lorentz transformation, the time axis t' of such a system must be inclined to the t-axis at the same angle as the angle between the axes x and x'.

The areas separated by light cones have interesting and important philosophical interpretations. To begin with, let us assume that no causal interaction can propagate faster than the speed of light. Electromagnetic interactions certainly meet this condition, and as for other types, we do not know of any cases where this assumption is broken. The only way to circumvent this principle, which we can call the principle of the locality of the causal relationship, appears in quantum mechanics – we will discuss this in the following parts of the book. For now, however, we will assume that if two events are linked by a causal relation, then it must be possible to connect them with a signal propagating at a speed at most equal to the speed of light. This means that all events that could potentially affect our chosen event *A* are either inside or on the surface of the "lower" light cone (Fig. 5.9). Similarly, the events that can be physically affected by event *A* are arranged in the "upper" light cone. For this reason, the lower cone is sometimes called the "absolute causal past" of event *A*, and the upper cone its "absolute causal future". The term "absolute" indicates that these concepts are independent of the frame of reference, since light cones do not change when moving from one frame of reference to another. The term "causal" implies that we are only interested in the potential interaction with the selected event, and not in the mere fact of temporal precedence or succession.

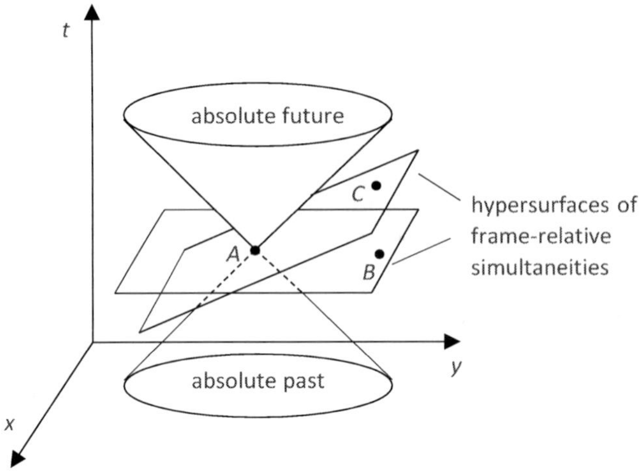

Fig. 5.9 Division of space-time into absolute causal past and future

How does the concept of an absolute causal past/future relate to the ordinary concepts of past and future? The standard understanding of the past, for example, is that it includes everything that has happened before the present moment. In a spacetime diagram, this would therefore be the part of spacetime with the time coordinate smaller than the coordinate of the selected event *A* (the part below the plane of simultaneity passing through *A*). But in different frames of reference, this area will look different, because the plane of simultaneity itself is transformed ("tilted"). The light cone includes all those events that are common to all frame-relative pasts. One can therefore alternatively characterize the absolute causal past for *A* as containing those events for which all observers agree that they occur earlier than the selected point *A*.

Why can't events occurring outside the light cones originating from point *A* simply be treated as simultaneous with *A*? Unfortunately, it is not that simple. Consider two events *B* and *C*, as in Fig. 5.9. Although both events are spatially separated from *A*, event *C* is in the absolute future from *B*. It is impossible for two events simultaneous with a third not to be mutually simultaneous. Therefore, it is not simultaneity that occurs between *B* and *A* and *C* and A. This relation can be called quasi-simultaneity at most: it is simultaneity relativized to a certain system, and for pairs of events B, A and C, A these are two different systems. Quasi-simultaneity does not have one important property of simultaneity – namely, it is not transitive.

> There is an important reason why we should not accept lightly the existence of interactions (signals) propagating faster than the speed of light. The presence of such signals could open the way to backward causation, with all its consequences (grandfather's paradox). If event *C* from the previous example could act on the spatially separated event *A*, and event *A* could in turn send a signal to the also spatially separated event *B*, then it would be possible for *C* to influence its absolute past. We could then construct a situation in which *C* was caused by *B*, and on the other hand *C* would send a signal to *A* initiating the elimination of event *B*. Therefore, we would have a contradiction: event *C* exists (because it is assumed to be caused by an earlier event *B*) but at the same time it does not exist (because *B* is eliminated by a causal sequence occurring backwards in time).

5.5 Relativistic effects on diagrams

We already know how to graphically represent Lorentz transformations, which result in the "tilt" of both the time and spatial axes. There is one more important element of space-time geometry that we have not yet introduced. It has to do with how to mark the appropriate units of length for time and space on the axes in a moving system. Let us remember that the geometry of spacetime is not Euclidean, so we cannot be guided by the "ordinary" distance between points. In particular, if, for example, one centimeter on the t-axis represents a one-second interval, then on the t'-axis one second will certainly not correspond to the same centimeter. This is true even in Galileo's classical space-time. One second in a moving system is determined by a horizontal simultaneity line coming from a point with coordinate 1 on the t-axis, and therefore the Euclidean length of a one-second segment in a primed system will be greater than in a non-primed.

In Minkowski spacetime determining the unit of time (and space) is a little more complicated. To do this, we need to use the space-time interval (5.4), which is always the same in every frame of reference and thus gives us the opportunity to compare measures of length in different frames. On the x and t axes of the initial system, let us select the points corresponding to the units $t = 1$ and $x = 1$. Let us now consider how to determine all the points whose space-time interval calculated from the origin of the system (0, 0) is the same as the interval from the selected unit points $t = 1$ and $x = 1$. This is not difficult: for example, the squared interval between the point on the x-axis with coordinate 1 and the origin of the system equals -1, so all points having the same interval must satisfy the equation $t^2 - x^2 = -1$. Similarly, in the case of point $t = 1$, the equation will be $t^2 - x^2 = 1$. We know from geometry that these equations describe hyperbolas converging in infinity to "diagonals" (i.e. light lines), as in Fig. 5.10.

Now consider the intersections of both the hyperbolas of the fixed interval with the t' and x' axes respectively. It is not difficult to see that these points will represent the units of time and space sought in the new system. For example, the (squared) interval between the intersection of the hyperbola with the t'-axis and the origin calculated in a moving system is simply the square of its time coordinate (the space coordinate x' is zero), but since this interval in any system is 1, the time coordinate of this point is also equal to 1. Similarly, in the case of a point on the x'-axis, its spatial coordinate must be equal to 1. The fact that the points determined in this way have a different Euclidean distance from the origin of the system is irrelevant. On the other hand, thanks to the introduced units, we can easily illustrate the relativistic effects of time dilation and length shortening.

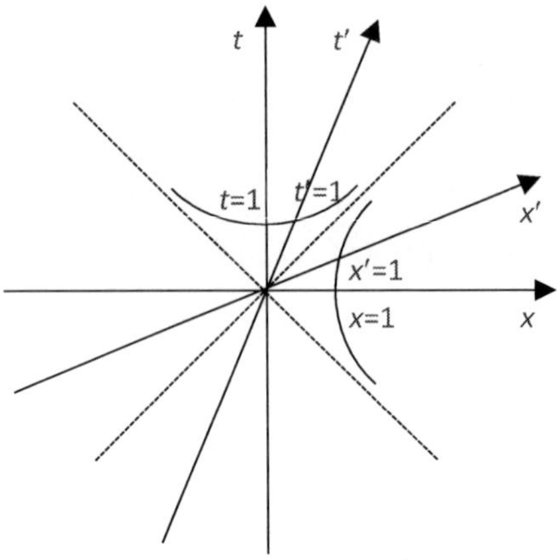

Fig. 5.10 Determination of units in a moving system

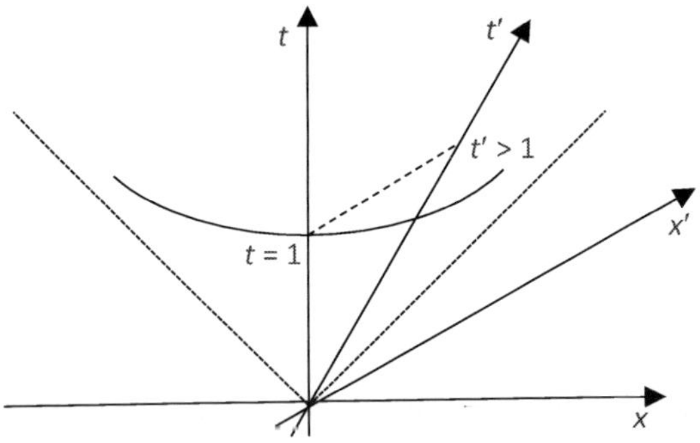

Fig. 5.11 Time dilation in the diagram

A point on the t-axis with a unit coordinate will have its time coordinate in the moving system determined by a line parallel to the x'-axis (the simultaneity line in this system), so it can be seen that this coordinate will be greater than 1 in the primed system (Fig. 5.11). To illustrate the shortening of length, consider two parallel lines, one of which passes through the zero point (it coincides with the time axis t') and the other goes through the point with the coordinate $x' = 1$ (Fig. 5.12). These lines represent the temporal evolution of the beginning and end of a rod of length 1 at rest in the moving system. The length of the rod measured in the "stationary" system will be given by the coordinate of the intersection of the second line with the x-axis (remember that the measurement of the position of both ends of the rod is to take place simultaneously at $t = 0$), so again we can see from the diagram that it will be shorter than one.

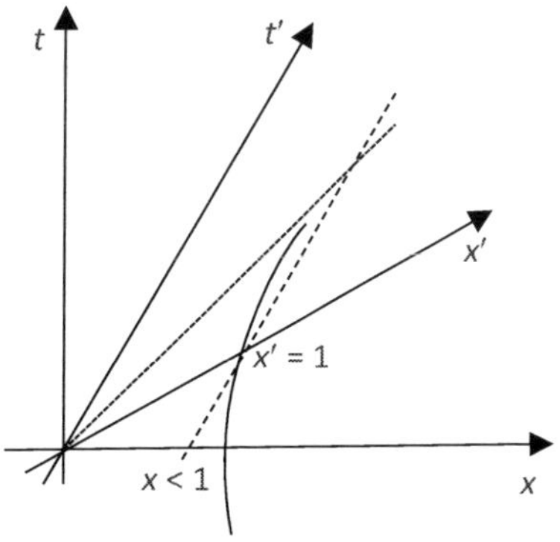

Fig. 5.12 Length contraction in the diagram.

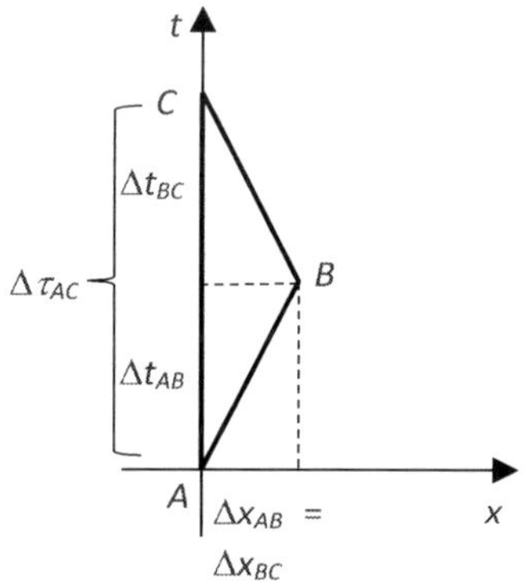

Fig. 5.13 Spatiotemporal diagram illustrating the twin paradox

Finally, let us move on to the analysis of the famous twin paradox. As we will see in a moment, this is not a paradox in the sense of reasoning leading to contradiction, but only a derivation of a very unintuitive consequence from the principles of relativity. Let us assume that the twin brother stayed at home and his sister set off on a space journey, from which she returned years later. As it turns out, the duration of the journey measured by the sister's own time will be shorter than the time of her own brother, i.e. the sister will become younger than her twin brother. Let us draw a diagram of the sister's journey from the perspective of the brother's frame of reference. Point A represents the departure event, point B is the destination of space travel and the point of turning back, and point C indicates the safe return to Earth (Fig. 5.13). Let us calculate the spatiotemporal intervals between points AB, BC and AC in the brother's system. The interval between A and C will obviously be equal to the time interval $\Delta I_{AC} = \Delta \tau_{AC}$, since the spatial distance of these events is zero. (Remember that the spatiotemporal interval between any events separated by a time-like is always equal to the proper time, i.e. measured in a frame in which these events have the same spatial location.) The other two intervals are given as follows:

$$\Delta I_{AB}^2 = \Delta t_{AB}^2 - \Delta x_{AB}^2,$$

$$\Delta I_{BC}^2 = \Delta t_{BC}^2 - \Delta x_{BC}^2$$

and of course we have

$$\Delta\tau_{AC} = \Delta t_{AB} + \Delta t_{BC}.$$

Since the spatiotemporal interval for a given object is simply equal to its proper time, we can write

$$\Delta\tau_{AB}^2 = \Delta t_{AB}^2 - \Delta x_{AB}^2$$

$$\Delta\tau_{BC}^2 = \Delta t_{BC}^2 - \Delta x_{BC}^2,$$

where $\Delta\tau_{AB}$ and $\Delta\tau_{BC}$ are the travel times from A to B and from B to C as measured by the sister. From the above equations we immediately get the inequalities:

$$\Delta\tau_{AB} < \Delta t_{AB}$$

$$\Delta\tau_{BC} < \Delta t_{BC}$$

and therefore

$$\Delta\tau_{AC} > \Delta\tau_{AB} + \Delta\tau_{BC}.$$

Conclusion: the travel time measured by Earth's clocks will be longer than the time recorded in a space rocket. So the sister will be younger than her twin brother when she returns. It seems paradoxical, but only because we have never had to deal with distant space travel. There is nothing contradictory in the assumption that time can "flow" differently in different frames of reference. So we should not insist on our intuitions acquired in contact with slow-moving objects at a short distance from us (compared to the cosmic scale). However, it may be worth explaining what the source of the "asymmetry" between brother and sister is. What distinguishes the situation of the traveler from that of the homebound brother, since, as we know, motion is relative and we can just as well say (from the perspective of the flying rocket) that it was the brother who went on a space trip together with the entire Earth and the solar system?

The answer is simple and visible to the naked eye on a space-time diagram: the trajectory of the sister is described by a broken line, and the trajectory of the brother by a straight line, no matter what (inertial) frame of reference we describe the whole situation in. At some point, the sister has to change frames of reference – from moving at a constant speed in the direction away from the Earth, to moving in the opposite direction. On the other hand, the "lazy" brother remains in the same frame all the time. The geometry of Minkowski spacetime shows that the length of the path along a polygonal chain (measured, of course, by the spatiotemporal interval) is shorter than the length along a straight line. It seems absurd, but only because our intuitions grew on the basis of Euclidean geometry, in which the so-called triangle inequality applies. Let us emphasize that the effect of faster aging for the brother is independent of the chosen frame of reference. For example, we can describe the whole situation from the vantage point between the Earth and a rocket, from the perspective of which both the brother and the sister are moving – first in opposite directions, and then the sister starts to catch up with the brother. Another possibility is to transform into a reference frame associated with the rocket in the first phase of the journey. Then, at first, the sister is stationary and the brother rushes in the opposite direction, after which the sister begins to fly towards the brother at twice the speed. Regardless of how we describe the whole situation, it is always a fact that the brother does not change the frames of reference, while the sister changes them.[3]

5.6 Some philosophical consequences of special relativity

Special relativity is one of the most intensively analyzed physical theories by philosophers (along with quantum mechanics, perhaps). It is impossible to include in this textbook presentation the wealth of all aspects of this theory important to the philosopher. However, it is necessary to pay attention to the most frequently cited philosophical consequences of relativity. First of all, let us focus on a problem that has already arisen in the analysis of Newtonian mechanics, namely the question of the ontological status of time and space. It is indisputable that the theory of relativity definitively refuted the absolutist position on the status of time and space conceived as separate entities. As we have seen, the temporal and spatial relations between events are not absolute in the physical sense, i.e., they are different in different frames of reference. Two simultaneous events in one system will generally not be simultaneous in a system moving with respect to the former. Similarly, the spatial distance between events changes as you move from system to system, even if the events are simultaneous in one system. Therefore, both time and space must change when the frame of reference is changed. The temporal locations (moments) of two events that we consider identical in one system may not be identical in another (the same applies to spatial locations, which was already true even in classical physics). In other words, each frame of reference has its own time and its own space.

It follows that in the ontological version of the dispute over the nature of time and space, the scales also tip in favor of relationalism. Time cannot be a substance independent of the existence of material objects, since in every frame of reference, determined by means of material measuring devices, time takes a different form. The same is true of space. Can we then declare Leibniz the winner in his dispute with Newton? The matter is not so simple, because the problem can be reformulated so that the question is not about the status of time and space, but about the inseparable whole that is space-time. The fundamental space-time relation, i.e. the relation of coincidence (occurring at the same time and in the same place) is obviously independent of the frame of reference. Therefore, it can be argued that a structure composed of space-time points is not ontically dependent on material objects occupying these points. Special relativity tells us only that the "breakdown" of space-time into independent time and complementary space is not determined in an unambiguous, objective way. However, the status of space-time as a whole is still an open question.

---

[3] Of course, all this assumes that we are talking about inertial systems all the time. If we allowed non-inertial frames, whose velocity can change, then we could choose a frame of reference that is "permanently" associated with the sister's rocket, and then the kinematic behavior of the brother would look exactly the same as that of the sisters in the Earth's system: the brother would move in one direction and then turn back. However, the rocket system is not physically "equivalent" to the Earth-bound system, as the rocket accelerates when it turns. Equivalence can only be restored when gravity is taken into account – accelerating systems are equivalent to "stationary" systems in a gravitational field. As it turns out, however, the gravitational field has exactly the same distorting effect on time as being in a non-inertial frame of reference – we will talk about this in the chapter devoted to the general theory of relativity.

Substantivalism regarding the question of the status of space-time, on the other hand, remains vulnerable to Leibniz's charge of violating the principle of the identity of indiscernibles (or the principle of sufficient reason, as discussed in Chapter 2). The substantial interpretation of space-time allows for the possibility of alternative scenarios differing only in the identity of the space-time points and areas occupied by given material objects, without violating the space-time relations between these objects, and therefore also being empirically indistinguishable. On the other hand, Newton's bucket argument, countering Leibniz, does not apply directly here, as the existence of apparent forces in non-inertial systems (e.g. rotating) cannot be simply explained by the existence of absolute space-time – as we remember, it is necessary to assume the existence of absolute *space* with respect to which rotation takes place. However, absolute space is, as we have already said, manifestly incompatible with the principles of relativity.

Thus, although it is theoretically possible to defend absolutism and substantivalism in relation to space-time on the basis of special relativity, relationalism seems to be in a much better position. This situation will change radically with the transition to general relativity, where new, strong arguments for substantivalism will appear. Interestingly, there will also be a new argument for relationalism, which is in a sense a modernized and mathematically advanced version of Leibniz's argument from the shift. The dispute over the ontological status of space-time is – as we can see – far from being resolved, although there has undoubtedly been a huge theoretical progress related to the development of physical theories.

Special relativity also has important implications for one of the more controversial problems in the philosophy of time—the problem of the passage of time and the related division into past, present, and future. The experience of the passage of time is one of the most basic experiences of human life. Despite this fundamentality, it is very difficult to try to make it sufficiently precise. The passage of time is a kind of change that is profoundly different from other changes such as movements in space. Motion occurs with respect to time, while the passage of time itself is a change of time from the future to the present and past. Some believe that a "second-order" time is needed to describe the passage of time, which leads to a regress (second-order time requires third-order time to describe its changes, and so on). Critics of the concept of the passage of time also point out that with a literal understanding of this concept, the question can be asked how quickly time passes. However, the speed of the passage of time does not make a clear physical sense (in what units would this speed be measured: seconds per second?)[4]

Special relativity provides a new argument in the dispute over the ontological status of the passage of time. The key assumption of the supporters of the objectivity of this concept is the thesis about the existence of a division into present, past and future events, independent of the observer. The present for a given observer includes all events that are simultaneous with the act of perception (consciousness). However, we already know that the simultaneity relation depends on the state of the observer's motion. Two observers moving relative to each other will consider different events to be simultaneous, and thus the "presents" for these ob-

[4] However, the "seconds per second" criticism is sometimes refuted – for example, the philosopher of physics Tim Maudlin uses the situation of currency exchange as an example. We can talk about the conversion factor of the "euro to dollar" dimension, but we can also consider the dollar to dollar conversion factor, with a constant value equal to one. Let us add that in the theory of relativity there is a concept of a "four-velocity vector", the length of which in each system is one (details will be given in the next section). It can be argued that this concept represents the speed of the passage of time.

servers will be different. There are infinitely many planes of simultaneity passing through a given point event. Which of these planes determines the "real" present? The relativization of simultaneity to a frame of reference strongly suggests that the objective present has no deeper physical meaning.

Supporters of the dynamic nature of time have some strategies to defend their position. First, it can be maintained that although empirically we are not able to choose one of many planes of simultaneity, objectively there is one such distinguished plane that defines events that are really happening "now". Such a strategy leads to the separation of the empirical from ontology, i.e. to the recognition that there are some objective facts that cannot be known by us empirically (at least for the time being). Some argue that it is possible to distinguish a certain privileged frame of reference by referring to facts outside the theory of relativity itself – e.g. to the cosmological facts concerning the distribution of matter on a global scale in the universe. Finally, there is a strategy based on replacing the notion of the global present with the notion of the local present, limited to a given physical body and its history. Each body, and therefore also each observer, would have their own past, future and present. Such a solution could be called "temporal relativism" or even "temporal solipsism". Although the ontological debate on the passage of time remains unresolved, it should be emphasized that the theory of relativity significantly limits the acceptable interpretations of this phenomenon.

5.7 Relativistic dynamics

For most philosophers, the special theory of relativity "ends" with the problem of time and space. For a physicist, however, the new theory of space-time is just the beginning. The essential goal of any physical theory is to predict the future behavior of physical systems. This goal is accomplished by formulating laws of motion that enable us to calculate the future state of a system based on the knowledge of the present state. As we remember, the cornerstone of classical mechanics is Newton's law of motion $F = ma$, which gives us the theoretical possibility of calculating the trajectories of physical bodies subjected to appropriate forces. Unfortunately, Newton's law is in obvious conflict with the principles of relativity. This is easy to show: just imagine a body on which a constant force acts for a sufficiently long time. In this case, the body could reach any arbitrarily high speed, which is inconsistent with the principle of the maximality of the speed of light. The relativistic law of motion must be modified.

The easiest way to introduce the new laws of motion is simply to write the appropriate equations and show that they avoid the above difficulty. For example, applying a constant force according to the new equation will lead to the effect that the rate of increase in the velocity of the body will decrease exactly in such a way that the limiting but impossible to achieve speed is the speed of light. Most popular introductions to special relativity are content to introduce and describe new equations of dynamics. However, let us try a longer, but perhaps more interesting way, starting with the introduction of new mathematical concepts. The basic concept used in this section and in later chapters will be the concept of *a four-vector*.

A well-known type of vectors, used in physics, are vectors in a three-dimensional Euclidian space. As we know from elementary geometry, when choosing a given coordinate system, we can represent each vector using a triplet of numbers, determining the lengths of the components of this vector along the axes of our system. It is therefore not surprising that in four-dimensional Minkowski spacetime, vectors will have four, not three, components: one temporal and three spatial. To distinguish them from "ordinary" vectors, we will call them four-vectors. Since the geometry of four-dimensional spacetime is different from Euclidean geometry, the four-vectors will also be correspondingly different from the three-vectors in Euclidean space. This difference is revealed in the way of transformation. When moving from one coordinate system to another, we need to modify the numbers representing the vector accordingly. However, the length of a given vector should not change – the length is of course calculated by the appropriate invariant of a given geometry.

Consider the simplest example of a four-vector – a four-vector of displacement, connecting the origin of a coordinate system with any point in space-time. Until now, we have symbolized the coordinates of points as (t, x, y, z). Currently, we will adopt the notation using indices from 0 to 3: (x^0, x^1, x^2, x^3). The zeroth coordinate will be the time coordinate, and the coordinates from 1 to 3 will be the three spatial coordinates. We will generally symbolize the four-vector of displacement as X^μ, where $\mu = 0, 1, 2, 3$. Using the known formulas for Lorentz transformations (5.3) we can write what the components of the same vector will look like in a new frame of reference, moving with velocity v in the direction of the x_1 axis (all the while using the convention $c = 1$):

$$(X')^0 = \frac{X^0 - vX^1}{\sqrt{1 - v^2}}$$

$$(X')^1 = \frac{X^1 - vX^0}{\sqrt{1 - v^2}}$$

$$(X')^2 = X^2$$

$$(X')^3 = X^3$$

In general, we will assume that each four-vector transforms its components in the above way.[5] In turn, the "length" of the four-vector will be given by a familiar formula based on the spatiotemporal interval:

$$(X^0)^2 - (X^1)^2 - (X^2)^2 - (X^3)^2$$

The length of the four-vector does not change when the coordinate system is changed (it is an invariant of the accepted spacetime transformations).[6]

Now let us introduce an important new type of four-vector: the so-called four-velocity. The standard definition of velocity is well known: it is a vector whose components are given by the time-derivatives of the components of the displacement vector: $v_x = \frac{dx}{dt}, v_y = \frac{dy}{dt}, v_z = \frac{dz}{dt}$. Similarly, the components of the velocity four-vector will be given by differentiating the components of the displacement four-vector X^μ with respect to time. However, in special relativity, there are many "types" of time (since time is different in each coordinate system).

---

[5] The above equations describe only the transformation of four-vectors due to the so-called Lorentz boost, i.e. the transition to a system moving at a certain speed. In general, Euclidean transformations of three-dimensional space, such as translations and rotations, should be added to the principles of transformation. The group of all transformations that preserve the equations of special relativity is called the Poincaré group.

[6] Another important quantity that is preserved during the transition between coordinate systems is the dot product of four-vectors. Let A^μ and B^ν be two four-vectors. The product $A^\mu B^\mu$ is defined as $A^0B^0 - A^1B^1 - A^2B^2 - A^3B^3$. Transforming the components of both vectors, according to the formula above, we can prove that the result of the product will be the same in both systems.

In order for the velocity four-vector to transform in the right way, the assumed time must be invariant, and therefore we have only one acceptable choice – it must be proper time τ, i.e. time measured in the system in which a given body rests. Thus, the definition of four-velocity will be as follows:

$$U^{\mu} = \left(\frac{dX^0}{d\tau}, \frac{dX^1}{d\tau}, \frac{dX^2}{d\tau}, \frac{dX^3}{d\tau}\right).$$

The coordinates U^1, U^2, U^3 (which we can abbreviate as U^i, bearing in mind the convention that the Latin indices range over the numbers 1, 2 and 3) are "ordinary" components of velocity, but calculated using proper time. To calculate the four-velocity of a given body in a selected system, we need to convert proper time to the time t measured in this particular system. The formula of "exchange" is familiar to us from previous considerations: $\tau = t\sqrt{1 - v^2}$. Inserting this expression into the above definition, we get the following formula for the spatial components of the velocity four-vector:

$$U^i = \left(\frac{dX^1}{dt}\frac{1}{\sqrt{1 - v^2}}, \frac{dX^2}{dt}\frac{1}{\sqrt{1 - v^2}}, \frac{dX^3}{dt}\frac{1}{\sqrt{1 - v^2}}\right)$$

or simply (v^i is the i-th component of the ordinary velocity of the body with respect to a given system):

$$U^i = \frac{v^i}{\sqrt{1 - v^2}}.$$

But how to interpret the "zero" component of the four-velocity? Converting proper time τ into time t relative to a given system, and remembering that X^0 is none other than the system's time coordinate, we get a simple equation:

$$U^0 = \frac{1}{\sqrt{1 - v^2}}.$$

The zero component of the four-velocity in a given frame of reference is, in a sense, the coefficient of transformation between proper time and the time measured in that frame. In general, a four-vector velocity can be represented as a unit vector tangent to the "world line" of a given body (the trajectory of this body in four-dimensional space – cf. Fig. 5.14).[7] In a system in which a given body rests, the three spatial components of the four-velocity will be equal to zero, but the time component remains non-zero (it is equal to 1). It is obvious that even if the world-line of a body is perpendicular to the plane of simultaneity in a given system, the tangent vector to this line will not cease to exist, although in this system it will have only one component – the time component U^0.

[7] Please, check for yourself that the "length" of the four-velocity vector calculated using the spatiotemporal interval is always equal to 1. Note: $v^2 = \left(\frac{dx^1}{dt}\right)^2 + \left(\frac{dx^2}{dt}\right)^2 + \left(\frac{dx^3}{dt}\right)^2$.

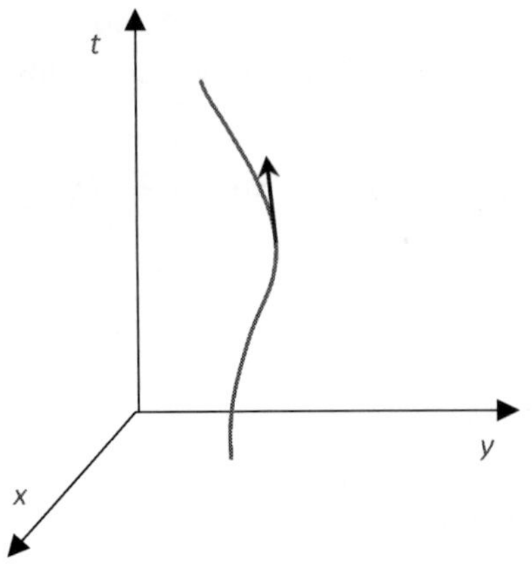

Rys. 5.14 Four-vector velocity tangent to the world-line

We can now move on to the concept of momentum, fundamental for the laws of dynamics. Let us remember that momentum is connected to ordinary velocity by the formula: $p = mv$. The same relationship applies in the theory of relativity, so we can write:

$$P^{\mu} = mU^{\mu}.$$

This results in the following relationships between the spatial components of the four-momentum and the ordinary three-velocity:

$$P^{i} = \frac{mv^{i}}{\sqrt{1-v^{2}}},$$

where $v^{i} = \frac{dx^{i}}{dt}$. Note that as the velocity v increases, the momentum increases at a rate much greater than it would result from simple proportionality. In particular, when v approaches the speed of light (i.e. 1), the momentum increases to infinity. This is fundamental to the principles of relativistic dynamics. Remember that Newton's second law of motion can be represented by the equation:

$$F = \frac{dp}{dt},$$

which means that the rate of increase of momentum is equal to the acting force. Applying this principle to the relativistic momentum, we get the consequence that the action of a constant force will not cause velocity to increase to infinity. The momentum will increase at a constant rate described by the equation above, but the correlated velocity v will approach the limit velocity $c = 1$ at an increasingly slower rate, but never reaching it (because the momentum in such a situation would have to be infinite). Thus, the required modification of the laws of motion was accomplished by modifying the concept of momentum.

We still have the remaining task of interpreting the null component of the four-vector momentum. As you can see, it amounts to:

$$P^{0} = \frac{m}{\sqrt{1-v^{2}}}.$$

What is the physical meaning of this formula? Let us try to guess it by considering an approximation of the above formula for low velocities. For this purpose, we will use the following mathematical approximation. For small values of v, the fraction with the root in the denominator can be represented with good accuracy as follows:

$$\frac{1}{\sqrt{1-v^2}} \approx 1 + \frac{v^2}{2}.$$

Using the above estimation, we get the following formula:

$$P^0 = m + \frac{mv^2}{2}.$$

The second component of the sum on the right side of the equation is recognized as the kinetic energy of a body moving with velocity v. But what about the first ingredient? In order to preserve the consistent units, let us reintroduce the speed of light c, which we have omitted for convenience as equal to unity. In this notation, the right side of the equation will look like this:

$$mc^2 + \frac{mv^2}{2}.$$

I'm sure that every reader will now recognize the first ingredient even faster than the second, "classic" one. Of course, this is Einstein's famous formula for so-called relativistic rest energy, $E = mc^2$. Together with the second term, the whole expression describes the total energy. Thus, the zero component of the four-vector momentum turned out to be nothing but the total energy of the body.[8]

However, we must remember that we have made a big conceptual leap here. Looking for an interpretation for the component P^0, we first made an approximation for low speeds. Usually, such a transition reproduces the well-known equations of classical physics. For example, we remember that in the limit of low velocities the Lorentz transformation formulas turn into the classical Galilean transformation. Similarly, relativistic effects (time dilation, length contraction) disappear when considering low velocities. However, in the above case, the expression for P^0 clearly does not turn into the well-known formula for kinetic energy. Regardless of how low velocities we are dealing with, the expression defining P^0 will contain a term that does not appear in classical physics (let us add: a term with a huge numerical value, proportional to the square of the speed of light). By interpreting the expression mc^2 as a new, previously unknown form of energy, we thus admit that discrepancies between classical and relativistic physics also appear in the area of low velocities and small distances known to us from everyday experience. So why didn't we notice this difference earlier?

An exact answer to this question would require delving into the concept of energy itself and its relationship to observational data, which we do not have space for here (although we have already touched on this problem in earlier chapters when discussing the principle of conservation of energy and the equivalence of mechanical and thermal energy). However, we can see that the rest energy of the body is not observable by itself. Its presence can only be revealed when we are dealing with a process in which there is a difference between the total mass at the beginning and at the end of the process. In this case, the loss of mass Δm must be

---

[8] For this reason, the momentum four-vector is also called the energy-momentum four-vector.

compensated by kinetic energy of the value of Δmc^2. We do not observe such processes in the macroscopic world, but the subatomic world provides us with plenty of examples of the conversion of rest energy into kinetic energy. The most spectacular example is, of course, nuclear reactions, including the fission of heavy atomic nuclei (or the fusion of light nuclei). The mass of the fission products is slightly less than the mass of the initial nucleus, and this difference is released as kinetic energy – either violently in the form of a nuclear explosion or slowly in the form of heat release in a nuclear reactor.

In the following box, I discuss a separate case of photons that requires special treatment. If you are not particularly interested in this case, this is the end of the introduction to special relativity for inquisitive philosophers. The next two sections are intended for people who are slightly more familiar with higher mathematics – they deal with certain conceptual and formal issues related to the relationship between the classical theory of electromagnetism and the theory of relativity.

It is not difficult to see that many equations of special relativity "collapse" in the case of objects moving at the speed of light. (According to the current state of knowledge, the only such objects are photons, i.e. quanta of light. For a time, it was thought that neutrinos also belonged to the same category of objects as photons, but this assumption was refuted experimentally.) It turns out that the concept of four-velocity, for example, cannot be applied to photons. This is because the proper time of photons is zero. Remember that the space-time interval between two points connected by a light ray is zero, and proper time is nothing more than space-time interval. Therefore, differentiation with respect to the variable τ, for which $d\tau = 0$, does not make mathematical sense. This problem can be expressed in another way: there is no unit vector tangent to the world line of the light ray, for a simple reason: all vectors tangent to it have a length (calculated in the Lorentzian way) equal to zero. However, the velocity four-vector must have a unit length, so photons cannot have a defined four-velocity.

On the other hand, four-momentum does not have to be identical to one – its length is, as it is easy to calculate, equal to m. Therefore, mathematically it is possible to assign photons a four-momentum, but then we have to assume that $m = 0$. Hence we have a simple argument that photons are massless. The four-vector momentum for a photon is of the form (E, p_1, p_2, p_3), where $E = p$ (or, going back to ordinary units, $E = cp$; p is, of course, the total momentum). The energy of a single photon is given by the formula $h\nu$ where ν is the frequency and thus the momentum of the massless photon is $\frac{h\nu}{c}$. Finally, let us add that in the general case the relativistic relation between energy and momentum results from the fact that the square of the length of the four-momentum is equal *to* m^2, i.e. $E^2 - p^2 = m^2$, and therefore $E = \sqrt{m^2 + p^2}$, or in any units of the speed of light c, $E = \sqrt{m^2c^4 + c^2p^2}$. When $m = 0$, this formula becomes the equality $E = cp$ valid for photons.

5.8* Relativistic theory of electromagnetism: Lorentz force

In this section, we will deal more closely with the issue of formulating a relativistically correct (i.e., Lorentz-invariant) version of Maxwell's theory of electromagnetism. In a sense, this task has already been accomplished, since Maxwell's theory, even in its original form, is essentially invariant with respect to the Lorentz transformation (this was the reason for its problems with Galileo's principle of relativity). It is instructive, however, to present Maxwell's theory in its full relativistically invariant form by deriving it from the fundamental principles of relativity. This will also enable us to accomplish a proper unification of electrical and magnetic phenomena, which we discussed in the previous chapter. Besides, not all laws of electromagnetism are actually in line with the principles of special relativity. For example, the formula for the Lorentz force, despite its name, is not Lorentz invariant and must be replaced by a relativistically correct formula.

Our approach will be quite abstract – we will derive the laws of electromagnetism from the principle of least action, which we introduced when discussing alternative approaches to Newtonian mechanics. The abstract nature of this approach will consist in the fact that we will have to guess the correct form of action so that the laws derived from it agree with the empirically confirmed laws of Maxwell's theory. What is crucial, however, is that the action we adopt (and consequently also the Lagrangian) should have an overtly relativistically invariant form, which will also guarantee the invariance of the whole theory. Let us begin with the question of how to describe in the Lagrangian formalism the behavior of a charged body in an electromagnetic field. As we remember from chapter 2 (section 2.8), the actual trajectory of a material body subjected to the appropriate forces can be determined by calculating the quantity called action for each possible trajectory from the starting point to the end point and choosing the one for which the action takes an extremal (usually minimal) value. The action can be represented as the integral from the start to the end point from a quantity called the Lagrangian, which in standard applications of classical mechanics is the difference between the kinetic energy and the potential energy of a body. In turn, the condition of minimizing the action is mathematically equivalent to a certain equation called the Euler-Lagrange equation (2.2). Finally, the trajectory of the particle is determined by solving this equation for a given Lagrangian.

Now our goal will not be so much to determine the specific trajectory of the particle, as to find the appropriate equation of motion that must be satisfied by the particle in the electromagnetic field – i.e. the relativistic variant of the formula for the Lorentz force. However, for the time being, we do not have a suitable Lagrangian – we will want to construct it by adopting an appropriate form of action. Suppose that the action describing the behavior of a particle is the sum of two components: the free action, which describes a particle on which no force acts, and the additional action, which describes the interaction of the particle with the field. Both components of the action must be relativistically invariant – as we already know from the previous section, this means that they must be given in the form of scalars, which do not depend on the adopted coordinate system. The free action is best represented by the proper time τ, which, as we know, is invariant with respect to the Lorentz transformation (remember the convention $c = 1$ from the previous sections):

$$\mathcal{A}_{free} = -m\int_a^b d\tau = -m\int_a^b \sqrt{1 - v^2}dt$$

The other part of the action must include information about the field. Let us assume that the electromagnetic field is described, in accordance with the formalism of the theory of relativity, by a four-vector $A_\mu(t, x)$, which is obviously a function of location in space-time. The dot product of such a vector with the infinitesimal displacement vector dx^μ is a scalar, and thus is invariant with respect to the choice of reference frame. Let us therefore try to take the following form of the action describing the interaction of a particle with a field, using Einstein's summation convention, which requires summation over repeated indices (e is the charge of the particle):

$$\mathcal{A}_{int} = e\int_a^b A_\mu(t,x)dx^\mu$$

I expect that many readers will vent their frustration at this point – how do you know that this is what the right action should look like? On what basis can we take this and not another form of mathematical expressions describing a given phenomenon? Especially for a philosopher, accepting far-reaching theses without proper justification – empirical or mathematical – may seem very suspicious. Philosophers are accustomed to questioning even the most seemingly obvious statements, such as "the table I see in front of me really exists." My response to these legitimate objections is that we should not treat the above constructions as logical deductions from certain axioms, but rather as the use of trial and error to reproduce certain already known results. Such a method is called reverse engineering – having a ready-made construct in the form of Maxwell's theory of electromagnetism we want to work backward and find a way to build it from scratch. Our goal is to obtain the known laws of electromagnetism – or similar laws to the known form of the laws of electromagnetism – using the fundamental principles of mechanics (including the principle of least action) and maintaining relativistic invariance. If we fail at the first attempt, we will try another approach. Of course, you really know that it has to work – after all, I didn't come up with this method on my own, but I learned it from others who tested it well.

Hoping that these remarks have dispelled at least some of the doubts, I will now return to the task of deriving the law governing the interaction of particles with the field. In fact, the matter is rather simple. In the expression for the action $\mathcal{A}_{int}$ describing a particle interacting with the field, we will formally replace the integration variable dx^μ with the time variable dt by substituting $dx^\mu = \frac{dx^\mu}{dt}dt$. As a result, we get the form of a Lagrangian describing the interaction with the field A_μ (this is an expression which must be integrated with respect to the time variable to calculate the action):

$$\mathcal{L}_{int} = eA_\mu\frac{dx^\mu}{dt} = eA_\mu\dot{x}^\mu.$$

Remembering that $x_0 = t$ and $\frac{dt}{dt} = 1$ we can write an expression for the total Lagrangian (let me remind you that the Latin indices run over three spatial coordinates 1, 2, 3):

$$\mathcal{L} = -m\sqrt{1+\dot{x}^2} + eA_0(t,x) + e\dot{x}^n A_n(t,x). \tag{5.5}$$

I hope that the minor formal transformations did not make the reader's head spin. Now we need to recall the "terrible" Euler-Lagrange equation (2.2):

$$\frac{d}{dt}\left(\frac{\partial\mathcal{L}}{\partial\dot{x}^p}\right) = \frac{\partial\mathcal{L}}{\partial x^p}$$

The only thing left to do is to differentiate the Lagrangian expressed in (5.5) with respect to the variables $\dot{x}^p$ and x^p, remembering that A_μ depends on x^p but not on $\dot{x}^p$. I hope that more mathematically skilled readers will be able to do it themselves, but if you have problems with it, take the result "on faith":

$$\frac{\partial \mathcal{L}}{\partial \dot{x}^p} = \frac{m\dot{x}^p}{\sqrt{1-\dot{x}^2}} + eA_p$$

$$\frac{\partial \mathcal{L}}{\partial x^p} = e\frac{\partial A_0}{\partial x^p} + e\dot{x}^n\frac{\partial A_n}{\partial x^p}.$$

I know, it's getting a little hot. A little more patience and we will see the light at the end of the tunnel. For now, let us insert the above into the Euler-Lagrange equation:

$$\frac{d}{dt}\left(\frac{m\dot{x}^p}{\sqrt{1-\dot{x}^2}}\right) + e\frac{dA_p}{dt} = e\frac{\partial A_0}{\partial x^p} + e\dot{x}^n\frac{\partial A_n}{\partial x^p}.$$

We still have to deal with the derivative $\frac{dA_p}{dt}$. It is a total derivative, so it takes into account both the explicit dependence of $A_p(t,x)$ on t and the implicit dependence, due to the dependence of spatial coordinates x on t. Let us remember that the method of minimizing action is based on the search for the trajectory of the body, and the trajectory is nothing more than the dependence of spatial coordinates on time in the form of a functional relationship $x^p(t)$. From calculus, we know that the total derivative with respect to a given variable x is equal to the sum of the partial derivatives with respect to each variable times its derivative with respect to the variable x. Maybe I will save the reader a few tedious transformations and write the final result (however, I urge the more courageous ones to try their mathematical skills – it really gives a lot of pleasure when you get the right result).

$$\frac{d}{dt}\left(\frac{m\dot{x}^p}{\sqrt{1-\dot{x}^2}}\right) = e\left(\frac{\partial A_0}{\partial x^p} - \frac{\partial A_p}{\partial t}\right) + e\dot{x}^n\left(\frac{\partial A_n}{\partial x^p} - \frac{\partial A_p}{\partial x^n}\right) \tag{5.6}$$

Now we can take a closer look at the result. On the left side of the equation, we should recognize the expression known from the previous section – it is a relativistic formula for the rate of change of momentum (more precisely, its p-component), or in short, the relativistic force acting on a particle. The right side of the formula contains two terms – one independent of velocity and the other dependent on it. Let's try to refresh our memory a bit and write the standard formula for the total electric and magnetic Lorentz force:

$$\boldsymbol{F} = e\boldsymbol{E} + e\boldsymbol{v} \times \boldsymbol{B}$$

Comparing the two terms, we can conclude that the first parenthesis to the right of the equality (5.6) represents the electric field:

$$E_p = \frac{\partial A_0}{\partial x^p} - \frac{\partial A_p}{\partial t} \tag{5.7}$$

There is a slight problem with the second expression, because there are two indices – one index p consistent with the index on the left, and the other "involved" with the index of the velocity components $\dot{x}^n$. I suggest that the reader writes down the entire sum using the "ordinary" notation for spatial coordinates x, y and z instead of x^1, x^2, and x^3. For example, let x^p be the z coordinate. We will then get the following expression:

$$\dot{x}\left(\frac{\partial A_x}{\partial z} - \frac{\partial A_z}{\partial x}\right) + \dot{y}\left(\frac{\partial A_y}{\partial z} - \frac{\partial A_z}{\partial y}\right).$$

This expression is similar to the already known formula (4.10) for the components of the cross product of two vectors. The product $\boldsymbol{A} \times \boldsymbol{B}$ is a vector whose component, e.g. in the z direction, is given by the appropriate combination of vector components $\boldsymbol{A}$ and $\boldsymbol{B}$:

$$A_x B_y - B_x A_y$$

and similarly for the other components. To obtain an analogous form for the above expression, we need to introduce a vector $\boldsymbol{B}$ with the following components x and y:

$$B_x = \frac{\partial A_z}{\partial y} - \frac{\partial A_y}{\partial z}$$
$$B_y = \frac{\partial A_x}{\partial z} - \frac{\partial A_z}{\partial x}, \tag{5.8}$$

and also for the component in the z direction (I omit the relevant derivation, but it works similarly to the above):

$$B_z = \frac{\partial A_y}{\partial x} - \frac{\partial A_x}{\partial y}. \tag{5.9}$$

Thus, the second expression in the formula for the relativistic Lorentz force can be represented as the p-component of the product of the velocity vector and the newly introduced vector $\boldsymbol{B}$: $\boldsymbol{v} \times \boldsymbol{B}$. This is exactly the same as in the original, non-relativistic expression for the force coming from the magnetic field. That is, vector $\boldsymbol{B}$ should be interpreted as the strength of the magnetic field. Its relationship to the four-vector A_μ is derived from the above equations (5.8) and (5.9): $\boldsymbol{B}$ is the cross product of the nabla vector ∇ and the three-vector A_n defined by the three spatial coordinates of the four-vector A_μ: $\boldsymbol{B} = \nabla \times \boldsymbol{A}$. In other words, $\boldsymbol{B}$ is the rotation of the magnetic potential $\boldsymbol{A}$.

So much effort to derive a formula that we practically knew at the beginning of our encounter with the theory of electromagnetism! However, some progress has been made. First, we made sure that the formula for the Lorentz force, after the necessary corrections, is fully consistent with the principles of relativity. Second, we obtained details of the mathematical relationship between the electric and magnetic field vectors and the four-vector A_μ. This four-vector is none other than an electromagnetic four-potential, the zero component of which represents the electric potential (scalar) and the other three components represent the magnetic vector potential. The electric potential A_0 (with a minus sign) turns into the well-known electrostatic potential V (whose gradient is equal to the field $\boldsymbol{E}$) when the magnetic potential is constant. Although it is possible to write the formula for the Lorentz force using only the four-potential A_μ, without the need to introduce vectors $\boldsymbol{E}$ and $\boldsymbol{B}$, it should be remembered that this potential is given only up to the gauge transformation, and therefore does not fully reflect physical reality. It can be proven that adding partial derivatives of any scalar S of the form $\frac{\partial S}{\partial x^\mu}$ (in other words, the gradient of S) to the components of the four-vector A_μ does not change the value of the action $\mathcal{A}$, and therefore has no effect on the physical situation.

There is one more issue left – namely the introduction of an electromagnetic field tensor, which unifies seemingly separate electric and magnetic fields. This tensor appears naturally when we rewrite the formula for the Lorentz force (5.6) in its full four-vector form. Again, I encourage the reader to derive this formula from the version for three-vectors (where the equation for the zero component is simply assumed, as it follows from the equations for the other components):

$$m\frac{d^2x^\mu}{d\tau^2} = e\left(\frac{\partial A_\nu}{\partial x^\mu} - \frac{\partial A_\mu}{\partial x^\nu}\right)u^\nu$$

where u^ν is the four-velocity of the particle (cf. the previous section). The expression in parentheses with two indices μ and ν is interpreted as a new tensor – the electromagnetic field tensor (also known as the Faraday tensor):

$$F_{\mu\nu} = \frac{\partial A_\nu}{\partial x^\mu} - \frac{\partial A_\mu}{\partial x^\nu}. \tag{5.10}$$

Note that the components of this tensor will be the corresponding components of the electric and magnetic fields. For example, let's take the F_{10} component. Given the formula (5.7) expressing the relationship between the field E_p and the potential A_μ, we get $F_{10} = E_x$. In turn, e.g. $F_{21} = -B_z$ (on the basis of equations (5.8) and (5.9) defining the components of the magnetic field vector). Let us also note that the tensor $F_{\mu\nu}$ is antisymmetric: $F_{\mu\nu} = -F_{\nu\mu}$, which immediately implies that the "diagonal" elements of the tensor $F_{\mu\mu}$ are equal to 0. The complete form of the tensor $F_{\mu\nu}$ in the matrix form is given below:

$$\begin{pmatrix} 0 & -E_x & -E_y & -E_z \\ E_x & 0 & B_z & -B_y \\ E_y & -B_z & 0 & B_x \\ E_z & B_y & -B_x & 0 \end{pmatrix} \tag{5.11}$$

Finally, let us show that the tensor $F_{\mu\nu}$ does not uniquely determine the four-vector of the electromagnetic potential A_μ – we are dealing here with gauge freedom as to the choice of the appropriate potential. Let there be a given potential A_μ. Consider a new, "re-scaled" potential:

$$A_\mu' = A_\mu + \frac{\partial S}{\partial x^\mu}$$

where S is any scalar. Calculating the form of the new electromagnetic tensor, we get:

$$F_{\mu\nu}' = \frac{\partial A_\nu'}{\partial x^\mu} - \frac{\partial A_\mu'}{\partial x^\nu} = \frac{\partial A_\nu}{\partial x^\mu} - \frac{\partial A_\mu}{\partial x^\nu} - \frac{\partial S}{\partial x^\mu \partial x^\nu} + \frac{\partial S}{\partial x^\nu \partial x^\mu}$$

And since the double differentiation with respect to the variables x^μ and x^ν is commutative, we get back the old tensor: $F_{\mu\nu}' = F_{\mu\nu}$. So adding a gradient of any scalar to A_μ has no physical consequences – it is a gauge transformation.

5.9* Derivation of Maxwell's equations

"Rationalist" philosophers tend to derive even the most obvious theses and statements from basic principles about which there can be absolutely no doubt. This method of proceeding was chosen, for example, by Descartes, who tried to justify all our empirical knowledge (and thus tried to refute the skeptical doubts that came to his mind) on the basis of the first principle of *Cogito ergo sum*. Spinoza and many others set themselves a similar task. Although physicists generally do not share radical skepticism about the reliability of sensory experience, in some cases they follow a similar direction, axiomatically deriving certain laws of nature from the first principles instead of experimental generalizations. This is an interesting task for many reasons – the possibility of such deductions potentially gives us insight into the most universal regularities of the physical world of an almost metaphysical nature. No philosopher interested in a general reflection on reality should be indifferent to such attempts.

We will now show how the four basic laws expressed in Maxwell's equations, describing electromagnetic phenomena, can be derived from some first principles. In fact, in the previous section we have already taken an important step in this direction. On the basis of the principle of least action and the principle of invariance with respect to the Lorentz transformation, we have justified the law of interaction of electric charges with the electromagnetic field. Moreover, we have also derived the interdependencies between electric and magnetic fields on the one hand, and the electromagnetic potential on the other. It turns out that with the help of these relationships alone, two of Maxwell's four equations can be mathematically proven – namely those equations in which there are no electric charges or currents (these equations are called homogeneous). On the other hand, the other two equations – inhomogeneous – can be derived assuming an additional premise in the form of the principle of conservation of electric charge. Let us first show the details of these deductions, and then consider their metaphysical interpretation.

Let us start with presenting in a synthetic way the relations between vectors $\boldsymbol{E}$ and $\boldsymbol{B}$ and the four-vector of electromagnetic potential A_μ, which we have previously written down in terms of coordinates (formulas 5.7 – 5.9). It is easy to check (please do it yourself) that these relations can be expressed by single formulas in terms of entire vectors, not only their components:

$$\begin{gathered}\boldsymbol{E} = -\left(\frac{\partial \boldsymbol{A}}{\partial t} - \boldsymbol{\nabla} A_0\right) \\ \boldsymbol{B} = \boldsymbol{\nabla} \times \boldsymbol{A}\,.\end{gathered} \tag{5.12}$$

We will now use two mathematical equalities that can be easily proven using the definition of the scalar and vector product (the first of which follows immediately from the fact that the cross product of two vectors is perpendicular to each of them):

$$\begin{gathered}\boldsymbol{\nabla} \cdot (\boldsymbol{\nabla} \times \boldsymbol{A}) = 0 \\ \boldsymbol{\nabla} \times (\boldsymbol{\nabla} S) = 0.\end{gathered} \tag{5.13}$$

Applying the first equality in (5.13) to the expression for the magnetic field vector $\boldsymbol{B}$ from (5.12), we immediately get one of Maxwell's equations (characterizing the magnetic field as sourceless):

$$\boldsymbol{\nabla} \cdot \boldsymbol{B} = 0\,.$$

Calculating the rotation of the electric field expressed in the first of equations (5.12), we obtain:

$$\nabla \times \boldsymbol{E} = -\left(\frac{\partial \nabla \times \boldsymbol{A}}{\partial t} - \nabla \times (\nabla A_0)\right).$$

The rotation from the gradient of A_0 disappears, according to the second equation in (5.13), and thus only one term on the right side of the equality remains, which we recognize as a time derivative of the vector of magnetic field. In this way, we obtained the familiar Faraday law:

$$\nabla \times \boldsymbol{E} = -\frac{\partial \boldsymbol{B}}{\partial t}.$$

At this point we are ready to show how to express two homogeneous Maxwell's equations in the compact tensorial form, presented earlier on p. 138. In order to derive the appropriate formula, we have to rewrite the vectorial equalities above in terms of the individual components of the electric and magnetic fields, and then to replace these components with corresponding components of the electromagnetic field tensor $F_{\mu\nu}$ as given in the matrix (5.11). Starting with the equation $\nabla \cdot \boldsymbol{B} = 0$, we first rewrite it as:

$$\frac{\partial B_x}{\partial x} + \frac{\partial B_y}{\partial y} + \frac{\partial B_z}{\partial z} = 0$$

and then replace numbers B_x, B_y and B_z with the components F_{23}, F_{31} and F_{12} (while at the same time expressing the coordinates x, y, z in the universal form as x^1, x^2, x^3). The result will be:

$$\frac{\partial F_{23}}{\partial x^1} + \frac{\partial F_{31}}{\partial x^2} + \frac{\partial F_{12}}{\partial x^3} = 0.$$

Similarly, let us write Faraday's equation for the x-component:

$$\frac{\partial E_z}{\partial y} - \frac{\partial E_y}{\partial z} = -\frac{\partial B_x}{\partial t}$$

and substitute F_{30} for E_z, F_{02} for $-E_y$ and F_{23} for B_x (note that $t = x^0$). Here is what we will get:

$$\frac{\partial F_{23}}{\partial x^0} + \frac{\partial F_{30}}{\partial x^2} + \frac{\partial F_{02}}{\partial x^3} = 0.$$

Observe that an interesting pattern emerges. In each derived equation we use a selected triple of indices (1, 2, 3 in the first, and 0, 2, 3 in the second), and we take the partial derivative with respect to the first index from the component of the electromagnetic tensor numbered by the remaining two indices, then the derivative with the respect to the second index from the appropriately numbered component (keeping the same order of indices), and then the third derivative in the same order of indices. It can be easily verified that the same pattern occurs with the remaining two components for the equation expressing Faraday's law. Thus all four equations can be synthetically captured in the following formula:

$$\frac{\partial F_{\mu\nu}}{\partial x^\sigma} + \frac{\partial F_{\nu\sigma}}{\partial x^\mu} + \frac{\partial F_{\sigma\mu}}{\partial x^\nu} = 0$$

where σ, μ, ν are any three distinct indices.

Yet this result is surprising. We remember that our goal was to derive Maxwell's laws from certain fundamental principles describing the physical world. The above reasoning shows that two of Maxwell's four equations can be derived from equalities (5.12) and some mathematical theorems. But what are these equalities? Let us recall that we introduced them on the occasion of the analysis of the formula for the relativistic Lorentz force, as a kind of abbreviations for the appropriate expressions that appeared there. So it may seem that the formulas (5.12) are nothing more than definitions of vectors ***E*** and ***B***. We simply agree to appropriately name certain mathematical combinations of the four-vector A_μ. But if this is so, then Maxwell's homogeneous laws turn out to be mathematical consequences of some linguistic conventions, and thus have the character of *a priori* analytic propositions, true by virtue of meaning, which actually say nothing about the world. We are again dealing with the problem that arose during our analysis of Newton's laws of mechanics – the fundamental laws of nature take on the status of definitions.

However, we remember that on closer examination, the issue of the analyticity of the Newtonian laws of motion turned out to be more complex. We have shown that conventional elements are combined there with unmistakable empirical components that could be false. It can be argued that now we are dealing with a similar situation. The key to solving the puzzle is the fact that many definitions contain a factual element in the form of a statement about the existence of the defined object (these are so-called non-conservative definitions: cf. the entry in the box on p. 38). If, for example, we decide to define a real number *a* using the equation $a^2 = -1$, then the condition of existence will not be met, and therefore the definition is incorrect.[9] In the case of formulas (5.12), we are not concerned with mathematical existence, but rather with physical existence. More precisely, we can argue that these equations already assume the physical existence of a corresponding quantity – the electromagnetic four-potential A_μ. To put it another way: equation (5.12) implies that for the experimentally defined quantities ***E*** and ***B*** there is a corresponding physical object that satisfies the appropriate mathematical equations. However, it does not have to be this way! For example, it is an empirical fact that the magnetic field vector can always be represented in the form of the rotation of another vector. If, for example, there were magnetic charges, there would exist magnetic fields with non-zero divergence, and such fields could not be represented as the rotation of a different vector. An analogous argument can be applied to the equation characterizing the electric field vector.

So what about our rationalistic ideal of knowledge? What additional non-analytic (i.e., synthetic) principles do we need to add to the laws of mathematics and analytic definitions of concepts in order to obtain the necessary assumption about the existence of an appropriate electromagnetic potential? The matter is not simple. Some physicists believe that such a missing axiom is the principle of least action, quoted earlier. It can be expressed in the form of a methodological rule: for each type of physical phenomenon, one should look for an appropriately formulated action in the form of an integral along the trajectories of a certain expression called the Lagrangian, so that minimizing this action would reconstruct the actual behavior of the physical system under consideration. Such a rule, however, does not tell us what exactly this action or the associated Lagrangian should look like. We remember from the

[9] The fact that this definition is non-conservative is seen in that it implies a non-trivial (and in fact false) consequence "There is a real number x such that $x^2 = -1$", which does not contain the defined term "*a*".

previous section that the key element of our reasoning, leading to the relativistic formula for the Lorentz force (and consequently to Maxwell's homogeneous equations), was to adopt a suitable form of action $\mathcal{A}_{int}$ describing the interaction. However, this choice was not justified by any general premise of an axiomatic nature. Rather, it was dictated by the fact that it was "right", i.e. that the formula for the Lorentz force obtained should have a form resembling the previously known formula. Through reverse engineering we guessed the correct expression for the Lagrangian, which allowed us to consequently introduce electric and magnetic fields as mathematical functions of the potential. An important element of this strategy, however, was the prior assumption of at least an approximate truth of the principles of electromagnetism known from experience.

I am sorry to disappoint philosophical rationalists, but it seems that without our previously acquired empirical knowledge of the interaction of electric charges with electric and magnetic fields, we would not be able to derive the laws of electromagnetism from the "first principles". Despite this, we obtained an interesting result. The combination of the principles of relativity with the previously known regularity concerning the action of electromagnetic fields on electric charges allowed us to derive two laws describing the behavior of these fields themselves. The interaction of fields with charges and certain relationships between fields, including the fact that magnetic charges do not exist, have been shown to be related. This is surprising, though perhaps not as much as the possibility of deriving the entire theory of electromagnetism from the first principles. Just think: the phenomenon of electromagnetic induction, described in Faraday's law, can theoretically be derived from the relativistic generalization of the formula for the Lorentz force.[10] Does this mean that the experimental studies of Faraday's and others were worthless? I leave this question for the reader to consider.

We still have to deal with two remaining Maxwell's equations containing reference to charges and current (inhomogeneous equations). What about them? Here it turns out that the relations between the fields $\boldsymbol{E}$, $\boldsymbol{B}$ and the potential four-vector alone are insufficient. We need to go back to basics again and start with the principle of least action. This time, however, we will apply this principle not to describe the motion of a particle in an electromagnetic field, but to derive how the field itself evolves. So we change the perspective – the electromagnetic field itself becomes the subject of physical changes, and not a simple provider of forces acting on physical objects. (Thus we treat fields as legitimate entities inhabiting the physical world.) But how to describe the evolution of the field in the language of trajectory? For this purpose, the concept of action must be slightly modified. In the case of a single particle, the action was described as an integral whose integration variable was time, and the limits of integration were determined by the initial and final states. In the case of fields, the situation is different. A field assigns a certain quantity to each point in spacetime – a scalar or a vector. The "trajectory" of the field will simply be an assignment of a field value to each point in some enclosed area of space-time. The problem we want to solve is the question of what the value of the field inside a given area will be, assuming that the field at the boundary of that area is

[10] This result may not be so surprising if we recall the discussion in section 4.6, where we showed that the effect of inducing the electromotive force in a circuit can be explained by the Lorentz force alone when moving to another frame of reference. Similarly, the fact that a magnetic field can only change the direction of motion of a charged particle, and not its speed, contains implicit information about the non-existence of magnetic charges.

known. To do this, we need to calculate the appropriate action and minimize its value. Therefore, we must integrate the Lagrangian in the entire volume of the area under consideration, i.e. the integration variables will be all four coordinates: one temporal and three spatial:

$$\mathcal{A} = \int \mathcal{L}\, dtdxdydz.$$

The Lagrangian $\mathcal{L}$ will generally be a function of both the value of the field φ and the derivatives of the field with respect to each coordinate: $\frac{\partial\varphi}{\partial t}, \frac{\partial\varphi}{\partial x}, \frac{\partial\varphi}{\partial y}, \frac{\partial\varphi}{\partial z}$. The Euler-Lagrange equation for fields will therefore have to take into account all these Lagrangian parameters:

$$\sum_{\mu=0}^{3} \frac{\partial}{\partial x^{\mu}} \frac{\partial \mathcal{L}}{\partial \frac{\partial \varphi}{\partial x^{\mu}}} = \frac{\partial \mathcal{L}}{\partial \varphi}.$$

It may look a little scary, but the structure of this equation is basically similar to the structure of the original Euler-Lagrange equation for a single particle. Let us write down the original equation for the x coordinate of the particle (we have analogous equations for the other coordinates y and z):

$$\frac{d}{dt}\left(\frac{\partial \mathcal{L}}{\partial \dot{x}}\right) = \frac{\partial \mathcal{L}}{\partial x}.$$

For a field, the equivalent of the x-coordinate (also called the particle's degree of freedom) is the value of the field φ. Therefore, it would seem that on the left side of the equation we should have the partial derivative of the Lagrangian with respect to $\dot{\varphi}$ (i.e. the time-derivative $\frac{d\varphi}{dt}$ of φ). However, since the field φ now depends on four variables x^{μ}, and not on single time, as was the case with the position x of the particle, it is not surprising that we have to calculate the partial derivatives of the function $\mathcal{L}$ with respect to all four rates of change of the field regarding each of the coordinates and their rate of change with respect to this coordinate, and sum up the results.

All that remains is to find the right Lagrangian to derive the missing Maxwell equations. Piece of cake! Unfortunately, we have to pull the rabbit out of the hat again. We have to consider two cases: the situation when there are no charges and currents, and the case with charges and currents. In the first situation, the adopted Lagrangian will have the following form:

$$\mathcal{L} = -\frac{1}{4} F_{\mu\nu} F^{\mu\nu}. \tag{5.14}$$

What can I say? Let's forget about any attempts at an *a priori* justification of this formula and simply try to analyze it a bit. As you can guess, the function $\mathcal{L}$ was chosen so that the result would be correct. First of all, let us recall Einstein's summation convention once again: the abbreviated notation $F_{\mu\nu}F^{\mu\nu}$ means that here we have the sum of the corresponding number of components of the electromagnetic field tensor: $F_{00}F^{00} + F_{01}F^{01} + \cdots$. The only difference between a tensor with upper indices ("contravariant") and a tensor with lower indices ("covariant") is that all components with zero indices have opposite signs ($F^{0\nu} = -F_{0\nu}$ and $F^{\nu 0}$

$= -F_{\nu 0}$).[11] You can calculate, taking the matrix form of the tensors $F_{\mu\nu}$ and $F^{\mu\nu}$ in (5.11), that as a result of this long sum we will get a pretty simple result:

$$\mathcal{L} = \frac{1}{2}(E^2 - B^2).$$

It is an expression invariant with respect to the Lorentz transformation (as a scalar). However, the most important thing now is to identify the Lagrangian field variable φ and its partial derivatives so that we can apply the generalized Euler-Lagrange equation. What is the field in our case? It is known that neither the electric nor the magnetic fields are suitable, as they are not invariant with respect to the Lorentz transformation. So we need to use the four-vector of electromagnetic potential A_μ again. Since the tensor $F_{\mu\nu}$ was introduced as a function of the potential (5.10), we can easily represent the above Lagrangian (5.14) in terms of A_μ:

$$\mathcal{L} = -\frac{1}{4}\left(\frac{\partial A_\nu}{\partial x^\mu} - \frac{\partial A_\mu}{\partial x^\nu}\right)\left(\frac{\partial A^\nu}{\partial x^\mu} - \frac{\partial A^\mu}{\partial x^\nu}\right).$$

Note that $\mathcal{L}$ does not depend directly on the potential A_μ itself, but only on its partial derivatives. This means that in the Euler-Lagrange equation, the right side will be zero. The left side, on the other hand, must be calculated by differentiating the above expression with respect to the appropriate derivatives. I will save the reader the tedious calculations (although they are not that complicated) and write the final result in the synthetic form of the tensor equation:

$$\frac{\partial F^{\mu\nu}}{\partial x^\nu} = 0.$$

It may be a little hard to believe, but this equation contains two inhomogeneous Maxwell equations, assuming that there are no charges and no electric current. For example, for $\mu = 0$, the above equation becomes the following equality:

$$\frac{\partial E_x}{\partial x} + \frac{\partial E_y}{\partial y} + \frac{\partial E_z}{\partial z} = 0$$

which is easily recognized as setting the divergence of the $\boldsymbol{E}$ vector equal to zero. On the other hand, for $\mu = 1, 2, 3$ the above equation will give us the corresponding components of the vector equation expressing Ampère-Maxwell's law for the case when there is no current (the current density $\boldsymbol{j}$ is zero).

Let us take a moment to catch a breath before the final stretch. We will now want to derive the general inhomogeneous Maxwell equations in a situation where we have both non-zero electric charges and non-zero current. An additional assumption needed to derive these equations will be the principle of conservation of electric charge, which is one of the fundamental laws of nature. So far, no phenomenon has been observed that would break this rule. But how to express it mathematically? First of all, we need to clarify what exactly it means to conserve the charge. One possible ("global") interpretation would be that the total electric charge in the universe does not change. However, such a principle does not exclude a situation in which an electric charge in a certain volume (e.g. on Earth) would suddenly disappear,

---

[11] A precise definition of contra- and covariant tensors can be found in the chapter on general relativity.

and immediately the same charge would appear in a different volume (e.g. on the Moon). The total charge would be conserved, but in a laboratory on Earth we would observe an inexplicable disappearance of the charge.[12] Such a phenomenon would break the principle of conservation of charge in the local version, which can be expressed in the claim that any change in charge in a given volume must be "compensated" by an appropriate flow of current through the boundaries of that volume.

The local version of conservation of charge will be presented mathematically in the form of a rule that equates the rate of change of charge density at a given point (or infinitesimal area) with the "outflow" of current from this point (area). As we remember from section (4.9), the outflow of a given field is mathematically expressed by its divergence. Thus, the equation that is sometimes called the continuity equation will be as follows:

$$\frac{\partial \rho}{\partial t} = -\nabla \cdot \boldsymbol{j}.$$

This equation can be rewritten in the four-vector form, typical of the theory of relativity. Let us introduce the current density four-vector J^μ as follows:

$$J^\mu = \left(\rho, j_x, j_y, j_z\right).$$

Thus, the zero component of the four-vector J^μ is the charge density, and the other components are the same as the current density components. The continuity equation can now be written in an elegant form (remember that $t = x^0$):

$$\frac{\partial J^\mu}{\partial x^\mu} = 0.$$

The last step in our arduous journey to deriving Maxwell's inhomogeneous equations will of course be to guess the new Lagrangian. To the expression (5.14) defining the Lagrangian for the sourceless situation, we will add a term representing the interaction with the sources of electric and magnetic fields, i.e. containing the density of charge and current. It will have the form of a dot product of two four-vectors (which secures Lorentz invariance):

$$\mathcal{L}_s = J^\mu A_\mu.$$

Note that the addition of this term to the Lagrangian (5.14) does not change the value of derivatives with respect to the rate of change of the four-vector A_μ (i.e. with respect to $\frac{\partial A_\mu}{\partial x^\nu}$), because there are no derivatives of A_μ in the new term. Thus, the left side of the Euler-Lagrange equation will not change. On the other hand, the right side of the equation will no longer be zero, because the Lagrangian depends on A_μ. Its partial derivative with respect to A_μ is of course J^μ. So finally our Euler-Lagrange equation gives the following result:

---

[12] One may wonder whether such a phenomenon would not violate the principle of charge conservation in the global version as well. We remember that in the special theory of relativity, the choice of the plane of simultaneity is to some extent a conventional matter. In one frame of reference, the event of the "disappearance" of the charge on Earth and its "appearance" on the Moon are simultaneous, and therefore globally the charge does not change. However, consider another frame of reference in which the appearance of charge on the Moon occurs later than its disappearance on Earth. In such a system, we will have the missing charge for a while – it is no longer on Earth, and it has not yet appeared on the Moon.

$$\frac{\partial F^{\mu\nu}}{\partial x^{\nu}} = J^{\mu}.$$

As you can guess, this is the full form of Maxwell's inhomogeneous equations (please expand this equation into components to see that for $\mu = 0$ it gives us Gauss' differential law, and for the remaining indices Ampère-Maxwell's law in its full form with current).

Wait a minute, but haven't we said earlier that in order to derive Maxwell's complete equations we will need the conservation of charge (continuity equation)? At what point did we use this equation? It turns out that the continuity equation is needed to argue that the action determined by the new Lagrangian is invariant with respect to the gauge transformation, i.e. it does not change when the four-vector potential A_μ is replaced by an "equally good" potential $A_\mu + \frac{\partial S}{\partial x^\mu}$. We will not prove this – let's leave this matter to more mathematically proficient readers to verify on their own.

Let us summarize the results obtained in this section in order to highlight some general regularities. It turns out that the two types of Maxwell's equations – homogeneous and inhomogeneous – differ significantly in terms of their grounding in certain general principles. Homogeneous equations (i.e. the equation expressing the absence of magnetic charges plus Faraday's law) can be derived from the Lagrangian describing the interaction of charged particles with a magnetic field, assuming the existence of appropriate relationships between electric and magnetic fields and the four-vector potential. On the other hand, the inhomogeneous equations (Gauss' law and Ampère-Maxwell's law) require the use of the Lagrangian formalism to describe the evolution of the field itself. The appropriate Lagrangian for the situation of the absence of sources is written in the form of a function of the four-vector potential, while in the general case we add to this Lagrangian a term containing information about the density of charge and current (four-vector of current density). In the case of both homogeneous and inhomogeneous equations, we formulate the appropriate Lagrangians in the version invariant with respect to the Lorentz transformation and with respect to the gauge transformation. The invariance with respect to the gauge transformation in the case of homogeneous equations is guaranteed by the mathematical form of the Lagrangian, while in the case of inhomogeneous equations it requires the assumption of the local conservation of electric charge (continuity equation).

Questions and problems

1. What influence did the experiments of Fizeau and Michelson-Morley have on the ether hypothesis?

2. Discuss in detail the structure of the Michelson-Morley experiment. Provide possible explanations for its negative outcome. Why have they been rejected in favor of a new theory of time and space?

3. How can simultaneity be defined for spatially distant events? Show that Einstein's signal-based definition leads to the conclusion that the simultaneity of events depends on the adopted frame of reference.

4. Compare the Lorentz transformations with the Galilean transformations. Under what assumptions can the former be reduced to the latter?

5. Show how Lorentz transformations justify the existence of two relativistic effects: time dilation and length contraction. What is the proper time for an object?

6. Compare the relativistic length contraction resulting from the Lorentz transformation with the Lorentz-FitzGerald conjecture to explain the negative result of the Michelson-Morley experiment. Why are these two effects different? According to the theory of relativity, do objects in motion undergo physical "flattening"?

7. Is it true in relativity that if a body moves with a certain velocity with respect to one system, and this system moves with a different velocity with respect to us, then the velocity of the body with respect to us will be the sum (in general vectorial) of these two velocities? What does the situation look like if the moving body is a photon?

8. Explain the concept of space-time interval and the distinction between space-like, time-like, and null intervals. For each of these concepts, give two definitions of it – one based on the possibility of connecting events with a light signal, and the other on the existence of appropriate reference frames in which simultaneity or collocation occurs.

9. Discuss the concept of a light cone and the concepts of the absolute (causal) future and absolute causal past based on it. How do they relate to the concepts of the relative (frame-dependent) past, present and future?

10. What is quasi-simultaneity? Are two events quasi-simultaneous with a third event always quasi-simultaneous with each other?

11. Let A be the common part of all areas of space-time representing the relative (frame-dependent) future of a certain event. What area will A be? Consider an analogous question for the case of relative pasts.

12. Depict on a space-time diagram the time and space axes for a system moving relative to a given system in the direction of increasing values of the x coordinate and in the direction of decreasing values of x. How to determine the points on these axes corresponding to the unit value of the time and space coordinate?

13. Present the reasoning leading to the "twin paradox", using the condition of the invariance of the spatiotemporal interval and the concept of proper time. What physically distinguishes the astronaut twin from the twin remaining on Earth?

14. Two twin brothers flew out of Earth in opposite directions in space rockets, after which the rocket of one of the brothers turned around and began to fly at twice the speed towards the other brother. Which of the brothers will be older when they meet?

15. What are the consequences of special relativity for the philosophical dispute over the status of time and space between absolutism and relationalism? Do these consequences also apply to the dispute regarding the status of space-time?

16. Does the theory of relativity refute the belief in the existence of the objective passage of time? Consider possible attempts to defend this belief.

Further reading

A very good summary of the special theory of relativity can be found in Chapter 1 of the well-known textbook: B.F. Schutz, *A First Course in General Relativity*, Cambridge University Press, Cambridge 2009.

Some historical aspects of the development of special relativity in the context of the classical theory of electromagnetism are discussed in Part VI of the book already cited above: J. T. Cushing, *Philosophical Concepts in Physics*, Cambridge University Press, Cambridge 1998.

This chapter owes much to the elegant introduction to the mathematical foundations of relativity in the second volume of the series "Theoretical Minimum. What You Need to Know to Start Doing Physics": L Susskind, A. Friedman, *Special Relativity and Classical Field Theory. Theoretical Minimum*, Basic Books, New York 2017.

CHAPTER 6. GENERAL RELATIVITY

What is special about special relativity (SR)? As the name suggests, this theory only applies to a particular case and therefore requires an appropriate generalization. Special relativity describes mechanical phenomena in distinguished reference frames, namely inertial frames. We remember that the Lorentz transformations, which are the foundation of SR, connect time and space coordinates measured in one system with coordinates in another system, moving at a constant speed relative to the former. But what if the other system starts to accelerate? The special theory of relativity does not answer the question of what the measurements of time and space will look like in such a frame. For this, a new theory is needed, called general relativity (GR). Its foundation is the postulate that all reference frames should be treated equally, regardless of the state of motion in which they are, or even whether the adopted coordinates are rectilinear or curvilinear. One of Einstein's goals in trying to formulate a new generalized theory of space and time was to make sure that the equations of this theory will "look" the same in every imaginable frame of reference. By reference frame we mean any assignment of four numbers (coordinates) to each event in space-time. The only condition imposed on such an assignment (called a map in the language of mathematics) is the requirement that it respects the "proximity" relation between points (mathematically the proximity relation is called "topology"). The point is that a map imposed on a manifold composed of space-time points should not make "jumps" when moving from a point to an adjacent point, and thus should be consistent with the topology of space-time.

But how to express fundamental laws of physics using such a general concept of a coordinate system? It turns out that the key to solving this problem is gravity. It may seem surprising, but as we will soon see, there is an amazingly close analogy between the gravitational field and accelerating reference frames. The general theory of relativity pursues two seemingly different goals at the same time. It equates the status of all reference frames and formulates a completely new theory of gravitational interactions. Gravity in this theory is something like a disturbance of space and time, not a mere interaction between bodies endowed with mass. To formulate a new theory of gravity precisely, it will be necessary to use subtle geometric concepts describing the curvature of time and space. These concepts will enable us to clarify yet another important difference between special and general relativity. A special theory describes space-time that is "flat" – not in the sense of being a two-dimensional surface, but in the sense of not being curved (twisted or bent). Straight lines in Minkowski spacetime are "genuinely" straight. In contrast, the geometry of general relativity is curved, similar to the geometry on the surface of a sphere. In such geometry, even the straightest possible lines have their internal curvature. Curvature in the theory of relativity is the result of the presence of bodies endowed with mass, and its observable effect is the behavior of bodies in the gravitational field, which move as if a force was acting on them. However, it is not the force that causes the appropriate curvature of the paths of bodies in the gravitational field, but the geometry of time and space itself.

6.1 Einstein's equivalence principle

However, let us proceed slowly by first discussing Einstein's equivalence principle. We remember from earlier chapters that Galileo argued for the principle of relativity – no physical phenomena should distinguish between systems moving relative to each other at a constant speed. Galileo's famous argument with the ship shows that we are not able to distinguish by means of "internal" observations whether we are below the deck of a ship moving at a constant speed or a ship at rest. On the other hand, if the ship begins to accelerate – e.g. by rocking on the water – this will cause observable effects in the form of moving objects under the influence of apparent forces. Einstein noted that Galileo's argument from the observational indistinguishability of inertial systems can be generalized to the case of accelerating systems if gravity is taken into account. To do this, consider two situations: in one, a space rocket is moving far away from gravitational sources with a constant acceleration equal to that of the free-fall acceleration on Earth $a = g$, and in the other, the rocket is stationary on the Earth's surface (Fig. 6.1). In both cases, the behavior of objects inside the rocket will be the same – all bodies will fall at a constant acceleration of g. Therefore, an astronaut enclosed in a rocket will not be able to determine which situation is taking place based on local observations.

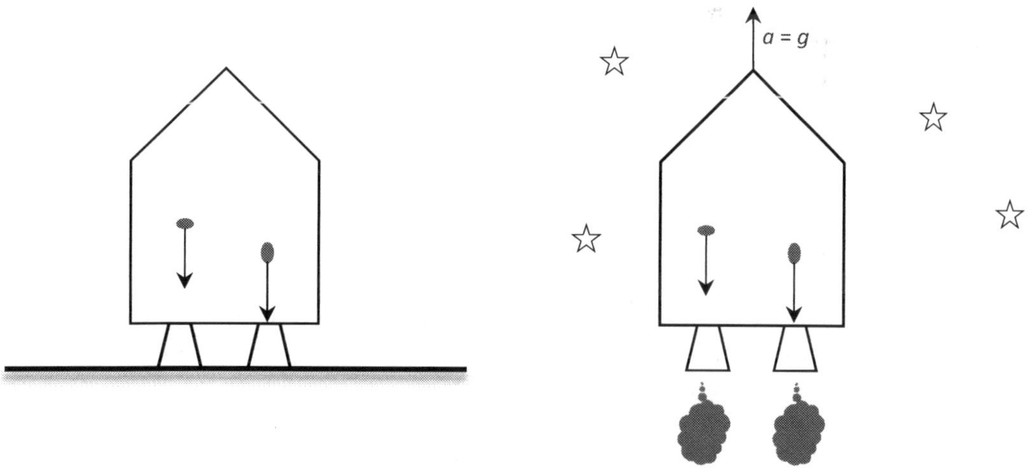

Fig. 6.1 Einstein's equivalence principle

Let us note two important aspects of Einstein's equivalence principle. First, this principle follows from the empirical fact of equality between inertial mass and gravitational mass. As we remember from the chapter devoted to classical mechanics, mass plays a dual role in it:

as a factor of proportionality in the second law of motion $F = ma$, and as a measure of gravitational "charge" for a given body, appearing in Newton's law of universal gravitation. Only assuming that the two masses are identical can we conclude that all bodies in the gravitational field will fall with the same acceleration, and therefore their behavior will be indistinguishable from that in the accelerating system. Secondly, the thesis about the equivalence of the accelerating system with the system resting in the gravitational field should be restricted with the caveat "locally". In fact, there is an observational way to distinguish between the two situations if we include bodies that are far apart. In the case of an accelerating system, the direction of the apparent force is the same regardless of the position of the bodies under consideration (it is opposite to the direction of acceleration of the entire system). In contrast, in a central gravitational field from a body such as the Earth, both the direction and the value of the gravitational force (gravitational acceleration) change depending on the position of the body. Bodies separated horizontally will *de facto* get a little closer to each other, and bodies separated vertically will move away (this effect is often described as caused by the action of the so-called tidal forces, which result in a corresponding deformation of the extended bodies in the gravitational field – see Fig. 6.2). Thus, observational equivalence is valid only for small areas for which we can ignore tidal effects. In this respect, Einstein's principle of equivalence differs from Galileo's principle of relativity – the latter applies globally to all phenomena in a given system, regardless of their mutual spatial separation.[1]

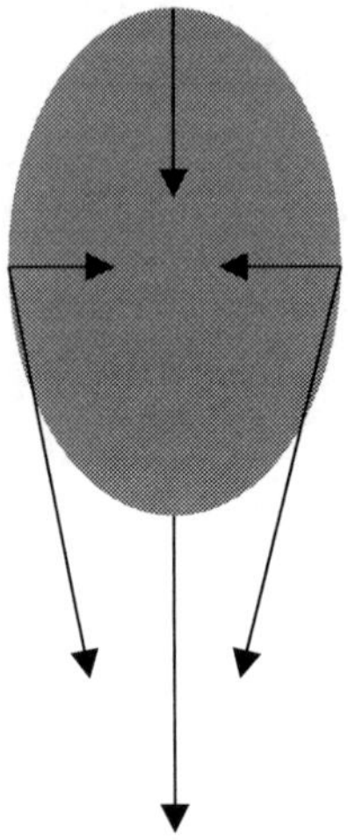

Fig. 6.2 Tidal forces in central gravitational field

Another example of local empirical indistinguishability is the case of free fall in a gravitational field. People in a falling elevator are weightless due to the cancelling-out of the gravitational force with the apparent force (the same applies to the less dramatic situation of orbiting the Earth in a space station). Therefore, locally such a situation is indistinguishable from remaining in an inertial frame away from all sources of gravity. This fact seems to

[1] The question arises: what about the inertial forces that arise in rotating systems, such as centrifugal or the Coriolis force? Is it possible to "eliminate" them in an analogous way to the above, by moving to a system resting in a gravitational field? The answer to this question is yes, but it requires a transition from the classical theory of gravity to the theory of gravity as a curvature of space-time. In the classical theory of gravitation, for example, there is no place for forces dependent on the velocity of bodies, such as the case of the Coriolis forces.

suggest the following generalization of the concept of an inertial frame in the case of a gravitational field: an inertial system can be interpreted as any system in free fall in a gravitational field. In other words, free fall will be the equivalent of the rectilinear and uniform motion of a body devoid of all forces, described by Newton's first law of motion. Of course it remains true that the space-time trajectory of a body falling freely in a gravitational field is not a straight line (it is curved in space or space-time, due to the acceleration that occurs). However, this difference between classical free motion and free fall in the gravitational field can be "minimized" by introducing the concept of a geodesic and geometries with intrinsic curvature.

A possible relationship between gravity and the curvature of spacetime can be seen using the equivalence principle. On the one hand, we know that gravitational field is empirically equivalent to the acceleration of the corresponding non-inertial frame. Accelerating systems, on the other hand, are characterized by coordinate axes, which are curved from the perspective of a non-accelerating system. As we can see in Figure 6.3, spacetime points whose spatial coordinates in a uniformly accelerating system are the same (e.g., $x' = 0$, $x' = 1$, and so on) form parabolas rather than straight lines. Therefore, it can be suggested that gravity has something to do with curvature. However, this loose thought should be properly specified conceptually and mathematically. In the next section, we will try to explain a little more clearly the concept of the intrinsic curvature of space (and space-time) and its relationship to the kinematic behavior of bodies.

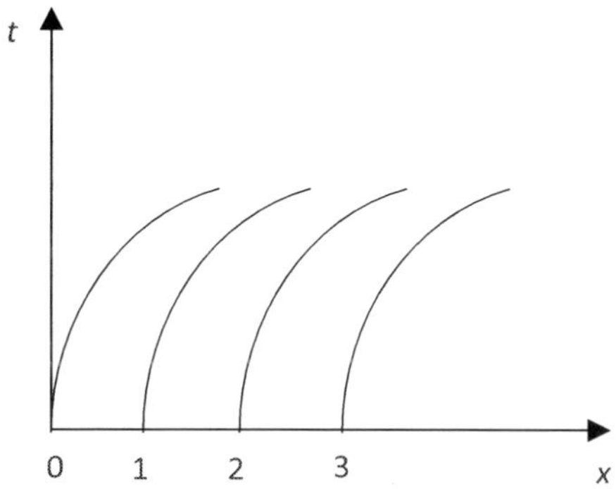

Fig. 6.3 Points with the same spatial coordinates in an accelerating system

6.2 Curvature of space(-time)

Popular approaches to general relativity often refer to the example of geometry on the surface of a sphere. This is a very useful example for pedagogical purposes, but it suffers from significant limitations. First, the geometry of a sphere is two-dimensional, while physical space has three dimensions. Second, and more importantly, a sphere is a spatial object, while relativity takes into account the geometric properties of the entire four-dimensional space-time. Only the curvature of space-time, with its non-negligible temporal aspect, reveals the essential properties of the behavior of bodies in the gravitational field. With this warning in mind, however, let us use the example of the sphere as an illustrative, albeit simplified, tool.

Let us choose any two points on the sphere. How do you define the distance between these points without leaving a two-dimensional surface? A natural approach, used on a daily basis in geography and air transport, is to minimize the length of paths between these points. The distance will be the length of the shortest possible path from one point to another. For obvious reasons, the shortest path will not be a straight line – you can see that the shortest distance will be achieved by moving along fragments of great circles, i.e. circles formed by intersecting two selected points on a sphere and the center of the sphere with one plane. The great circles on a sphere represent the "straight" (or straightest possible) lines in a given geometry. Such lines are called "geodesics". Let us now consider two points at the equator and draw from them two geodesics at right angles to the equator (Fig. 6.4). Let us imagine that two ants are moving along both lines. Although at the beginning of their journey their paths could be considered parallel, at the moment of reaching the pole, the paths converge. From the ants' point of view, it looks as if there was an attractive force between them, which finally led to the meeting. However, in reality there is no force, only a curvature of geometry that causes lines that originally looked parallel to touch at a distant point.

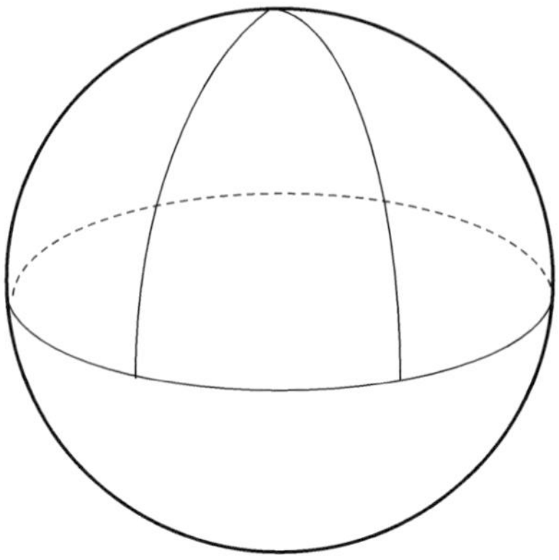

Fig. 6.4 Geodesics perpendicular to the equator converge at the pole

It is perhaps worth mentioning one of the basic assumptions of Euclidean geometry, which is the so-called fifth postulate (the axiom of parallels). The geometry we learn in school is based on the rather intuitive thesis that exactly one line parallel to a given line can be drawn through one point. It turns out that on a sphere this rule is broken, if we only interpret the concept of straight lines as referring to great circles. For a given line (geodesic) on the sphere there is no line parallel to it. This is due to the fact that every two great circles on the sphere must intersect. The geometry of the sphere is non-Euclidean – it belongs to the category of so-called Riemannian geometry with positive curvature. There are also non-Euclidean geometries with negative curvature (the so-called Lobaczewski geometry), for which there is more than one line parallel to the given passing through a selected point. Lobaczewski's two-dimensional geometry can be illustrated by the example of a "saddle" (mountain pass).

Returning to our example of ants on a sphere, we can say that it provides us with an illustrative argument for the fact that gravitational interactions can be tried to be explained purely geometrically using the concept of curvature. It should be emphasized that this type of direct geometric interpretation is possible only in the case of gravity, and not in the case of other interactions (e.g. electromagnetic). This is because only gravity is universal, i.e. it acts on all bodies equally. Thanks to this fact, we can interpret gravity as a property of (space-time) independent of what test body occupies a given area. The basic idea of this approach is that free bodies in a gravitational field move along "straight" lines (i.e., geodesics), as if they were not subjected to any force. However, the geodesics themselves are characterized by an intrinsic curvature due to deviations from Euclidean geometry. As we will see, this curvature depends on the masses present in a given area.

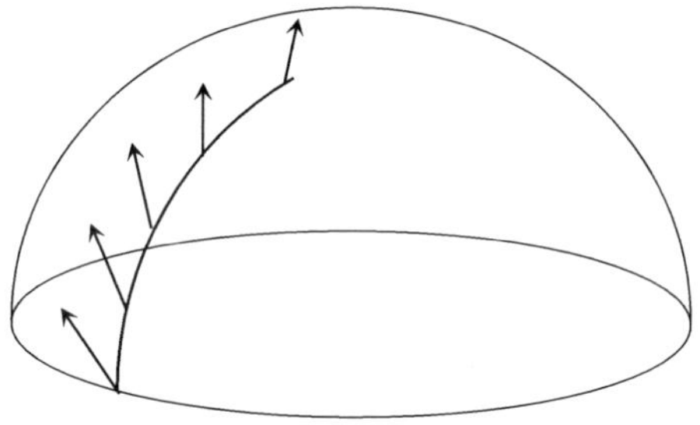

Fig. 6.5 Parallel transport of a vector on a sphere

Let us take a closer look at the concept of curvature, which for now we have treated quite loosely, relying on intuitions. How do you know if a two-dimensional surface is curved? It turns out that this is not a trivial question. To illustrate the problem, let us compare two cases: the surface of a sphere and the surface of a cylinder. Intuitively, both surfaces look curved. However, mathematically only the first of them satisfies the formal conditions of curvature. The geometry of the cylinder surface is completely Euclidean: two lines perpendicular to a third will not intersect each other. Figuratively speaking: the surface of a cylinder can be "straightened up", creating an ordinary two-dimensional plane (just as a newspaper rolled up can be flattened out). On the other hand, analogous straightening in the case of a sphere is impossible without changing the intrinsic geometry. These intuitive considerations can be expressed a little more precisely with the help of the important concept of *parallel transport*. Let us consider any point on a sphere and draw from it a vector tangent to the sphere in a certain direction. Mathematically, the following operation can be defined on a selected vector: we connect two points with any line, and then "transport" the vector from the starting point along this line to the end point, maintaining its direction (defined by the so-called infinitesimal concept of parallelism). In other words, we divide a given curve into very small (infinitesimally small) segments and try our best to maintain the same direction of the vector on each segment (Fig. 6.5).[2] Of course, from the external perspective, the tangent vector to the surface of the sphere will change direction, but relative to the sphere itself, the direction remains unchanged, thanks to the fact that we will maintain parallelism in every small step.

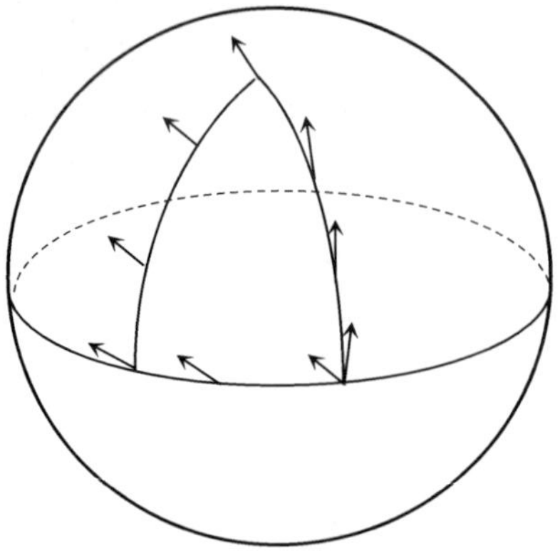

Fig. 6.6 Parallel transport of a vector along a closed curve on a sphere. The end vector does not coincide with the initial.

With the help of the concept of parallel transport, we can define the mathematical notion of curvature. For this, we need to consider a closed curve on a given surface. By making a parallel shift of a given vector along this curve, we compare the start and end vectors at the same point. If both vectors coincide, we are dealing with a flat surface. If, on the other hand, the end vector is pointing in a different direction than the initial vector, it means that the considered surface is characterized by curvature, the magnitude of which depends on the magnitude of the angle between the two vectors (the greater the angular divergence between them, the greater the curvature). Let us apply this concept to the case of a sphere. Consider a closed line in the shape of a triangle, the sides of which are, respectively: a segment of the equator and two lines joining at a certain point. Starting from a point on the equator and moving a given vector parallel along the first line, and then returning along another line and the equator, we get a vector pointing in a different direction from the original (Fig. 6.6). On the other hand, an analogous operation on the surface of the cylinder will not bring about a change of direction (it is best to notice this by "unrolling" the surface of the cylinder – then you can immediately see that the surface must be flat).

An interesting case of curvature is the surface of a cone. Since a cone can be unrolled without deformation to a flat form (imagine cutting the cone along a line coming from the vertex – the so-called generating line of the cone – and then straightening it, then we will get a flat surface with a cut-out triangle), in general its surface is characterized by zero curvature. However, there is one exception, namely the top of the cone. To see that we are dealing with a non-zero curvature, consider a closed curve around the apex (e.g. a circle). On the straightened surface of the cone, we can easily see that by choosing a given vector at the starting point and moving it along our curve, we will get another vector

[2] Formally, this preservation of direction (or infinitesimal parallelism) is expressed by an operation called covariant derivative. We will mention this concept in more detail in the sections marked with an asterisk.

at the end of the path, (remember that the end point of the path on the cut is the same as the starting point – see Fig. 6.7) We can show that the angle between the start and end vectors is equal to the angle at the tip of our cone. The direction of the transported vector is changed only in the case of closed paths encompassing the apex – for all other closed curves, the transport will not change the direction of the vector.

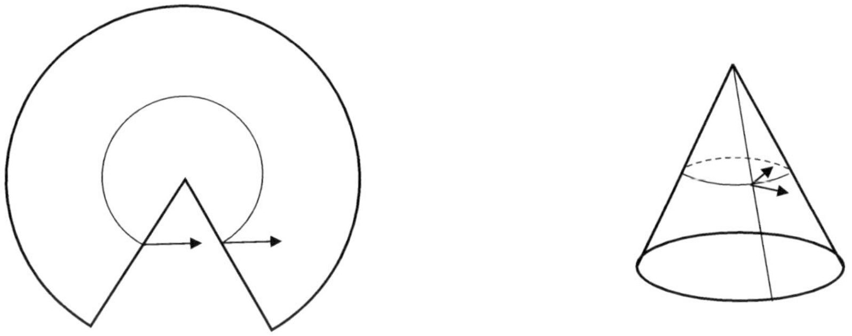

Fig. 6.7 Parallel transport of a vector along a curve containing the vertex of a cone. The drawing on the left was created by cutting the cone along its generatrix and stretching it out on a plane. Because the start and end vectors form different angles with the cut line, when re-glued, these vectors will not coincide with each other (figure on the right), indicating the existence of a curvature at the apex of the cone.

The mathematically strict definition of the concept of curvature applies this concept to a given point. To achieve this, we need to modify our criterion based on the parallel transport of vectors along a closed curve – namely, we need to take the limit of closed curves shrinking to a point. Of course, in this case, the magnitude of the angular divergence alone will be useless as a measure of curvature at a point, because the difference between the start and end vectors during parallel transport will typically converge to zero for smaller and smaller "loops". The solution is to take the curvature measure R as the ratio of the angular difference to the area enclosed by the curve.[3] In this case, the curvature will be the ratio of two numbers converging to zero, which can turn out to be any number. It is not difficult to notice that in the special case of a cone and closed curves encompassing its vertex (see the entry in the box), this ratio will converge to infinity, because the angular difference between the initial and final vectors is constant, regardless of the curve in question, and the area enclosed by the curve obviously converges to zero for incrementally smaller curves. Thus, the curvature at the apex point of the cone is infinite (the apex is "sharp" rather than "smooth").

[3] It is perhaps worth adding that the full definition of curvature also includes a sign, which can be positive or negative. To achieve this, we compare the direction of "rotation" of the selected vector during parallel transport with the direction of the curve's rotation. If both of these directions are the same, the curvature is positive (as on the surface of a sphere). If, on the other hand, the transferred vector rotates in the opposite direction to the direction of rotation, we are dealing with a negative curvature, as on the surface of a saddle.

The case of two-dimensional curved surfaces can be generalized to any number of dimensions, so that the concept of curvature can be applied to, for example, four-dimensional space-time. In general, however, the curvature will then be represented not by the number R (scalar), but by a mathematical object called a tensor. We have already encountered tensors in the chapters devoted to electromagnetism and special relativity – in simple terms, they are objects represented in each coordinate system by strings of numbers (tensor components). Each tensor has a number of indices that determine its *rank*. A tensor with one index is simply a vector, and the number of its components is equal to the dimension of the given space. A tensor with a number of indices k will have k^d components, where d is the dimension of space. It turns out that the curvature tensor (also called the Riemann tensor) in general must have four indices: $R_{\alpha\beta\gamma\delta}$, and therefore it is characterized by $4^4 = 256$ components in a four-dimensional spacetime. However, not all components of the Riemann tensor are independent from each other (the number of independent components actually reduces to 20).

The Riemann tensor is related to another very important tensor that plays a fundamental role in the geometry of spacetime. It is a metric tensor that defines an infinitesimal distance at a given point in space. It can be used to precisely determine the distance between any two points in space (in mathematical language, the distance function between points is often called metric, hence the name metric tensor). It is worth taking a closer look at it. Let us start with the known formula for the distance between two points in Euclidean geometry when the difference between the coordinates of these points is known. Suppose our space has three dimensions, and the coordinates of any point are given in the form of three numbers x, y, and z. The coordinate differences between two infinitesimally close points (they are referred to as *differentials*) are denoted by dx, dy and dz. According to the Pythagorean theorem, the distance between such points is given by $ds^2 = dx^2 + dy^2 + dz^2$. In general, however, the expression for the infinitesimal distance (displacement) can contain any products of coordinate differentials: $dxdy$, $dzdy$, etc. The formula describing all possible variants of the form of the infinitesimal displacement can be written as:

$$ds^2 = \sum_{i,j=1}^{3} g_{ij} dx^i dx^j.$$

Instead of writing $x, y, z,$ we now use a more universal form of coordinates: x^1, x^2, x^3. The coefficients g_{ij} are the components of the metric tensor, which, as you can see, has its rank equal to 2. In the case of Euclidean geometry expressed in Cartesian coordinates, the metric tensor will have very simple components: $g_{11} = g_{22} = g_{33} = 1$, and all other components are equal to 0. This is sometimes expressed by a useful symbol δ_{ij} called the Kronecker delta, which takes the value of one when $i = j$, and zero when $i \neq j$. It is worth remembering that the metric tensor will have different forms in different curvilinear coordinate systems. However, as long as the space under consideration has zero curvature, it is always possible to find a coordinate system in which the metric tensor takes the form of the Kronecker delta. On the other hand, if a given space is characterized by a non-zero curvature, there is no coordinate system in which the metric tensor would take the simple form δ_{ij}.

To illustrate, let us give the form of a metric tensor specifying the geometry on a sphere. The standard coordinates on a sphere are the so-called spherical coordinates, defined by two angles θ (latitude) and φ (longitude). The angle θ runs from $-\frac{\pi}{2}$ to $\frac{\pi}{2}$, and the angle

φ from 0 to 2π. It can be calculated using trigonometry that the differential distance ds^2 on a sphere with radius r is given by the following formula depending on the angles φ and θ:

$$ds^2 = r^2 d\theta^2 + r^2 \sin^2\theta\, d\varphi^2$$

Thus, the metric tensor has the following components: $g_{11} = r^2$, $g_{22} = r^2 \sin^2\theta$, and for the "non-diagonal" components: $g_{12} = g_{21} = 0$. It is different from the Kronecker delta δ_{ij} and cannot be reduced to it in any coordinate system.

Let us also mention that the concept of parallel transport can be used to alternatively define geodesics, which we have previously characterized as the shortest lines connecting given points. Let us consider any line and draw a vector tangent to it at each point. If it turns out that the tangent vectors to a given line are parallel to each other, this line will be called geodesic. The condition of tangential parallelism can be written in the form of an equation that equates the covariant derivative of the tangent vector with zero. This can be formally written as

$$\nabla_\mu V^\mu = 0,$$

where ∇_μ is the covariant derivative (also called affine connection), and V^μ is the vector tangent to the curve. Since the covariant derivative of a given vector represents its variation along a given curve at a given point, the geodesic equation means that the vectors tangent to the geodesics do not change their direction (remain parallel to each other).

Let us now move on from purely geometric considerations to physics. In the chapter on special relativity, we said that the motion of a particle is described by a trajectory in four-dimensional space-time (this is called the world-line of this particle). We also introduced the concept of a four-vector velocity, which is a unit vector tangent to the world line of a particle, and its spatial components are components of velocity with respect to proper time. From the above considerations, we know that if the trajectory of the particle is a geodesic, the vectors tangent to the trajectory do not change their direction, i.e. their derivative is equal to zero. Thus, for a particle moving along a geodesic, its four-velocity U^μ does not change as it moves along the trajectory:

$$\nabla_\mu U^\mu = 0. \tag{6.1}$$

The above equation can be interpreted as a generalization of Newton's first law of motion. This means that free particles, i.e. not subjected to any forces other than gravity, move along geodesics. This observation is the key to describing the behavior of objects in geometry with curvature. According to the principle of inertia of classical mechanics, free bodies move along trajectories that are straight lines.[4] In geometry characterized by curvature, this principle is preserved, except that the role of straight lines is played by geodesics. Gravity manifests itself only in the curvature of geometry, not in the form of an acting force.

An illustration of these considerations can be the description of the planet's orbit around the Sun. Figure 6.8 shows the space-time trajectory of the planet, which takes the shape of an extended spiral (of the three spatial dimensions, we have chosen two that are sufficient to

[4] Remember that this is about trajectories in space-time! The straightness of a space-time trajectory means that a given body will not accelerate.

describe the planet's motion, as it occurs on a plane). We interpret the curvature of the planet's world line purely geometrically, treating the trajectory as a geodesic in geometry with non-zero curvature. Metaphorically speaking, it is not the Sun that directly causes the variability of the planet's path, but space-time itself, which "pushes" the planet in a given direction. Let us also note that the path of the planet in four dimensions does not deviate significantly from the "true" straight line (remember that under the assumption *of* $c = 1$, one unit of time will correspond to a huge spatial distance covered by light during this time). Since the speed of the planet in rotation is incomparably lower than the speed of light, its world line will be almost vertical and locally it will differ very little from the straight line. This means that the curvature of space-time in areas distant from the Sun is small, and the geometry "locally" differs in an almost imperceptible way from Euclidean.

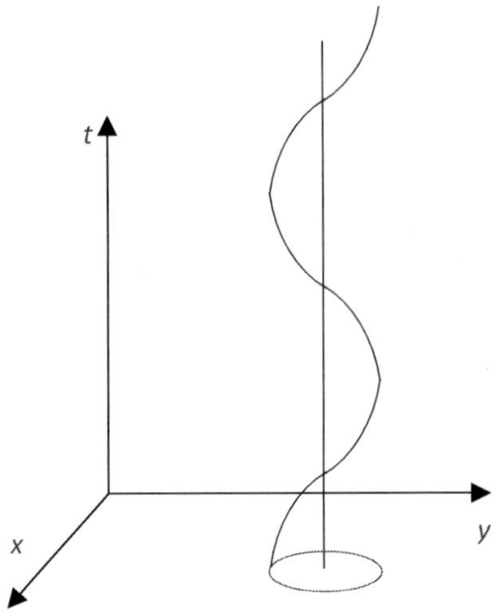

Fig. 6.8. Space-time trajectories of the Sun and the planet orbiting it. After an appropriate selection of units, the spiral trajectory of the planet will "stretch" to the form of an almost straight line.

6.3 Einstein's equation

No popular introduction to general relativity can do without discussing the basic equation of this theory, i.e. the Einstein field equation. Its exact formulation requires quite advanced mathematics – we will sketch it a little more precisely in the sections marked with an asterisk. However, the basic physical meaning of this equation is rather simple and not difficult to grasp. In the classical Newtonian approach, gravity is produced by objects endowed with mass, i.e. by matter. Since we are now replacing gravitational forces with a "perturbation" of space-time, or more precisely, its curvature, it seems natural to expect that matter will cause the curvature of space-time. Einstein's equation clarifies this general idea. In the most general outline, the idea of this equation can be expressed as follows:

Geometry = Matter

We have yet to specify exactly what property of matter is responsible for which property of geometry. Unfortunately, the situation is more complicated than in Newton's theory, in which the only source of gravity is mass. In general relativity, this is still (partially) true, but in addition to mass, energy also plays the role of sources of the gravitational field, as it is essentially one and the same quantity. Furthermore, the curvature of space-time is influenced not only by energy itself, but also by how it changes over time. More precisely, in the Einstein equation there are quantities that determine the flow of energy in different directions. For example, a stream of cosmic dust carries a certain amount of energy, for which we can calculate how much of it flows through a unit area per unit of time (this is analogous to the concepts of electric and magnetic field fluxes, discussed in the chapter on electromagnetism). In addition, the momentum of objects also contributes to the curvature of space-time, along with the information about how much such momentum flows through a given surface. Although in typical situations the contribution from momentum is incomparably smaller than the energy input (which is expressed by the formula mc^2, and therefore is numerically enormous), this element should also be taken into account in the precise equation.

A synthetic presentation of all quantities representing matter and its contribution to the curvature of space-time is given in the form of the so-called energy-momentum tensor, denoted by $T^{\mu\nu}$. It is a second-order tensor, and thus has sixteen components, which we can represent in the form of a table with four rows and columns. Individual rows contain quantities characterizing, respectively, the density and flux of energy and the densities and fluxes of the individual components of momentum. Einstein's equation, on the other hand, can be rewritten in the following form:

$$G^{\mu\nu} = kT^{\mu\nu},$$

where k is a constant. The expression on the left denotes a quantity that characterizes the geometry of spacetime. For a long time Einstein grappled with various possibilities of defining the mysterious tensor $G^{\mu\nu}$. He was guided by two tips. First, the above equation should reproduce Newton's familiar law of gravitation in the case of small masses and low velocities. The point is that Newton's theory of gravity is empirically well confirmed by the data available to us, and therefore the new theory should also have the same confirmation. The only problem that remains is how to make such a limit transition from GR to Newton's theory, since they use completely conceptually different methods of describing gravity – in the form of forces or potentials in Newton's approach, and in the form of metric properties of space-time in GR. What is needed here is a certain "bridge law", combining the old approach with the new one. Einstein showed that in the case of small masses and velocities, the role of the gravitational potential of Newtonian theory can be played by the temporal component of the metric tensor g_{00}. This gives us a suggestion that in a limiting situation, the tensor $G^{\mu\nu}$ should approach an appropriate function of the component g_{00} (more precisely, a combination of the second derivatives of g_{00}, called the Laplacian – see the asterisked section for details).

At this point, it is worth mentioning a very important principle used in physics and discussed by philosophers, called the principle of correspondence (originally formulated by the Danish physicist Niels Bohr, whom we will meet later in the next chapter). The correspondence principle states that any new theory that replaces the old should roughly "reduce" to the old theory in the area where that old theory was empirically well confirmed.

An example of the application of this principle is the limit transition from the Lorentz transformation in SR to the Galilean transformation, assuming that the velocities of bodies are much smaller than the speed of light c (and additionally that the considered spatial distances are also small). We are currently considering another example of an application of the correspondence principle – the GR equations should recover the SR equations for the case of small masses, as well as transition into the equations of Newton's classical theory of gravity with the additional assumption of low velocities. Yet another example of the principle of correspondence in action concerns quantum mechanics.

The second tip takes into account the principle of conservation of energy and momentum. Since momentum and energy cannot disappear in a given region of spacetime, the tensor $T^{\mu\nu}$ must satisfy an equation called the continuity equation. Einstein suggested that the geometric tensor $G^{\mu\nu}$ should satisfy the same equation in virtue of the laws of mathematics. This assumption would automatically ensure that the principle of conservation of energy and momentum is met. More precisely, Einstein's equation, which is obviously not a mathematical truth, would nevertheless mathematically imply the law of conservation of energy and momentum in such a situation. Einstein showed that both imposed conditions are satisfied by a combination of three quantities: the Ricci curvature tensor $R^{\mu\nu}$ (mathematically related to the Riemann tensor), the metric tensor, and the curvature scalar R (this combination is called, of course, the Einstein tensor). So Einstein's equation "in its full glory" looks like this:[5]

$$R^{\mu\nu} - \frac{1}{2} g^{\mu\nu} R = kT^{\mu\nu}. \tag{6.2}$$

So we already have two fundamental principles on which the general theory of relativity is based: Einstein's equation, which determines how matter affects geometry, and the principle of motion, which states that bodies move along geodesics (briefly expressed in the form of the equation $\nabla_\mu U^\mu = 0$). Hermann Weyl presented the quintessence of GR in the neat phrase "matter tells space-time how to curve, space-time tells matter how to move". It may come as a surprise, however, that the relationship between gravity, motion and space-time properties is not at all a unique feature of general relativity that distinguishes it from Newton's classical theory. It turns out that Newton's theory can also be "geometrized" in the manner of GR. By writing the classical equation of motion for a particle in the gravitational field in the language of differential geometry, the gravitational potential can be included in the covariant derivative, thus obtaining a geodesic equation of the same form as the GR equation. However, the curved geometry of Newtonian theory will be different from the geometry of GR. The main difference is that GR is a generalization of special relativity with its Minkowski geometric structure, which is different from the Newtonian-Galilean spacetime structure. Thus, the general theory of relativity is based not on two, but on three pillars. These are: Einstein's equation (6.2) linking the curvature of geometry with the distribution of matter, the equation of motion of bodies in curved spacetime (6.1) and the local Minkowski geometry in the limit of weak fields.

[5] This equation was independently derived by David Hilbert, a German mathematician. Some believe that it should therefore be called the Einstein-Hilbert equation.

6.4 Empirical consequences of general relativity

The general theory of relativity explains the observable facts about gravitational interactions in an extremely elegant way – first of all, their unique feature, which is universality (producing the same gravitational acceleration at a given point regardless of the mass of an object present). It also realizes the ideal of independence of fundamental physical facts from the arbitrarily accepted frame of reference in which we describe these facts. It should also be added that GR eliminates the troublesome non-locality (action at a distance) present in the original Newtonian theory of gravity. The gravitational interaction between distant bodies occurs through the "distortion" of space-time that obeys the principle of locality. However, it must also be admitted that GR is formally a very complicated theory. Its basic equations require the use of very advanced mathematical techniques, far beyond the Newtonian and Leibnizian calculus. Solving Einstein's equations for even the simplest situations is an extremely difficult task and, in many cases, practically impossible to accomplish due to the degree of mathematical complexity. The question then arises whether the replacement of Newton's classical theory of gravity with Einstein's theory is justified by empirical facts. Are there empirical data that unequivocally indicate that the general theory of relativity correctly describes reality? It turns out that there are many spectacular observations and experiments which reinforce the belief that Einstein's theory of gravity is closer to the truth than its classical predecessor.

An important group of empirical tests of general relativity concerns the behavior of light rays in the vicinity of objects of significant mass. In the classical theory of gravity, light does not interact with the gravitational field. As we showed in the chapter on special relativity, photons do not have a rest mass and therefore, according to Newton's law of gravitation, the force of the gravitational interaction on the photon is zero. As a result, the light rays should always travel in straight lines, regardless of the surrounding gravitational field. However, the situation is different within the general theory of relativity. According to the laws of motion operating in GR, all objects, including photons, move along geodesics in space-time. As we know, geodesics near massive objects are curved, following Einstein's equation. Thus, the light ray in the vicinity of such masses should also be visibly curved.

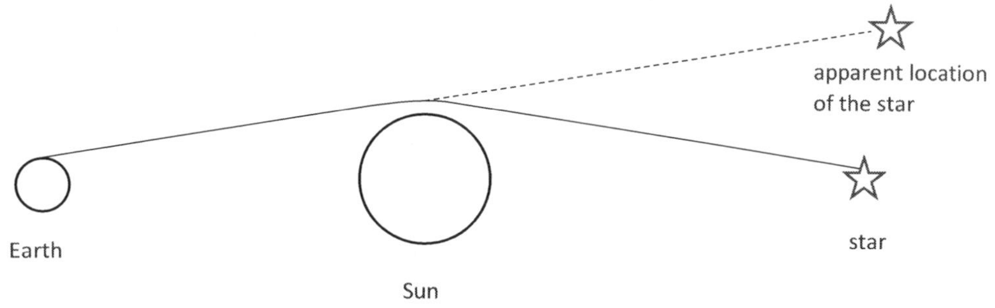

Fig. 6.9 The effect of deflection of a light ray near the Sun

An object with the greatest mass in our system is, of course, the Sun. This led to the idea that the strongest effect of curvature should be observed when a light ray passes near the Sun. An English astronomer, Arthur Eddington, fascinated by Einstein's theory, came up with an idea of empirically testing this effect. He chose a convenient time for a total solar eclipse (which at that time happened to be visible in Africa, so Eddington had to organize an expedition to this continent). During the eclipse, which blocks the Sun's brightness, Eddington recorded on photographic film the apparent position of stars near the Sun. He then compared them with their "normal" position when far away from the Sun (Figure 6.9). It turned out that the stars visible near the Sun undergo a slight shift, which corresponds to the deflection of a light ray passing by the Sun (this is an effect similar to the deflection of a ray running through hot air). Thus, the prediction of the theory of relativity has been confirmed (we are dealing here with a classic example of *experimentum crucis*, described in the first chapter).

Another example of the "anomalous" (from the point of view of the classical theory) behavior of light in the gravitational field is provided by an experiment performed in the 1960s on the surface of the Earth (the Pound-Rebka-Snider experiment). General relativity predicts that a light wave traveling in a gravitational field away from the source of gravity (in the direction of the weakening of the field) will decrease its frequency (move the wavelength in the direction of the red color – so-called red shift). In turn, the frequency of a wave propagating towards a given mass will increase accordingly (shift towards blue). This can be justified quasi-classically by noting that the energy of a single photon depends on its frequency ν ($E = h\nu$), and since the photon will lose some of its energy to overcome the gravitational field (or gain it), its frequency must be reduced (or increased) accordingly. However, a more correct explanation of this effect refers to the "influence" of mass-endowed bodies on space-time, and therefore also on time. Depending on the distance from a given mass, time will flow slightly differently (slower near a large mass), and since frequency is the number of vibrations per unit of time, this will have an effect on the change in radiation frequency.

The near-Earth effect described above is extremely small. However, there is an experimental method that allows its detection. It is based on the phenomenon of emission and absorption of gamma radiation of a strictly defined frequency by atomic nuclei. A given atomic nucleus can radiate a photon as a result of the transition of its nucleon (proton or neutron) from one energy level to another. The same nucleus can absorb the photon again, but it must have exactly the same energy (it must exactly match the energy gap between the ground and excited states of the nucleus). If the photon has a slightly different energy, absorption will not occur. An experiment using this effect has a very simple structure: we place two atomic nuclei of the same kind at different heights in the Earth's gravitational field. A nucleus located above, for example, radiates a photon, which is then to be absorbed by the second nucleus. Under normal conditions, such absorption should occur unhindered, because the photon has exactly the energy it needs. However, in the gravitational field, this energy will change, so the photon will "miss" the target. The result of the experiment shows that photons undoubtedly change energy and frequency when crossing the gravitational field.

The most spectacular consequence of general relativity regarding the influence of matter on the behavior of light is the possibility of the existence of objects called black holes. Let us recall once again the situation of the deflection of a light ray near the Sun. What happens if we consider a body the size of the Sun but much more massive? It is quite obvious that the curvature of the light ray will then be correspondingly greater. By increasing the mass of a body, theoretically we should achieve a situation in which the path of the ray closes around this body. This means that the light will not be able to leave the region near such a supermassive object – it will "orbit" around its center indefinitely. Such an object is, of course, a black hole. For a given size of a star, it is possible to calculate what its mass must be for it to become a black hole.[6] For a long time, black holes were considered purely theoretical constructs, mainly due to the impossibility of observing them directly (since, as we assumed, light is not able to "break out" from such an object). However, it is possible to confirm the existence of black holes indirectly. For example, in a so-called binary system, in which two stars orbit each other, one of the stars may be a black hole. In such a situation, we will notice a shining star orbiting a certain point in a tight orbit, where no other object is visible. Another observable effect of a black hole is when it begins to absorb material from a neighboring object, accelerating its jet to incredible speeds. Such a jet will then emit high-energy gamma rays before it is completely absorbed. Observations of similar effects allow for an identification of a large number of black holes in the universe, which is a strong confirmation of general relativity.

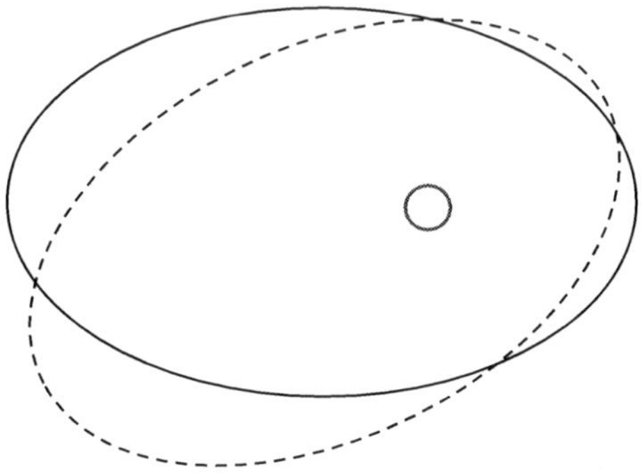

Fig. 6.10 The phenomenon of perihelion precession in Mercury's orbital motion around the Sun

The empirical successes of the general theory of relativity were sealed by the case of the so-called anomalous precession of Mercury's perihelion. The orbiting along the ellipse of Mercury – the closest planet to the Sun in the Solar System – is combined with a slow rotation of the entire orbit, and thus also of its closest position to the Sun, i.e. the perihelion (Fig. 6.10). Such a motion can be explained on the basis of Newtonian mechanics by the disturbing influence coming from other planets, above all from Jupiter, the largest planet in our system. However, the exact calculations made on the basis of the equations of the classical theory of gravity deviated slightly from the observed speed of perihelion rotation. The difference was not great – 43 arcseconds per century (one arcsecond is equal to 1/3600 of a degree). Let us add that precession itself takes place at a speed of 5600 seconds per hundred years, so the

---

[6] Or, alternatively, for a given mass – for instance the mass of the Sun – we can calculate how small the radius of a star must be in order to turn into a black hole. This radius is known as the Schwarzschild radius, and the formula expressing it is as follows: $\frac{2GM}{c^2}$, where G is the gravitational constant, c – the speed of light, and M – the mass of an object. From this it follows that in order to turn our Sun into a black hole, we would have to "squeeze" it to the size of a ball with a 3-kilometer radius.

discrepancy between theory and experiment was less than 1%. However, this problem bothered astronomers for quite some time. There have been suggestions that the detected incompatibility is the result of the influence of a new, unknown planet. It was not an *ad hoc* solution at all. A few decades earlier, a similar problem occurred regarding the mismatch between the observed orbit of the planet Uranus and the predictions of the theory. Although it had been suggested that Newton's beautiful and so far flawless theory may need to be corrected, astronomers John Adams (British) and Urbain Leverriere (French) posited the existence of a new planet that is disrupting Uranus' trajectory. They then calculated where the planet should be located to explain the empirical data they observed. The German astronomer Johann Galle used these calculations and found the planet he was looking for in the sky, named after Neptune, the Greek god of the sea. This episode was quickly recognized as the greatest, spectacular success of Newton's theory. Few people suspected that this theory had only a little more than half a century of peaceful existence before it was abandoned in favor of both theories of relativity.

Once a similar procedure had been successful, it was natural to repeat it for the case of Mercury. So the astronomers began to search for a hypothetical planet Vulcan, which was supposed to be so close to the Sun that the glare coming from our star made it impossible to observe it with the naked eye. However, despite the use of state-of-the-art instruments, the search proved fruitless. Today, the only use of the "planet Vulcan" is as the object of the musings of philosophers on the metaphysics of non-existent entities. The problem, however, remained until Einstein set about making the appropriate calculations using the GR equations. When the number 43 arcseconds appeared as a result of these calculations, it became clear that from then on Mercury would be a symbol of both the collapse of Newton's good old theory and the success of the young competitor. Let us add that discrepancies between Einstein's and Newton's theories appear only in strong gravitational fields, as in the case of the field acting on Mercury. The other planets of our system are located so far from the Sun that the differences in the predictions of the two theories are completely negligible, and you can easily use the equations of classical mechanics, e.g. to calculate the trajectories of space probes launched from Earth.[7]

6.5 Motion and space-time

Let us now turn to philosophical considerations on the nature of time and space and the relativity of motion. Let us recall that the idea of relativity of motion (according to which the motion of objects makes sense only in relation to other bodies) first occurred to Galileo in the form of the principle of relativity. However, this principle applied only to inertial systems, i.e. non-accelerating ones. The acceleration of the system creates additional inertial forces that enable us to distinguish between "true" motion and relative motion. The fact of the existence of inertial effects in accelerating systems was exploited by Newton in his famous bucket argument for the absoluteness of space. Newton argued that the only explanation for the appearance of centrifugal force in the case of rotational motion of a body is that the body is *really* rotating relative to absolutely stationary space. Newton's argument was countered

[7] However, it is worth remembering that if we want very high accuracy, as in the case of precisely determining the position of objects on the Earth's surface using satellite positioning (GPS), we must take into account the corrections resulting from the general theory of relativity.

by Mach, who pointed out that the famous scientist ignored the role of the rest of the universe. In the case of the relative rotation of a bucket with respect to the water, the presence or absence of a centrifugal force that changes the shape of the water determines whether the water rotates relative to the rest of the universe. Mach believed that every motion is relative, regardless of whether it is accelerated or not. A single body in an empty universe will not be affected by any inertial forces, because there are no other bodies relative to which it could accelerate.

What does the dispute between Newton and Mach look like on the basis of general relativity? At first glance, it may seem that GR is closer to Mach's position. This is due to Einstein's principle of equivalence, which locally equates accelerating systems with systems resting in the gravitational field. Therefore, it can be suggested that participation in accelerated motion is not an absolute property of a given body, because it can also be said that it is gravity that causes the observed effect in the form of inertial (apparent) forces. However, the matter is not simple. Einsteinian equivalence applies to two physically different situations, not two descriptions of the same situation. In one situation, we have an accelerating body without a gravitational field, and in the other, the same body in a gravitational field. But the gravitational field must have a source. Therefore, in the latter case, there must be some additional matter producing the corresponding gravitational field. Theoretically, we can distinguish between the two situations by carefully examining the environment of the test body. If we rule out the existence of external bodies that produce a gravitational field, we are forced to admit that our body is really accelerating.

This argument can be made more explicit by considering a hypothetical scenario of a solitary body in the universe. In the absence of gravitational field sources, the most natural (though not the only) solution to Einstein's equations is the Minkowski flat spacetime of special relativity. But, as we know, in SR there is a concept of absolute acceleration and the inertial forces associated with it. Thus, the general theory of relativity, of which SR is a special case, allows for the occurrence of apparent forces that allow us to determine whether the system is really accelerating (e.g. rotating), even if we are not able to relate such accelerated motion to external objects.

The problem of solitary rotating objects in GR was generally considered by Kurt Gödel, the famous logician and mathematician, whose long discussions with Einstein at the Institute for Advanced Studies in Princeton have become legendary. Gödel proved that there is a solution to Einstein's equations describing a single disk in which the lines of light emitted vertically in relation to the disk will outline a spiral shape. The most natural interpretation of this result is that it represents a rotating disk from the perspective of which the lines of light, which are really straight lines, appear in a spiral. Thus, GR allows for the existence of absolutely rotating objects in empty space! Mach would be very unhappy with such a turn of events. Of course, it remains an open question whether Gödel's solution makes physical sense. However, the very fact of allowing such a solution is significant. Many physicists have tried to improve the general theory of relativity in the spirit of Mach and relativism, so as to exclude all elements of absolute motion. One such recent attempt is so-called shape dynamics proposed by a British physicist Julian Barbour. However, it is too early to judge whether this approach will live up to the hopes placed in it.

Let us now recall another famous philosophical dispute between two outstanding minds of the era: Newton and Leibniz. As we remember from Chapter 2, the subject of this dispute was the ontological status of time and space. Leibniz defended the view that time and space are ontologically dependent on the physical objects within them, while Newton firmly argued that time and space are substances capable of existing independently. In a sense, Leibniz can be considered the winner of this dispute, since on the basis of special relativity neither time nor space have an absolute character – they are dependent on frames of reference, and therefore also on the physical objects that determine these systems. However, the dispute between Newton and Leibniz can be presented in a new form as concerning the status of all space-time, rather than its separation into time and space. Since, of course, in SR space-time with its Minkowski geometry is invariant with respect to system transformations, the question still remains open as to whether space-time is an independent substance, or whether its existence is dependent on objects and the relations between them, as Leibniz wanted it to be.

On the basis of special relativity, it seems that Leibniz has a good chance of success. Minkowski's space-time is in a sense an uninteresting "background" – it looks the same in every part of the universe, it is not subject to change or interactions, it is simply a convenient way to describe what happens to physical objects such as measuring rods or clocks. We remember from the previous chapter how the basic relativistic effects – length contraction or time dilation – were explained using appropriate measurement procedures. Many philosophers adopt an extremely empiricist, or even instrumentalist, attitude towards the theory of space-time present in SR. They treat this theory as a synthetic description of measurement procedures that use light rays to determine, among others, the simultaneity of events. In this approach, space-time does not seem to be the main subject of research, but only a reflection of measurements on physical objects.

General relativity is often, if somewhat misleadingly, referred to as "background independent" – it would be more appropriate to call it "background free". The point is that in all previous theories, such as classical mechanics or SR, space-time is a passive background, allowing the location of objects and events, but not participating in physical interactions. On the other hand, on the basis of GR, space-time takes an active role, because it is affected by material objects endowed with mass, and also affects physical bodies themselves, determining their motion along curved geodesics. However, some philosophers have doubts whether this is not an abuse of the concept of causal interaction. Bodies endowed with mass do not literally affect space-time in the sense of physical interaction, but simply there is a relationship between mass, energy, momentum and curvature. Similarly, following curved geodesics by free bodies is not a form of "pushing" but rather a certain regularity. We would not say for instance that empty Newtonian space "pushes" free bodies to continue their uniform and rectilinear motion.

This situation is radically changed within the framework of the general theory of relativity. First, the geometry of space-time becomes an important participant in the processes under consideration. Its properties may be modified. Space-time is subject to various deformations, expressed in the form of curvature, which can take an extreme form near singularities such as black holes. To this it should be added that such deformations have a significant physical impact on the behavior of bodies. The basic law of motion of GR, according to which free physical bodies follow geodesics, can naturally be interpreted in terms of the almost physical interaction or influence of space-time on the objects in it. A spectacular example of such an impact can be gravitational waves, i.e. propagating disturbances in space-time, carrying energy. Sufficiently strong gravitational waves could, for example, knock down the famous rock on the Gibraltar Peninsula. In the light of such facts, it is difficult to deny space-time full physical reality. A synthetic account of the relationships between objects, as Leibniz imagined space and time, would hardly be able to destroy a physical object such as the Rock of Gibraltar (but see the counterarguments in the box above).

However, Leibniz's supporters are not yet in a hopeless situation. They can reach for the hidden ace up their sleeve, which is Leibniz's argument from the shift. This argument must, of course, be adapted to the mathematics and physics of GR. In its original form, Leibniz's reasoning was carried out on the assumption of Galileo's flat space-time, broken down into instantaneous space and absolute time. We will now present a modified version of it, known as the *hole argument*. Let us start by mentioning the basic property of GR, which is its so-called general covariance. Behind this difficult name there is a fairly clear idea, explicitly formulated by Einstein: the equations of the theory should apply equally in every coordinate system, regardless of whether it is rectilinear, curvilinear, or moves relative to other systems, etc. Formally, such independence is encompassed in a generalized way of moving from one system to another, represented by so-called diffeomorphisms. A diffeomorphism is any transformation of a particular space that does not change the proximity relations between points (as we have mentioned, such relations are referred to as topology). For example, we can consider transformations that stretch a given space, deforming its shape, but not ones that tear it apart.

Formally, diffeomorphisms are smooth functions from N-dimensional real number spaces into the same spaces. These spaces represent the coordinates of the points of our initial space. (In the case of spacetime, the dimension N will of course be equal to 4.) Thus, a given diffeomorphism can simply be interpreted as a conversion of the coordinates of each point in spacetime to other coordinates. This is the so-called passive interpretation of transformation. It does not change anything in the physical situation, but only replaces one description with another, equally good one.

It turns out, however, that each passive transformation corresponds to a certain active transformation, making modifications in space-time itself. Consider a simple example to illustrate such a change. Let us select two space-time points A and B, whose coordinates in a certain system are (1, 0, 0, 0) and (2, 1, 0, 0). Select a diffeomorphism that converts the coordinates of point A to coordinates (2, 1, 0, 0), while changing the coordinates B to some other coordinates (it doesn't matter which ones).[8] However, such a passive transformation is matched by an active transformation, which does not move the coordinates, only the points themselves. That is, point A is moved to the place of point B. Of course, some other point will enter the "place" of point A (in the sense of appropriate coordinates), and in turn point B will move somewhere else.[9] The result will be a space-time that looks the same as the original one with the coordinates changed, but which is ontologically different.

[8] Such a diffeomorphism can be obtained, for example, by adding 1 to the time coordinate t and to the spatial coordinate x: $d(t, x, y, z) = (t + 1, x + 1, y, z)$.

[9] It can be objected – quite rightly so – that the active interpretation of the transformation given above is based on the erroneous belief that there are "places" where points are located and in relation to which these points can be moved. However, the points themselves are places – the question of the location of places is meaningless. Strictly speaking, the active version of the transformation is done not in relation to places, but in relation to physical fields or the distribution of matter. If, for example, point A in the original situation was occupied by some material object – e.g. a star – and point B was empty, then after the transformation point A will be empty, and B will go somewhere else, perhaps to some other object or physical field.

From Leibniz's point of view, such an active transformation does not make much sense, since the result is indistinguishable from the initial state. In principle, we have thus already reconstructed the good old argument from the shift in which the assumption of substantialism led to the assumption of the existence of two (in fact infinitely many) ontologically different, but empirically indistinguishable situations. However, in the case of GR, there is an additional "twist" to the shift argument. The Leibnizian shift had to be homogeneous – at all points in the universe we move matter by the same distance. On the other hand, diffeomorphic transformations in OTW are more flexible – some areas can be "stretched" more, others less. In particular, there is a diffeomorphic transformation of space-time, which shifts points only in a certain distinguished area (a "hole"), leaving the rest of space-time unchanged. We are therefore dealing with a particular form of indeterminism. There are two possible evolutions of the universe that are exactly the same over a period of time, but that differ from each other in the area of the hole. The difference is that the same points receive different physical and geometric properties. For example, a point that in one "world" was inside a neutron star will be devoid of all matter in another possible world. Therefore, its metric properties will also be different – in the first case, it will have a metric tensor (and curvature tensor) significantly different from the flat Minkowski tensor, and in the second case, it will have a metric tensor practically equal to the Minkowski tensor.

Of course, the differences described above are impossible to detect, because we cannot "color" individual points with markers in order to check what properties they have in the respective worlds. The indeterminism of the hole has no empirical consequences. Still, for a substantivalist, the change inside the hole is real, though unobservable. On the other hand, the Leibnizian relationist would say that space-time points are secondary to the coincidence of physical properties. It makes no sense to say that a point in the interior of a neutron star could be the same as a point in empty space. In every possible world with the same distribution of matter, there is one and the same point: the one that is determined by the specific physical conditions in the interior of the star. Thus, all the possibilities corresponding to the different diffeomorphic transformations are reduced to one and the same possibility.

However, is substantivalism doomed to a complete lack of determination of the ontological situation inside the hole area? It depends on what variant of substantivalism we adopt. The hole argument undermines the most radical version of substantivalism, according to which each point has an intrinsic "identity" or individuality, independent of all possessed properties. In the language of metaphysics, such a non-qualitative individuality is called *haecceity*, from the Latin *haecce,* meaning "this one" (we mentioned this concept in the box on p. 59). Having this *haecceity* allows us to talk about different alternative properties of the same object, e.g. about different possible locations of the same point in relation to material objects, such as our example neutron star. However, there are philosophers who question the legitimacy of introducing identity understood in this way. In an alternative approach, objects are characterized by the so-called essential properties, which determine the identity of each object. An essential property is one without which an object cannot exist. Philosopher of physics Tim Maudlin suggested that the essential properties of space-time points should be identified with their metric properties expressed by the metric tensor. With this assumption, a situation in which a point inside a neutron star would be in an empty region is metaphysically impossible, because it would then lose its essential property in the form of a strongly "warped" metric tensor. Hence, an active diffeomorphic transformation, although mathematically permissible, is excluded by virtue of a metaphysical assumption.

Thus, it seems that essentialist substantivalism (as well as a related version called *sophisticated*) is able to refute the hole argument, in contrast to haecceitistic substantivalism. Of course, as is usually the case in philosophy, there are various additional difficulties, counterarguments and defenses. One problem with Maudlin's conception is that it allows for a different form of indeterminism. Although our chosen point in the interior of a neutron star cannot assume other metric properties, it is not excluded that there would be an alternative possible world, in which a completely new, non-existent in reality point with the same properties would be found in the interior of the star (in the literature, such objects are called "moles" that impersonate legitimate objects). A solution to this problem may be, for example, to assume that essential properties are both necessary and sufficient to be a given object. However, the analysis of the consequences of such a position goes beyond the scope of this study. Readers interested in the deeper metaphysics of the hole argument are referred to the literature at the end of the chapter, while we will move from abstract philosophical considerations to no less abstract mathematics.

6.6* Fundamentals of differential geometry

6.6.1 Vectors and tensors

The general theory of relativity in its formal part is based on a mathematical theory called differential geometry. We have already encountered some concepts of this theory in the chapters devoted to the theory of electromagnetism and special relativity, but now we will look a little more closely and in a more orderly way at its basic concepts used in physics. Differential geometry uses methods of mathematical analysis to describe various geometric properties of spaces, such as curvature. In order to be able to apply mathematical analysis, we must be able to express geometric properties in the language of functions on real numbers. Probably most readers are familiar with analytic geometry, in which we can assign numerical coordinates to points with the help of a Cartesian coordinate system, and then use them to algebraically represent various geometric relations, such as in the equation of a line or a circle. However, analytic geometry is Euclidean, and we need more flexible (pun intended) geometry. The starting point of such geometry is the concept of a differentiable manifold – in short, a manifold.

Let us begin by introducing the concept of a map on a certain set of points M. In general, a map is any one-to-one function φ, which maps a subset $A \subseteq M$ into a subset (open[10]) of the N-dimensional space R^N (the Cartesian product of N sets of real numbers). The idea is very simple – the map assigns to each point in A its unique coordinates in the form of N real numbers. Of course, in the case of spacetime theory, the dimension N will be equal to 4, but differential geometry operates with arbitrarily multi-dimensional spaces. A set of maps C that are in a sense compatible with each other and that together cover the entire space M is called an *atlas*. A differentiable manifold will therefore be the set M together with its atlas: (M, C).[11]

[10] An open set is a set without the boundary – e.g. the interior of a circle with radius r without points on the circle.

[11] One may ask why we need to consider more than one map – isn't it enough to simply have one function φ assigning each point in M its coordinates? It turns out that in some cases this is impossible. For example, it is known from geography that there is no single flat map that covers the entire globe. We need at least two such maps – one with the North Pole, for example, without the South Pole, and

With the introduction of the atlas in the form of a series of maps, we can now represent numerical functions defined on a set of points M as functions from R^N to R, to which standard concepts from mathematical analysis apply: continuity, differentiability in degree d (existence of derivatives up to degree d). For example, any function α from M to R is "smooth" when, for every map φ in the C atlas, the function $\alpha \circ \varphi^{-1}$, whose domain is a subset of R^N and the codomain is R, is infinitely many times differentiable.

An extremely important concept, which allows us to introduce a whole range of mathematical characteristics with respect to a given manifold, is the concept of a space tangent to M at a given point p. Visually, we can imagine a tangent space as a piece of paper glued to a sphere at a certain point, for example. Tangent space is the "arena" in which objects such as vectors and tensors act. There are many equivalent ways to define the tangent space. The most elegant, though somewhat abstract, is the following definition. Let $S(p)$ be the set of all smooth functions to R defined in the neighborhood of a point p. On this set, we will consider all functions (in fact functionals – cf. section 2.7) ξ into real numbers that possess the formal features of derivatives (f_1 and f_2 are arbitrary functions from $S(p)$):

$$\xi(f_1 + f_2) = \xi(f_1) + \xi(f_2)$$

$$\xi(f_1 f_2) = f_1 \xi(f_2) + f_2 \xi(f_1)$$

$$\text{If } f = \text{const, then } \xi(f) = 0.$$

You may remember that the derivative of a function at a point satisfies all three of the above conditions: the derivative of the sum of functions is the sum of the derivatives, the derivative of the product is given by the above formula, and the derivative of a constant function is, of course, equal to zero. Each functional ξ satisfying the given conditions will be called a tangent vector at point p. The relationship between derivatives and tangents is already known from a basic calculus course (cf. the entry in the box on p. 43). In the abstract approach currently under consideration, there are an infinite number of "derivative" operations for a given point p, each of which determines a certain abstract direction tangent to a given manifold. In addition, the set of all such operations (tangent vectors) forms a vector space. This means that the sum of two tangent vectors is also a tangent vector (satisfies the above conditions), and the product of the tangent vector by a number is a tangent vector. The space of all tangent vectors at point p is called the tangent space and is denoted as T_p.[12]

---

the other vice versa. The condition of the agreement of two maps concerns their behavior in the areas of the "intersection" of their domains (where they overlap). The simplest requirement would be to assume that two maps $\varphi_1 : A \to R^n$ and $\varphi_2 : B \to R^n$ are compatible when they are identical in the area of $A \cap B$. However, this is too strong a condition, because specific coordinate values may change when the map is changed. Two maps are assumed to match when points considered "nearby" by one map are close in the other map. More precisely, this condition states that the compositions of the mappings $\varphi_1 \circ \varphi_2^{-1}$ and $\varphi_2 \circ \varphi_1^{-1}$, which are functions from subsets of R^n in R^n, are sufficiently "smooth" in the given domains (they are differentiable a certain number of times).

[12] I hope that as philosophers you will appreciate the subtle beauty of this construction, which at first glance may seem like an aberration. Here we define the purely geometric concept of a vector by means of the concept of a functional behaving like a derivative, taken from mathematical analysis (of real numbers). This may raise objections: in school we are used to looking at mathematical objects such as vectors or functions as separate types of entities. After all, a vector is not a function, but an arrow with a specific direction and length (if I remember the school definition correctly). However, it turns out that in higher mathematics, seemingly separate classes of objects often intersect with each

Let us add that tangent vectors belong to the category of objects called *contravariant*. There are also other types of vectors, referred to as *covariant* vectors (this nomenclature has no particular justification and is a certain historical accident). To introduce them, consider a new space T_p^*, called a dual space to the tangent space T_p. The dual space to a given vector space is generally defined as the space composed of all linear functionals operating on given vectors. Linear functionals form a vector space (the sum of two functionals is a functional as well as the product of a functional by a number). Again, we are dealing here with a situation where we identify as vectors mathematical objects that seemingly have little to do with geometry (these are functions that assign real numbers to the corresponding vectors from T_p).

The spaces T_p and T_p^* allow for a formal introduction of another very important category of objects, i.e. tensors. A covariant tensor is any linear function that assigns a number to n vectors in the tangent space T_p. In other words, a covariant tensor is a linear function (a functional) from the Cartesian product of n spaces $T_p \times T_p \times \ldots \times T_p$ to real numbers R. The number n is called the rank of the tensor. It is worth noting that a covariant tensor of rank 1 is simply a covariant vector. Contravariant tensors, on the other hand, are functions that operate on Cartesian products of dual spaces: $T_p^* \times T_p^* \times \ldots \times T_p^*$. Thus, in short, we can say that contravariant tensors and vectors "live" in the tangent space and act on elements of the dual space,[13] while covariant tensors and vectors inhabit the dual space and act on the elements of the tangent space.

The concepts of covariant and contravariant vectors and tensors that we have introduced are quite abstract (this is the method preferred by mathematicians). Physicists, however, like to think about these objects in a slightly more "down-to-earth" way, by referring to a specific coordinate system. Let $x^1(p), x^2(p), \ldots, x^N(p)$ be the coordinates of a point p in a certain map. Consider N functions from R^N to R, each of which assigns one selected coordinate x^i to every sequence of coordinates (x^1, x^2, ..., x^N) (we will denote this function using the same symbol x^i). Let us take any tangent vector ξ from the space T_p, defined as above. This vector, as a linear function, assigns a certain number ξ^i to each function x^i. The sequence of numbers $(\xi^1, \xi^2, \ldots, \xi^N)$ will be called the components of vector ξ in the coordinate system x^1, x^2, ..., x^N. As you can see, we now move from the abstract conception of tangent vectors as some functionals to the "concrete" and well-known characteristic in the form of N-tuples of numbers. It can be proven (we will not do this) that the operation of a vector-functional ξ on any function f of $S(p)$ will be given by the formula below. So indeed the vector ξ acts as a derivative (let's add, a directional derivative in the abstract "direction" defined by the components ξ^i).

---

other. A vector is not simply an arrow, but an object that "behaves" in a certain way, or more precisely, enters into specific relationships with other vectors. Vectors can be functions, functionals or other abstract objects.

[13] The explanation why contravariant vectors, which we initially introduced as elements of the tangent space, can be interpreted as functions (functionals) on the dual space T_p^* is a bit involved. This is because the dual space to the dual space, or T_p^{**}, is in a sense "identical" to the original space T_p (more precisely, in mathematics we speak of these spaces as being isomorphic). Thus, any element of the space T_p, which we know is a contravariant vector, can be also interpreted as an element of T_p^{**}, i.e. a linear functional on T_p^*. I leave it to the most inquisitive readers to think about what the relationship of "identity" between T_p and T_p^{**} might look like.

$$\xi(f) = \sum_{i=1}^{N} \xi^i \frac{\partial f}{\partial x^i}. \tag{6.3}$$

It should be emphasized that although in each coordinate system tangent vectors are represented by sequences of numbers, these numbers will change when moving from one system to another. Tangent (contravariant) vectors transform according to strictly defined rules. The transformation rule can be derived from formula (6.3). Let $y^1, y^2, \ldots y^N$ be the coordinates in a new coordinate system. These coordinates will of course be related by certain functional relationships with the "old" coordinates x^i. According to the chain rule, the derivative of a function f with respect to the x^i coordinate can be expressed as the product of the derivative with respect to the new coordinates times the derivative of the new coordinate with respect to the old coordinate:

$$\frac{\partial f}{\partial x^i} = \frac{\partial f}{\partial y^j}\frac{\partial y^j}{\partial x^i}$$

In the above formula, we applied a commonly used convention (the so-called Einstein summation convention), according to which repeating indices (in our case j) should be summed over all values. This simplifies the notation of the appropriate formulas (without this convention we would have to write the right side of the equation as $\sum_j \frac{\partial f}{\partial y^j}\frac{\partial y^j}{\partial x^i}$.)

Using the above formula, we can easily see that the formula for the transformations of the components of a vector when moving from one system (x) to another (y) has the form:

$$\xi^i_{(y)} = \frac{\partial y^i}{\partial x^j}\xi^j_{(x)} \tag{6.4}$$

where subscripts (x) and (y) denote the relativization to a given coordinate system. In textbooks on general relativity, one can often find the phrase that a contravariant vector is defined simply as an object that transforms in the above way. On the other hand, covariant vectors, which, as we remember, belong to the dual space T_p^*, have a different way of transformation. Let's start with the question of what the components of a covariant (dual) vector look like in a given coordinate system. To calculate them, we act with a given vector (which is a linear functional on the set of contravariant vectors) on vectors of the form (1, 0, ..., 0), (0, 1, ..., 0), etc. (Such vectors are called basis vectors, because we can express any other vector with their help.) The numbers obtained as a result of such operations will be the components of a given covariant vector – they are marked with symbols with lower indices (subscripts): ξ_j. It turns out that if we apply the transformation rule (6.4) for contravariant vectors to the definition of a covariant vector, the corresponding transformation rule will take the form:

$$\xi_i^{(y)} = \frac{\partial x^j}{\partial y^i}\xi_j^{(x)}. \tag{6.5}$$

This is, in a sense, the "reverse" of the rule (6.4). Moving from the system (x) to (y), we need to calculate how quickly the "old" coordinates change with respect to the new, and not the new coordinates with respect to the old ones, as was the case with contravariant vectors.

Similar transformation rules apply to tensors, both covariant and contravariant. To determine a tensor in a given coordinate system, N numbers are not enough – a whole array (matrix) is needed for this. For example, a second-order covariant tensor is a linear function that assigns numbers to pairs of tangent vectors. Therefore, in order to determine the way this tensor works on the N basis vectors (1, 0, ..., 0), (0, 1, ..., 0) etc., we need $N \times N$ numbers (each number gives the result for two basis vectors. To indicate that we are dealing with a rank 2 tensor, we use two indices, e.g. T_{ij}. In addition, we adopt the convention that the lower indices tensor denotes a covariant tensor ("inhabiting" the dual space and acting on the elements of the tangent space), and the upper indices denote the contravariant tensor (acting on the elements of the dual space). As with vectors, covariant and contravariant tensors differ in the way they transform when moving from one coordinate system to another. Here are the corresponding rules for covariant and contravariant second-order tensors:

$$T_{nm}^{(y)} = \frac{\partial x^r}{\partial y^n}\frac{\partial x^s}{\partial y^m} T_{rs}^{(x)} \tag{6.6}$$

$$T_{(y)}^{nm} = \frac{\partial y^n}{\partial x^r}\frac{\partial y^m}{\partial x^s} T_{(x)}^{rs} \tag{6.7}$$

Let us add that it is possible (in fact necessary) to consider "mixed" tensors, for which some indices are covariant and some are contravariant. This means that such tensors operate on appropriate combinations of covariant and contravariant vectors. Examples of such tensors can be: T_n^m (covariant in the n index and contravariant in the m index) or T_a^{bc} (contravariant in two indices and covariant in one). The transformation rules for such tensors will contain appropriate combinations of partial derivatives of one coordinate with respect to another.[14]

6.6.2 Covariant derivative, geodesic and curvature

Students of differential geometry learn different ways of operating on tensors – addition, multiplication, as well as more "exotic" operations, such as contracting (reducing the number of independent indices) or lowering or raising indices. Of course, we will not be able to discuss these issues – instead referring those interested to the literature. However, we should take a closer look at one important operation on tensors or vectors – namely the differentiation operation. It is very important that we are able to correctly differentiate, for example, a vector along a given curve. As you know, derivatives represent numerically how fast a given function changes with the change of the differentiation parameter. In particular, we can express the fact that a given vector does not change along a certain curve – this is the case when the derivative of this vector along this curve is equal to zero. This will allow us to introduce the concept of parallel transport, which as we know is crucial in geometry with non-zero curvature. On the surface, it may seem that the matter is simple. Consider any vector field $V_m(p)$ on some area of the manifold M written into components in some coordinate system (in some map). This means that $V_m(p)$ is in fact a function from R^N (coordinates of points) to R^N (components of a vector) that can be differentiated in a standard way. Therefore, we can calculate the values of partial derivatives $\frac{\partial V_m}{\partial x^n}$ at a given point p. They will create an array of numbers numbered with indices m, n. It seems, therefore, that we have obtained a well-known object – a second-order tensor.

[14] Try writing these rules for the tensors T_n^m and T_a^{bc}.

However, there is a problem. The set of numbers $\frac{\partial V_m}{\partial x^n}$ does not form a tensor, because these numbers do not transform in the right way when moving from one coordinate system to another. This can be verified by direct calculation, using known differentiation rules, such as the product differentiation rule, and remembering that when changing the system from (x) to (y), we need to change the components of the vector V_m, according to rule (6.5). The result is the following "transformation rule":

$$\frac{\partial V_m^{(y)}}{\partial y^n} = \frac{\partial x^r}{\partial y^m}\frac{\partial V_r^{(x)}}{\partial y^n} + \frac{\partial^2 x^r}{\partial y^n \partial y^m} V_r^{(x)}.$$

The first term of the sum on the right-hand side of the above equation represents the transformation that would be expected if the numbers $\frac{\partial V_m}{\partial x^n}$ were the components of a tensor. This can be clearly seen if we rewrite this term using the chain rule as $\frac{\partial x^r}{\partial y^m}\frac{\partial x^s}{\partial y^n}\frac{\partial V_r^{(x)}}{\partial x^s}$ – this is exactly the transformation rule of a second-order covariant tensor, written in the formula (6.6). Unfortunately, the rule is spoiled by the presence of the second component in the formula.

The solution is to adopt a new differentiation rule that will "compensate" for the occurrence of the term $\frac{\partial^2 x^r}{\partial y^n \partial y^m} V_r^{(x)}$. Notice that this term is given in the form of numbers with three indices n, m, and r, which "act" on the vector $V_r^{(x)}$. Thus, the new differentiation rule, which is called the covariant derivative, should look like this:

$$\nabla_n V_m = \frac{\partial V_m}{\partial y^n} - \Gamma^r_{mn} V_r. \tag{6.8}$$

Of course, you need to find the right form for the numbers Γ^r_{mn}, so that the whole operation creates a correctly transforming tensor. These numbers are called Christoffel symbols (they are also often called the affine connection). It should be emphasized that Γ^r_{mn} is not a tensor, which is understandable, since the numbers $\frac{\partial V_m}{\partial y^n}$ do not form tensor coordinates either. Only adding both components $\frac{\partial V_m}{\partial y^n}$ and $-\Gamma^r_{mn} V_r$ yields a correct tensor. It can be proven that the covariant derivative of a contravariant vector V^m will have an analogous but slightly different form:

$$\nabla_n V^m = \frac{\partial V^m}{\partial y^n} + \Gamma^m_{nr} V^r.$$

Covariant derivative ∇_n can be applied to any tensors, not just vectors. One of the most important applications of this operation is in the definition of geodesics. The idea is clear enough: a geodesic is a line for which the tangent vectors are parallel to each other according to the covariant derivative, i.e. the covariant derivative of the vectors tangent to a geodesic is equal to zero. We only need to define the tangent vectors to a given line. This is not a difficult task. First of all, let us specify what is meant by a curve in a manifold M: it is a smooth map from a segment of real numbers I into the set M (you can imagine that we number the points on a given line with real numbers). Let $\gamma(s): I \subseteq \mathrm{R} \to M$ be a smooth function defining some curve in M. The tangent (contravariant) vector to the curve γ at point p (for which $\gamma(s_0) = p$) is a linear functional γ^p such that

$$\gamma^p(f) = \frac{d(f \circ \gamma)}{ds}(s_o)$$

for any function f from the set $S(p)$ (the set of smooth functions from the neighborhood of the point p into real numbers). Again, we have a close affinity between the concept of tangency and a derivative – this time a derivative "along" a given curve.[15]

Let us calculate the components of the vector tangent to the curve γ at a given point p. As we remember, the components of a vector in a given coordinate system (x) arise as a result of applying this vector (which is a functional) to single coordinates (functions assigning one selected coordinate x^i to any N-tuple of coordinates). Using the above definition of the vector-functional γ^p we can see that its i-th component will be of the form $\frac{d(x^i \circ \gamma)}{ds}$. This expression is often simplified to $\frac{dx^i}{ds}$, with the presupposition that we differentiate the function of the x^i coordinate "along" a given curve γ parameterized by the variable s.

We will now want to calculate the covariant derivative of any vector V^m along a curve γ. This is done by taking the product of the covariant derivative with respect to the coordinates with the components of the tangent vector to the curve. The result is a covariant derivative operation ∇_γ along the curve γ, which looks like this:

$$\nabla_\gamma V^m = (\nabla_n V^m)\frac{dx^n}{ds} = \frac{\partial V^m}{\partial x^n}\frac{dx^n}{ds} + \Gamma^m_{nr} V^r \frac{dx^n}{ds}.$$

Notice that the first component of the sum can be simplified to the total derivative of the vector V^m with respect to the variable s (again, this is the chain rule in operation):

$$\nabla_\gamma V^m = \frac{dV^m}{ds} + \Gamma^m_{nr} V^r \frac{dx^n}{ds}.$$

Now suppose we want to study the variability of the tangent vector itself along the curve γ. To do this, we need to put $\frac{dx^m}{ds}$ in the above equation in place of V^m:

$$\nabla_\gamma \frac{dx^m}{ds} = \frac{d^2 x^m}{ds^2} + \Gamma^m_{nr} \frac{dx^n}{ds}\frac{dx^r}{ds}.$$

As we have already said, a geodesic is a line whose tangent vector does not change as you move along that line. This means that its covariant derivative along the geodesic is zero. So we have the so-called geodesic equation:

$$\frac{d^2 x^m}{ds^2} + \Gamma^m_{nr} \frac{dx^n}{ds}\frac{dx^r}{ds} = 0. \tag{6.9}$$

It is interesting that in this equation there is a second derivative of the position with respect to a certain parameter, which in the case of the world line can be treated as (proper) time, which gives acceleration. This brings to mind Newton's second law of motion if we identify the term $-\Gamma^m_{nr} \frac{dx^n}{ds}\frac{dx^r}{ds}$ with "force". It turns out that in geometry with zero curvature

---

[15] It turns out that every vector tangent to the manifold at a point (a vector in the tangent space T_p) can be represented as a vector tangent to some curve γ passing through p. Thus, we have an alternative and equivalent definition of a tangent space as the space of all vectors tangent to some curve passing through the point p.

(flat), such as Minkowski geometry, Christoffel symbols Γ_{nr}^{m} are equal to zero, and thus the geodesic equation simply turns into the formula $a = 0$, which indeed characterizes a straight line, according to the first law of motion. On the other hand, in curved geometry, the acceleration will not be zero – objects moving on geodesics will behave as if they were subjected to a force.

We are already able to express the concept of a "straight" line (geodesic) in the language of differential geometry. However, we still lack the very important concept of length. How can we mathematically represent the length of curves? A certain measure called a metric is needed for this. The most general way to introduce a metric is based on the concept of a metric tensor. Formally speaking, we start by introducing the dot product operation on tangent vectors. The dot (scalar) product, which we have already discussed, is known from the school course of geometry: it is usually presented as the product of the length of vectors times the cosine of the angle between them, or in the translation into components as the sum of the products of the individual components. In abstract terms, the dot product of vectors A and B can be any function in real numbers $g(A, B)$ that satisfies certain conditions – symmetry, linearity and (usually) the condition of positive definiteness. However, the latter condition can be omitted – e.g. in the case of Minkowski geometry, as we remember from the previous chapter, the product of two vectors can be a negative number (a metric based on this notion of a product is called pseudo-Riemannian).

Formally, a metric tensor is a second-order covariant tensor that assigns to each pair of tangent (contravariant) vectors their dot product. Consider a vector tangent to a given curve $\gamma(s)$, whose components in some system are $\frac{dx^i}{ds}$. The dot product of such a vector with itself (i.e. its length squared) can be generally represented in the form of a linear combination:

$$g(\gamma(s), \gamma(s)) = g_{ij}\frac{dx^i}{ds}\frac{dx^j}{ds},$$

where g_{ij} are some numbers. We will interpret these numbers as the components of the metric tensor in a certain coordinate system. Assuming conventionally that tangent vectors have a unit length, we can rewrite the above equation as:

$$1 = g_{ij}\frac{dx^i}{ds}\frac{dx^j}{ds},$$

And hence, applying ordinary algebra to the infinitesimal elements of ds, we get the most well-known formula for the metric tensor (the so-called linear element):[16]

$$ds^2 = g_{ij}dx^i dx^j.$$

The expression ds is interpreted as an infinitesimal displacement along a given curve. In this interpretation, the metric tensor tells us how such a displacement depends on the displacements along the individual coordinates. In Euclidean geometry represented in Cartesian coordinates, the linear element is obviously given by the Pythagorean theorem:

[16] It can be checked that the g_{ij} transforms in the right way for covariant tensors (6.6). Hint: the infinitesimal displacement dx^i can be represented in a new coordinate system as $dy^j\frac{\partial x^i}{\partial y^j}$.

$$ds^2 = \sum_i (dx^i)^2. \qquad (6.10)$$

From this it follows that the components of the metric tensor are of the form $g_{ij} = 1$ for $i = j$ and zero for the remaining cases.

With the help of the metric tensor, we can calculate the length of a given curve. The procedure is to determine the infinitesimal displacement at each point using the length of the tangent vector, and then sum up such displacements. Of course, we remember that the summation of infinitesimally small segments is done using the integration procedure. Therefore, the formula expressing the length of the curve between the two points p and q will have the following form:

$$\int_a^b |\gamma(s)| ds$$

where the integration limits *of a* and b are given by $\gamma(a)=p$, $\gamma(b)=q$, $\gamma(s)$ denotes the tangent vector to the given curve at the point parameterized by s, and the vertical lines denote the length. Taking advantage of the fact that the length of the vector is given by the root of its dot product with itself, we have the following definition of the length of the curve containing the metric tensor g_{ij}:

$$\int_a^b \sqrt{g_{ij}\frac{dx^i}{ds}\frac{dx^j}{ds}}\, ds$$

Parenthetically, let us mention that geodesics can alternatively be defined as lines for which the length calculated according to the above formula is minimal (assuming that the metric is Riemannian, i.e. non-negative). We remember from the chapter on analytical mechanics that the condition for minimizing an integral such as the above is the satisfaction of the Euler-Lagrange equation by the integrated function. It turns out (it is not difficult to verify) that the Euler-Lagrange equation for the above integral has the same form as the geodesic equation (6.9), which shows the equivalence of two concepts of geodesics – one based on the condition of minimizing length, and the other on the zeroing of the covariant derivative from the tangent vector.

The metric tensor plays a crucial role in describing the fundamental geometric properties of a given space. It can be used, for example, to determine whether a given geometry is curved or flat. This tensor also determines other important geometric parameters of space. For example, there is a relationship between the metric tensor g_{ij} and Christoffel symbols Γ^i_{jk}. This relationship results from the assumption of the so-called compatibility of the covariant derivative (affine connection) with the metric tensor. A covariant derivative is compatible with a metric tensor when its application to the metric tensor yields zero.[17] It turns out that for a given metric tensor, there is exactly one compatible covariant derivative, and therefore also one set of Christoffel symbols. The mathematical relationship between these symbols and the metric tensor is quite complicated – it can be checked in any textbook on general relativity.

---

[17] Intuitively, this means that the change of the metric tensor calculated alongside any curve is the "minimal" allowed by the intrinsic curvature of the geometry. If the geometry is flat, the metric tensor is obviously the same everywhere. When we move to a curved geometry, the metric tensor can change from point to point, but this change will be entirely due to the intrinsic curvature.

Now let us move on to another important concept, which is the curvature of space. More specifically, we are talking about the so-called intrinsic curvature, i.e. one that does not depend on the higher-dimensional space in which our space is "embedded". For example, the surface of a cylinder is curved if viewed from the perspective of a three-dimensional space, but it is not characterized by intrinsic curvature (all the laws of Euclidean geometry apply in it). The surface of the sphere, on the other hand, has a non-zero internal curvature. The internal curvature of space can be identified by examining the form of the metric tensor. It turns out that in a flat (non-curved) space, the metric tensor can be represented in the form of the Kronecker delta, where the diagonal elements g_{ii} of the metric tensor are all equal to unity (for now, we assume that the space is Riemannian, i.e. the dot product of vectors is positive definite). In other words, there is a coordinate system (Cartesian) in which the linear element has the form resulting from the Pythagorean theorem (6.10). It is worth noting that by changing the coordinates from Cartesian to curved (e.g. polar), we can obtain a form of the metric tensor that does not meet this condition (in which some diagonal terms will be different from 1). However, the only important thing is that in a certain system a metric tensor can be presented in the form of the identity matrix. On the other hand, spaces with intrinsic curvature (such as the surface of a sphere) do not have this property. The metric tensor in *each* coordinate system will be different from the identity matrix.

The most "synthetic" mathematical representation of the intrinsic curvature at a given point in space is given in the form of a tensor, called the Riemann curvature tensor. We will not deduce the form of this tensor in detail, limiting ourselves to an illustrative presentation of the appropriate procedure. The basic idea of this procedure is quite simple; we have already outlined it in general terms in section 6.2. The curvature in a given area is non-zero when we find a difference between the initial and end vectors by performing the operation of parallel transport of a given vector along a closed curve. Going a little more strictly, we choose two coordinates x^{μ} and x^{ν}, and consider a very small quadrilateral whose sides are given respectively by an infinitesimal shift δx^{μ} along the x^{μ} coordinate, an infinitesimal shift δx^{ν} along the x^{ν} coordinate, and then shifts in the reverse direction, so as to return to the starting point. We then select any vector V^{β} (strictly speaking, a vector field) at the starting point and try to calculate what the change of this vector along the four sides of the selected quadrilateral will look like. The general formula representing such an infinitesimal change will be as follows:

$$\delta V^{\alpha} = \delta x^{\mu} \delta x^{\nu} (...) V^{\beta}, \tag{6.11}$$

where the parentheses (...) symbolize some mathematical object, the exact form of which has yet to be calculated. Note that in order to maintain the consistency of the entire formula, this object must have four indices α, β, μ, ν (three will be "absorbed" by summing over the indices of the objects on the right side of the equation, and the fourth will be the same index as the index of the object on the left). Therefore, it can be assumed that the object we are looking for is a rank-four tensor. This is the so-called Riemann curvature tensor. It should also be expected that this tensor will be expressible in terms of covariant derivatives, because the infinitesimal changes of objects along given coordinates are usually given in the form of the derivatives times the change of this coordinate. Indeed, it turns out that the expression in brackets (...) can be presented synthetically in the form of the so-called commutator of covariant derivative operations along the coordinates:

$$[\nabla_{\mu}, \nabla_{\nu}],$$

whose definition is as follows: $\nabla_\mu \nabla_\nu - \nabla_\nu \nabla_\mu$. Note that this expression produces a zero value if the sequential operations of taking derivatives in different directions can be put in reversed order without any change, i.e. if they commute. It turns out that in a flat space (without curvature), the operations of covariant derivatives do not depend on the order in which they are performed. That is, the result of differentiating a given vector along the x^μ coordinate and then along the x^ν coordinate is the same as performing these operations in reverse order. It is easy to guess that the commuting of these operations entails that the change of the vector along the closed curve will be zero – we will match the starting point.

However, in curved space, covariant derivatives in different directions do not commute, i.e. their commutator is non-zero. This is the Riemann curvature tensor. Note that although in the above formula there are explicitly only two indices μ and ν, there are two additional indices "hidden" in it, resulting from the definition of the covariant derivative (6.8). By being consistent with the formula (6.11), we can write the Riemann tensor as $R_{\mu\nu}{}^{\alpha}{}_{\beta}$. You can also find a variant of this tensor with only lower indices $R_{\mu\nu\alpha\beta}$. created as a result of the operation of lowering the index. The Riemann curvature tensor can be represented in the form of an appropriate combination of Christoffel symbols, by substituting formula (6.8) into the expression (6.11) and expanding the entire commutator. The result is a rather complicated combination in which there are both partial derivatives of Christoffel symbols $\Gamma^{\alpha}_{\beta\gamma}$ and their products.

Finally, let us mention another important tensor – the so-called Ricci tensor. It is formed by "contracting" two indices of the Riemann tensor. Contraction consists in replacing two indices with one, using Einstein's convention requiring that all tensor components be summed over the values of repeating indices. In this way, from a tensor with four indices, we get a rank-two tensor:

$$R_{\mu\nu} = R_{\mu\alpha}{}^{\alpha}{}_{\nu}$$

The Ricci tensor plays a key role in the formulation of the basic law of general relativity, which links the geometric properties of spacetime with the distribution of mass and energy. In addition, there will be one more quantity in this law, resulting from the subsequent contraction of the other two indices of the Ricci tensor. Of course, this operation will result in an object with a zero number of free indices, i.e. a scalar. It is called the curvature scalar and is denoted by the letter R.[18] We will now proceed to discuss the mathematical and physical details of this law, expressed in so-called Einstein's equations.

6.7* Physics in curved space-time

The entire previous section was devoted to the introduction of mathematical concepts necessary to describe the geometry of curved spaces. These concepts include primarily: the concept of the metric tensor and mathematical objects based on it, such as the covariant derivative (including Christoffel symbols), geodesics (curves of the smallest possible length), the concept of parallel transport, and curvature tensors (Riemann and Ricci). In this section, we will

---

[18] Formally, the curvature scalar R is defined as R^{μ}_{μ}. Thus, it arises as a result of applying two operations to the Ricci tensor $R_{\mu\nu}$: first we raise one of the two indices, and then we contract it with the lower index.

introduce objects that describe not geometry, but the physical properties of objects located in space-time. More specifically, we will start by explaining the concept of the energy-momentum tensor, and then give a heuristic explanation and physical sense of Einstein's basic equation, combining physics with geometry. Before we do this, however, we must make the necessary modification of the geometric part of our theory. Until now, we have tacitly assumed that our space is essentially similar to Euclidean space, as long as its curvature is negligible. In particular, when the curvature of space is exactly zero, we assumed that the metric tensor would take the form of the Kronecker delta δ_{nm}, which is a matrix containing only ones on the diagonal and zeros everywhere else. In other words, in flat space, the infinitesimal distance ds is given in the form of the Pythagorean theorem, characteristic of Euclidean geometry.

However, as we remember from the chapter on special relativity, space-time has a different metric structure. Its invariant is not the Euclidean distance, but the spacetime interval, which takes into account the time coordinate, but in a way different from spatial coordinates (they enter with the opposite sign to the sign of the time coordinate):

$$ds^2 = (dt)^2 - (dx)^2 - (dy)^2 - (dz)^2.$$

So the metric tensor in this case will be in the form of a matrix with the numbers 1 and -1 on the diagonal (let us also remember the convention $c = 1$). Following the notation introduced in the previous chapter, according to which the time coordinate t is denoted by zero index x^0 and spatial coordinates are written as x^1, x^2, x^3, we can give the components of the metric tensor in flat Minkowski spacetime as follows:

$$g_{00} = 1;\ g_{11} = -1;\ g_{22} = -1;\ g_{33} = -1;$$

and zero in all other cases. This tensor is often denoted by the symbol $\eta_{\mu\nu}$.

The other definitions given in the previous section do not change. Let us now move on to the problem of the mathematical representation for the physical sources of the gravitational field, i.e. of the curvature of space-time, which is what gravity comes down to in GR. According to Newton's theory of gravity, the source of gravity is the mass of objects, thus a certain scalar. This approach must be doubly modified. First, as we know from special relativity, mass is equivalent to energy, as implied by the famous formula $E = mc^2$. Therefore, energy should also contribute to the formation of the curvature of space-time, which we observe as a gravitational field. However, there is a problem here. Energy is not an invariant of intersystem transformations. The energy of an object at rest is equal to mc^2, while from the point of view of a frame moving at velocity v, its energy increases by a factor of $\frac{1}{\sqrt{1-\frac{v^2}{c^2}}}$. To remedy this, we add momentum to the source of curvature. As we remember, in special relativity energy and momentum are related by the fact that both of these quantities occur in the so-called energy-momentum four-vector (energy is its first component, and the other three components represent the three components of momentum). The length of the energy-momentum four-vector P^μ (of course calculated in accordance with the Minkowski metric) is the invariant of the transformation.

However, it is not the length of the four-vector P^μ that is the quantity that will characterize the physical source of the gravitational field. The second modification of Newton's theory of gravity will be the assumption that the physical quantity "responsible" for the occurrence of the curvature of spacetime is a rank-2 tensor, denoted as $T^{\mu\nu}$. Why? The answer to this can be found in the detailed form of the components of this tensor. However, let us begin by recalling another important concept that appeared on the occasion of the analysis of electromagnetic phenomena in the language of special relativity (sec. 5.9). It is a four-vector of current density J^{μ}, which synthetically captures two sources of electromagnetic field: electric charge and current. This four-vector can be written in the form (ρ, j_x, j_y, j_z), where ρ is the charge density at a given point in space-time, and j_x, j_y, j_z are the corresponding components of the current vector along the *x*, *y* and *z* directions. More precisely, the component j_x can be represented as a flux of charge passing through some infinitesimal surface perpendicular to the *x*-direction per unit time. Since the area of such a surface is calculated by taking the product *dydz,* in general the current component j_x can be written as (Q is the charge):

$$j_x = \frac{dQ}{dydzdt}$$

and similarly for the other components:

$$j_y = \frac{dQ}{dxdzdt};$$

$$j_z = \frac{dQ}{dxdydt}.$$

Note that the charge density can be represented in an analogous way, because it is the charge divided by a unit of volume:

$$\rho = \frac{dQ}{dxdydz}.$$

We observe a certain pattern here. Each of the components of the four-vector J^{μ} is expressed as the quotient of the infinitesimal change in charge and the three-dimensional infinitesimal "surface" in four-dimensional space-time. The components representing the current flow involve surfaces made up of two spatial coordinates and one temporal coordinate, while the charge density component has a purely spatial surface in the denominator, with no temporal element. Thus the charge density ρ can be interpreted as the flux of charge flowing along the time coordinate *t*, in the same way that j_x denoted the flux of charge along the *x* coordinate.

An additional important element of the analysis of the four-vector current is the continuity equation, which expresses the generalized ("local") principle of charge conservation. This equation states that the change in the amount of charge in a given spatial volume must be "compensated" by the flow of current through the walls of that volume. Since charge density can be interpreted as a flux along the timeline, it simply means that the total amount of charge in a four-dimensional volume does not change, or that the total charge "outflow" in four dimensions is zero. The mathematical form of the continuity equation for charge, as we have previously discussed in section 5.9, is as follows:

$$\frac{\partial \rho}{\partial t} = -\nabla \cdot \boldsymbol{j}$$

In a homogeneous four-dimensional form, this equation takes the form:

$$\frac{\partial J^{\mu}}{\partial x^{\mu}} = 0.$$

The expression on the left is nothing more than a four-dimensional divergence, denoting the outflow of the four-vector J^{μ}.

An analogous procedure can be applied to energy and then to each component of momentum. In the case of energy, we can talk about both its density and the flow of energy through appropriate surfaces, perpendicular to the x, y and z axes. So energy will require four components. In addition, each of the three momentum components p_x, p_y, and p_z also requires four parameters. These are: the density of a given component and its flux in all three spatial directions. So we have a total of $16 = 4 \times 4$ parameters, which gives us a second-order tensor $T^{\mu\nu}$.[19]

Let us present in more detail the individual components of this tensor, called the energy-momentum tensor (another name is stress-energy tensor). The component T^{00} stands for energy density (energy per unit volume), i.e. the energy flux along the time axis t. The component T^{01}, in turn, is the energy flux along the x-axis. Similarly, T^{02} and T^{03} will represent energy fluxes along respective axes y and z. The tensor component T^{10} is the density of the p_x component of momentum, and the components T^{11}, T^{12}, and T^{13} are the fluxes of this component along the x, y, and z directions. The same will apply to the subsequent components of momentum p_y and p_z. This can be presented synthetically in the form of a table, in which the rows correspond to energy and momentum components, respectively, and the columns represent individual axes (temporal and three spatial). At the intersection of the corresponding row and column, we have a flow of a given quantity in the selected direction.

	Axis t	Axis x	Axis y	Axis z
Energy	T^{00}	T^{01}	T^{02}	T^{03}
Momentum p_x	T^{10}	T^{11}	T^{12}	T^{13}
Momentum p_y	T^{20}	T^{21}	T^{22}	T^{23}
Momentum p_z	T^{30}	T^{31}	T^{32}	T^{33}

In analogy with the continuity equation for charge, we can write the continuity equation for each of the four quantities included in the energy-momentum tensor. For example, the energy continuity equation (conservation of energy) will be:

$$\frac{\partial T^{00}}{\partial t} + \frac{\partial T^{01}}{\partial x} + \frac{\partial T^{02}}{\partial y} + \frac{\partial T^{03}}{\partial z} = 0.$$

Synthetically, the continuity equation for the energy-momentum tensor takes the form (remember the summation convention):

$$\frac{\partial T^{\mu\nu}}{\partial x^{\nu}} = 0.$$

However, the above equation applies only in spacetime without curvature. As we remember, in the general case, partial derivatives must be replaced by covariant derivatives. Thus, the general continuity equation for the energy-momentum tensor is as follows:

[19] Strictly speaking, for $T^{\mu\nu}$ to be a tensor, its components must transform in the right way. This is indeed the case, but we will not prove it.

$$\nabla_\nu T^{\mu\nu} = 0.$$

We already have basically all the components of Einstein's equation – the fundamental equation of general relativity. Let us now try to sketch the general way to reach this equation, starting with a reminder of the formal description of gravity in Newton's theory. In Chapter 2, we introduced Newton's well-known law of gravity, which describes the interaction between two point masses. In the general case, however, the distribution of masses can be arbitrary, not just point-like, and therefore the law of gravity must also take into account this possibility. The solution to this problem is to give an equation analogous to Maxwell's first equation in the differential version (4.7), discussed in Chapter 4. Maxwell's equation connects the divergence of the electric field $\boldsymbol{E}$ with the charge density at a given point. On the other hand, the electric field can be represented as a gradient of the electric potential (with a minus sign). Gradient divergence is an operation that differentiates a given scalar twice along each of the three spatial coordinates and adds the results. As a result of this operation applied to the gravitational field, we obtain an equation, called the Poisson equation:

$$\nabla^2 \varphi = -4\pi G \rho$$

where φ is the gravitational field potential, ρ is the mass density, and the operator ∇^2, as we have defined before, is the Laplacian, $\nabla^2 = \frac{\partial^2}{\partial x^2} + \frac{\partial^2}{\partial y^2} + \frac{\partial^2}{\partial z^2}$.

The idea is that the new equation describing gravity should approximately turn into the Poisson equation for small velocities and small masses. However, in the new approach, we no longer use the concept of the gravitational field and its potential, but only a geometric approach to the kinematic behavior of bodies in the form of the principle that all bodies move along geodesics. So we will have to find a way to "translate" the language of fields and gravitational potentials into the language of geometry. In the previous section, we introduced the description of geodesics by assuming that the covariant derivative of a vector tangent to a geodesic should be identically equal to zero (i.e. the tangent vector to the geodesic does not change its direction when parallel-transported). This leads to a geodesic equation containing Christoffel symbols and the corresponding coordinate derivatives with respect to the parameter characterizing a given curve. Currently, we are considering not just any curves, but the so-called world-lines for bodies, i.e. lines whose tangent vector is always a time-like vector (directed "upwards" in relation to the time axis). The world line of a given object is parameterized by its proper time τ (time calculated in the system in which the object rests). Therefore, the tangent vector to a given world line has the form $\frac{dx^\mu}{d\tau}$, and the geodesic equation will take the following form:

$$\frac{d^2 x^\mu}{d\tau^2} + \Gamma^\mu_{\alpha\beta} \frac{dx^\alpha}{d\tau} \frac{dx^\beta}{d\tau} = 0.$$

Consider the above equation only with respect to spatial coordinates, ignoring the time coordinate $x^\mu = x^0$(in the approximation of low velocities, proper time τ is practically equal to x^0, and thus the second derivative of x^0 with respect to τ is 0). The derivative $\frac{dx^0}{d\tau}$ is close to 1, while the velocities $\frac{dx^i}{d\tau}$ (where, according to the convention, the Latin index i ranges over the values of 1, 2, 3) are negligible in relation to $\frac{dx^0}{d\tau}$ – remember the assumption that $c = 1$. Thus, the geodesic equation transforms into the following relation:

$$\frac{d^2x^i}{dt^2} + \Gamma^i_{00} = 0.$$

It is, as you might have guessed, a variant of Newton's classical law of motion, where $\frac{d^2x^i}{dt^2}$ is acceleration and $-\Gamma^i_{00}$ represents force (divided by mass). Christoffel symbols in general are, as we have already mentioned, complicated functions of the metric tensor (they contain derivatives of the components of the metric tensor with respect to different coordinates). However, approximately when the metric tensor $g_{\mu\nu}$ deviates only slightly from the flat Minkowski spacetime tensor $\eta_{\mu\nu}$, the Christoffel symbol Γ^i_{00} can be written as:

$$\Gamma^i_{00} = -\frac{1}{2}\frac{\partial g_{00}}{\partial x^i}.$$

Since in Newtonian mechanics, the acceleration of a body in a gravitational field (i.e., force) is given in the form of the gradient of the potential:

$$\frac{d^2x^i}{dt^2} = -\frac{\partial \varphi}{\partial x^i},$$

finally, we have the relationship between gravitational potential and geometry, i.e. the metric tensor, which is valid for weak gravitational fields and low velocities:[20]

$$\varphi = \frac{1}{2} g_{00}$$

Inserting the above into the Poisson equation, and remembering that the T^{00} component of the energy-momentum tensor represents the energy density – i.e. mass – we get the first approximation of Einstein's equation, which combines physics with geometry:

$$\frac{1}{2}\nabla^2 g_{00} = 4\pi G T^{00}$$

The above equation indicates the key relationship for general relativity between the physical property of matter (represented by a component of the energy-momentum tensor) and the property of spacetime (a component of the metric tensor). However, this is not a satisfactory solution for the general case. The basic problem with this equation is that it contains the selected component, not the whole of the energy-momentum tensor. As we remember, the components of tensors change when the coordinate system changes. The fundamental equation of relativity should be invariant with respect to any coordinate transformations, so it should contain complete tensors, not their distinguished components. Of course, there is also the fact that the proposed equation is valid only in special reference frames (moving at a low speed) and for small mass-energies. Therefore, we should look for a general equation of the following form:

$$G^{\mu\nu} = kT^{\mu\nu} \tag{6.12}$$

of which the above equality is a special case.

[20] Of course, we can always add any constant to the potential (according to the gauge transformation). However, this does not change our considerations, because the derivative of a constant is zero.

In the above notation, $G^{\mu\nu}$ is a tensor representing the geometric properties of spacetime, and k is the corresponding constant. From our previous considerations, it follows that the tensor $G^{\mu\nu}$ should contain second derivatives of the metric tensor. Such a tensor could be the Riemann curvature tensor, which indeed satisfies such a condition (because it contains derivatives of Christoffel symbols, which in turn are defined by combinations of derivatives of the metric tensor). However, this tensor is of the fourth rank, so mathematically it cannot be equated with the energy-momentum tensor, which contains two indices. On the other hand, the Ricci tensor, which is a contraction of the Riemann tensor, could be used here. Unfortunately, the obstacle is the fact that the energy-momentum tensor must satisfy the continuity equations, expressing the principle of conservation of momentum and energy. Since the covariant derivative of the right side of the equation is zero, the same should apply to the left side of the equation. However, it can be calculated that the covariant derivative of the Ricci tensor $R^{\mu\nu}$ is not zero in general. It looks like this:

$$\nabla_\mu R^{\mu\nu} = \frac{1}{2} g^{\mu\nu} \frac{\partial R}{\partial x^\mu},$$

where R is the curvature scalar. Taking into account the aforementioned fact that the covariant derivative of a metric tensor is always zero[21], and that the covariant derivative of each scalar is the same as the ordinary partial derivative, we can represent the right side of the above equation as $\nabla_\mu \frac{1}{2} g^{\mu\nu} R$. So we have found an object whose covariant derivative is identically equal to zero. This is the so-called Einstein tensor:

$$G^{\mu\nu} = R^{\mu\nu} - \frac{1}{2} g^{\mu\nu} R.$$

Inserting this expression into formula (6.12), we finally get the famous Einstein equation, which is the quintessence of general relativity:

$$R^{\mu\nu} - \frac{1}{2} g^{\mu\nu} R = k T^{\mu\nu}.$$

Theoretically, the above equation contains everything we need to calculate the metric properties of spacetime, assuming a given distribution of masses (energy) and momentum. In practice, however, solving Einstein's equation (or rather equations, because it breaks down into many independent equations for individual components) is impossible due to the high degree of mathematical complexity. Therefore, various types of simplifying assumptions are used, e.g. in the form of imposed symmetry conditions.

In the case of small masses and velocities, the dominant component of the energy-momentum tensor is the component T^{00}, representing the energy density (there is a factor c^2 in it, which is obviously a huge quantity). It can also be shown that the left side of the equation, i.e. the Ricci tensor minus the metric tensor times R, converges in this case to the Laplacian from the zero component of the metric tensor: $\nabla^2 g_{00}$. Therefore, in order for Einstein's equation to become the equivalent of the classical Poisson equation, we must assume the following value of the constant k: $k = 8\pi G$.

[21] Zeroing the covariant derivative of the metric tensor is a condition for the compatibility of the affine connection with the metric tensor – we mentioned this earlier.

Let us also consider the question of what happens if the energy-momentum tensor is equal to zero in the entire region of spacetime. This, of course, corresponds to the situation of a completely empty universe, devoid of all matter. According to Einstein's equation, the left side must also be zeroed, which gives us the following relation:

$$R^{\mu\nu} = \frac{1}{2} g^{\mu\nu} R$$

On the surface, it seems that this equation can have solutions in the form of a non-zero Ricci tensor. However, this is not the case. By "multiplying"" both sides of this equation by the metric tensor $g_{\mu\nu}$, or more precisely, contracting both sides with this tensor, we get an equation containing only the scalar of curvature:

$$R = 2R.$$

Of course, this implies that $R = 0$, and therefore also the Ricci tensor is zero. Does this mean that space-time will be devoid of curvature? It may even seem, as Leibniz suggested, that in such a situation space-time will cease to exist altogether, since it is only an epiphenomenon of material objects. However, mathematics tells us otherwise. Zeroing the Ricci tensor, which is a contraction of the Riemann curvature tensor, does not imply the disappearance of the latter. There are non-zero forms of the Riemann tensor for which the Ricci tensor is zero. One such form represents "sourceless" gravitational waves. Thus, the mathematics of general relativity allows for the existence of curved spacetime even in the total absence of matter. Of course, it remains an open question whether such a possibility is physically meaningful. Most physicists seem to think that this is the case, but for obvious reasons we will never be able to subject this hypothesis to direct empirical verification.

Questions and problems

1. Discuss Einstein's equivalence principle, using examples of a free fall and a space rocket moving with constant acceleration. Compare Einstein's principle with Galileo's principle of relativity. What are the restrictions of the former that do not apply to the latter?

2. What are geodesics? Give an informal interpretation of two mathematical definitions of the concept of geodesic. Which lines on a sphere are geodesics?

3. Discuss the concept of intrinsic curvature using the notion of parallel transport. Give an example of a two-dimensional surface curved in three dimensions whose intrinsic curvature is zero, and one for which the intrinsic curvature is greater than or less than zero.

4. What principle in general relativity describes the kinematic behavior in the gravitational field of bodies not subjected to any other forces? What is the relation of this principle to Newton's first law of motion? What will the trajectory of a free body look like in a gravitational field?

5. Discuss the basic physical idea contained in Einstein's field equation. What property of space-time depends on the distribution of matter?

6. What two postulates led Einstein to adopting a specific form of the so-called Einstein tensor and the entire equation connecting the distribution of matter with the curvature of space-time?

7. Is the geometrization of the gravitational force a unique feature of general relativity, absent in classical physics? What is the essential difference between GR and the geometric version of Newtonian mechanics?

8. Discuss the most important empirical tests confirming the validity of the general theory of relativity.

9. Compare the historical problem of observed anomalies in the motion of the planet Uranus and its solution with the analogous problem of anomalies in the motion of Mercury. What do these examples teach us about empirical testing of general hypotheses in science?

10. Does the general theory of relativity implement Mach's idea of complete relativization of all motions, including accelerated motions?

11. What arguments from general relativity support the substantial nature of space-time?

12. Outline the idea of the hole argument for relationalism regarding the nature of space-time. Compare this argument with Leibniz's shift argument. Use the concepts of active and passive interpretation of transformations between coordinate systems.

Further reading

A classic textbook on the general theory of relativity for students of physics is the already cited B.F. Schutz, *A First Course in General Relativity*, Cambridge University Press, Cambridge 2009.

An accessible but in-depth discussion of spacetime in special and general relativity can be found in the book by a well-known American physicist: R. Geroch, *General Relativity from A to B*, The University of Chicago Press, Chicago 1978.

The following monograph contains a very elegant presentation of the mathematical and philosophical aspects of the theory of relativity: M. Friedman, *Foundations of Space-Time Theories*, Princeton University Press, Princeton 1983.

A thorough overview of the problems of time and space, from Newton to both theories of relativity, is contained in the monumental work: L. Sklar, *Space, Time and Spacetime*, University of California Press, Berkeley 1974.

More recent discussions of the status of space-time, including the hole argument and versions of substantialism, can be found in the article: O. Pooley, "Relationist and substantivalist approaches to spacetime", in: R. Batterman (ed.) R. Batterman (ed.) *The Oxford Handbook of Philosophy of Physics*, Oxford University Press, Oxford 2013, 522–586.

Again, I acknowledge the debt that I incurred while writing the asterisked sections of this chapter by using this book as a guide: L. Susskind & A. Cabannes, *General Relativity: The Theoretical Minimum*, Basic Books, New York 2023.

CHAPTER 7. QUANTUM MECHANICS

The quantum revolution in physics that took place at the beginning of the twentieth century resonated even more widely in philosophical circles than the emergence of both theories of relativity. There are many reasons for philosophers' interest in quantum mechanics. Firstly, this theory reaches deep into strictly ontological and epistemological issues, such as the problem of determinism and causality, realism or the problem of identity and persistence in time. Moreover, quantum mechanics is perhaps the only currently recognized theory in physics that contains a fundamental gap in its theoretical foundations, known as the measurement problem. Although many physicists question the importance or even existence of this problem, a significant number of theorists see the need to address this issue. As a result, quantum mechanics has received a significant number of so-called interpretations, the main purpose of which is to fill this fundamental gap, and at the same time solve lesser problems, such as the alleged existence of non-local interactions or radical indeterminism. These interpretations usually supplement the basic assumptions of the theory with some additional hypotheses or interpretative rules. The best-known interpretations of quantum mechanics include the many-worlds theory, Bohmian mechanics and the theory of spontaneous localization, although there are many other interesting approaches.

In this chapter, we will first of all trace the historical path that led to the creation of a new quantum theory, which radically broke away with the worldview present in classical mechanics, as well as in both theories of relativity. We will point to the experimental sources of quantum theory and its non-classical character. Next, we will discuss the most important concepts of the new theory, such as the concept of state, superposition of states, entangled states, and probability. We will also mention the fundamental law of state evolution in quantum mechanics, expressed in the Schrödinger equation. We will begin our analysis of the philosophical problems of quantum mechanics by presenting the famous EPR argument, formulated by Einstein and his collaborators. The purpose of this argument was to show that quantum mechanics, with its indelible use of probabilities, cannot be the "last word" of science. In other words, quantum mechanics is incomplete and must be replaced by a more accurate theory, based on the principles of determinism and predictability. The key assumption of the EPR argument is the thesis of locality, which we have already mentioned in the context of the theory of electromagnetism and special relativity. The principle of locality applied to the EPR case implies the existence of a deeper reality that current quantum mechanics is unable to describe. However, Bell's surprising theorem shows that the existence of such a deeper reality (expressed in so-called hidden variables) together with the assumption of locality leads to a conflict with the experimentally well-confirmed predictions of quantum mechanics.

After discussing Bell's theorem and its philosophical implications, we will move on to the central issue, which is the measurement problem in quantum mechanics. A popular approach to this problem takes the form of the famous paradox of Schrödinger's cat. The alternative interpretations of quantum mechanics that we will discuss in this chapter can be distinguished as different ways of resolving this paradox. We will also show how quantum mechanics describes systems composed of many particles of the same type, such as electrons or photons. It turns out that this description has very interesting implications for the issue of the identity and distinguishability of quantum objects. Therefore, we will again encounter the philosophical principle of the identity of indiscernibles, as well as the problem of identifying objects over time. As before, more inquisitive readers will find further details about the beautiful mathematics used in quantum theory, based on Hilbert spaces, vectors and operators.

7.1 From experiments to quanta

The emergence of quantum mechanics is inextricably linked to the rapid increase of our knowledge about the structure of matter. In the nineteenth century, atomism became the dominant approach in science (chemistry and physics), based on the search for the simplest components of matter, whose fundamental properties and interactions should explain the complexity of the observed material world. We have already encountered this approach when discussing the reduction of thermal phenomena to molecular processes. The fundamental assumption of this reduction was the thesis that the elementary components of matter (molecules, atoms) are subject to the laws of Newtonian mechanics. As we shall see in a moment, with an increasingly accurate analysis of the basic components of matter, this belief had to undergo a radical modification. The breakthrough here was the discovery of the internal structure of atoms. The concept of atoms as impenetrable "clumps" of matter was rejected in favor of the model of a small atomic nucleus and surrounding electrons. The discovery of the electron and the determination of its basic parameters (mass, charge) was another milestone on the way towards a new concept of matter.

However, the first indicator of the "non-classical" nature of the microworld was the analysis of electromagnetic radiation, and more specifically its interaction with matter. Almost every introduction to quantum mechanics begins with the presentation of an episode in the history of physics concerning the so-called black body radiation. In the context of this rather special issue, the concept of "quantum" was first introduced, which became the basis of the new theory. The problem facing physicists at the turn of the nineteenth and twentieth centuries was how to theoretically describe the thermal radiation emitted by an object at a certain temperature. We know from experience that such radiation has a certain frequency distribution – most of the energy is radiated at some intermediate frequency, and for higher and lower frequencies the radiated energy drops to zero. Trying to theoretically derive the formula for the distribution of such radiation, scientists assumed, in accordance with Maxwell's classical theory of electromagnetism, that a standing wave of a given frequency (a so-called vibration mode) can carry any value of energy between zero and infinity, depending only on its amplitude.

Unfortunately, the formula derived from this assumption had an absurd consequence: the energy radiated in waves of growing frequencies increased to infinity (the so-called ultraviolet catastrophe), which is of course physically impossible. The German physicist Max Planck found out, using trial and error, that if one assumes that the energy of a wave of a given frequency v can only take on distinguished values that are multiples of a certain constant *h* times the frequency *ν*, the formula derived from this assumption will agree perfectly with empirical observations. Planck called the portions of energy available for a given radiation "quanta", and the constant *h* is still referred to by his name: Planck's constant.

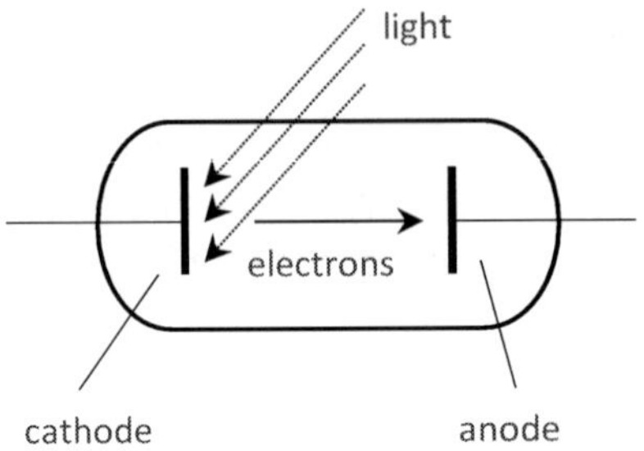

Fig. 7.1 Photoelectric effect.

Planck's hypothesis of radiation quanta can be considered *ad hoc* and its creator himself originally treated his idea as only a mathematical trick to solve a pressing problem. Soon, however, additional experimental facts appeared, indicating the existence of photons as individual energy carriers of electromagnetic radiation. One of such facts was the so-called photoelectric effect. It is based on the possibility found in the experiment of "knocking" free electrons out of metal by electromagnetic radiation (light). Two live electrodes are placed in a tube emptied of air (Fig. 7.1): a positive (anode) and a negative (cathode). Under normal conditions, no current flows through the tube, because there are no electricity carriers in a vacuum. However, when we illuminate the cathode, there is a flow of current, which is explained by the fact that the energy provided by the light radiation allows the electrons to escape from the cathode (overcome the energy barrier) and flow towards the positively charged anode. As experimentally demonstrated, the current increases with both the increase in radiation intensity and its frequency. This is explained by the classical theory of electromagnetism by the fact that both the intensity (proportional to the amplitude of the wave) and the frequency of the wave (the number of vibrations per second) determine the energy supplied by the wave. The higher the energy, the more electrons are knocked out and the greater the current. However, another fact was difficult to explain classically – namely, the observed existence of a threshold frequency, below which no current appeared, despite the increase in light intensity. The solution proposed by Einstein was to assume, following Planck, that the energy of radiation is transmitted in portions equal to $h\nu$. Each individual electron interacts with one quantum (photon), so for an electron to be released, the portion of energy $h\nu$ must be greater than a certain threshold value, regardless of how many such portions are in the beam.

Experimental facts about the behavior of light, or more broadly electromagnetic waves, reveal a fundamental interpretative problem. On the one hand, light exhibits indisputable features of wave propagation – it is subject to the phenomena of interference and diffraction, which are unambiguously related to the wave character. On the other hand, the empirical findings cited above show that under certain circumstances light behaves analogously to a swarm of particles. How can these mutually exclusive models be reconciled? This is one of the many challenges faced by the new theory of the microworld.

Other controversial empirical facts have been revealed during the study of the properties of the elementary components of matter, specifically electrons. One of the most spectacular experiments performed in the early years of the twentieth century was the Stern-Gerlach experiment, which investigated the interaction of the magnetic moment of an electron with a magnetic field. According to the classical theory of electromagnetism, rotating electrically charged bodies create a magnetic field – in short, they behave like small magnets with a specific orientation in space, parallel to the axis of rotation. This effect is described by the magnetic moment vector, which is a function of both the rotational speed (more precisely, angular momentum) and the charge. A body endowed with a magnetic moment, moving in a non-uniform magnetic field, will experience an acting force parallel to the direction of the force lines of the external field. The value of this force depends on the alignment of the magnetic moment vector in relation to the lines of force: it is greatest when the magnetic moment is parallel, and it drops to zero for the perpendicular direction. Therefore, by passing a stream of particles with randomly oriented magnetic moments through a properly shaped magnetic field, we should obtain the entire spectrum of deflections: from the maximum value "up", through the zero value, to the maximum value "down".

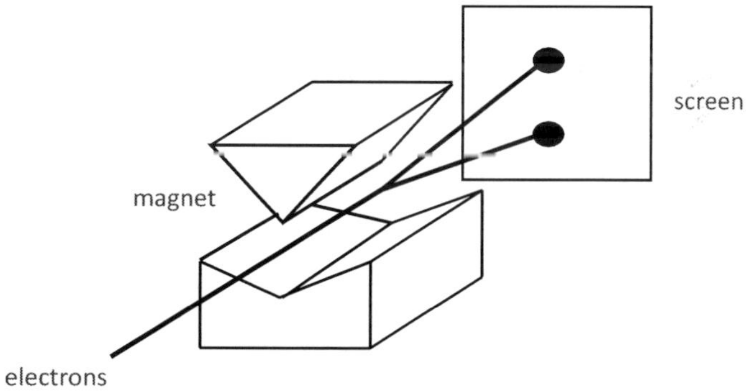

Fig. 7.2. Stern-Gerlach experiment

The aim of the Stern-Gerlach experiment was to investigate the interaction of the magnetic moment of an electron with a magnetic field. For this purpose, rather than free electrons, silver atoms were used, which have one "unpaired" electron in the highest orbit. Due to the "canceling-out" of the angular momenta of the remaining electrons, the total internal angular momentum of the silver atom is equal to the momentum of a single electron. In the following description, however, we will simply talk of single electrons. The stream of electrons was therefore passed between the poles of a suitably shaped magnet, recording their deflection on the screen (photographic plate). The result obtained differed significantly from the classical predictions – instead of a continuous distribution, only two maximum deflections "up" and "down" were observed (Fig. 7.2). It is as if all the electrons have aligned their magnetic moments parallel to the field lines of force, either up or down. However, this seems absurd. In addition, repeating experiments for different magnet orientations leads to exactly the same result. However, it is mathematically impossible for the components of a given vector in all directions to take the same maximum value.

The first conclusion drawn from the Stern-Gerlach experiment was that the internal angular momentum of an electron is a non-classical quantity that cannot be described in the language of vectors in three-dimensional physical space. To emphasize this new, non-classical character, the angular momentum of particles is referred to as spin. A detailed mathematical analysis of spin was given by Wolfgang Pauli (using matrices – the details will follow in the section with an asterisk), and later by Paul Dirac (using the concept of spinors). Moreover, spin exhibits a trait called *quantization*. This means that only specific, discrete values for a given quantity are acceptable, and not a continuous spectrum of them, as in the case of classical quantities. In the case of the spin of an electron in a given direction, there are exactly two permissible values, equal to $\frac{1}{2}\hbar$ and $-\frac{1}{2}\hbar$, where $\hbar$ is Planck's constant divided by 2π (such spin is often called fractional). Other particles admit more possible spin values, but still a limited, discrete numbers of them.[1] The quantization or discretization of physical quantities is a unique feature of quantum mechanics. It characterizes many other quantities, such as orbital angular momentum or energy in bound systems. The quantization of energy levels in an atom (in particular in the hydrogen atom) explains another important observational fact, namely the existence of discrete spectral lines in the radiation characteristic of a given type of atom.

However, this is not the end of the groundbreaking nature of the Stern-Gerlach experiment. An additional fascinating aspect of the quantum world was revealed by sequential spin measurements along different directions. Let us start with a simple case in which two magnets with the same orientation are placed one after the other (Fig. 7.3 on the left). After the electrons passed through the first magnet, one of the output beams (e.g., the lower beam) was blocked, and the upper beam was directed to the second magnet. The result of such an experiment is as expected – only the upper beam appears at the exit of the second magnet. The interpretation of this result seems obvious. The upper beam leaving the first magnet contains only electrons with spin directed upwards, and therefore the second measurement cannot reveal electrons with the oppositely directed spin. So far, everything is going as predicted. But let's see what happens if we position three magnets one behind the other in such a way that the first and third magnets have the same orientation, and the second is oriented perpendicular to the two (Figure 7.3 on the right). To fix the ideas, let's assume that the axes of the first and third magnets are parallel to the z-axis, and the second to the x-axis. So we are dealing with a sequential measurement of spins s_z, s_x, and then s_z again. The upper beam coming out of the first magnet is split again into two beams ("left" and "right") after the passing through the middle magnet. This is intuitive – within particles with spin s_z directed upwards, there should

[1] Note that not one spin is introduced here, but in fact an infinite number of spins, each defined with respect to one direction in space. Although these spins are typically referred to as "components", they are not components of a normal vector. However, you can in addition define the so-called total spin, the square of which is the sum of the squares of the three components in three perpendicular directions: $s_x^2 + s_y^2 + s_z^2$. As it turns out, we then get a quantity that takes only one value for all electrons, equal to $\frac{3}{4}\hbar^2$.

be about the same number of particles with spin s_x pointing in one direction as in the other. However, if one of these beams passes through the magnet along the z-axis again, we get a surprising result: two beams will appear again, an upper and a lower beam, of comparable intensity.

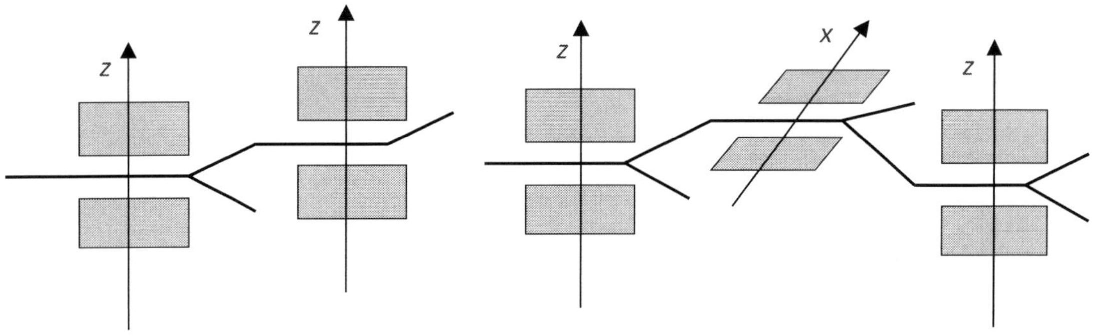

Fig. 7.3. Double and triple Stern-Gerlach experiments

This is puzzling. The particles entering the second magnet have their spin s_z equal $\frac{1}{2}\hbar$. After measuring the s_x spin and selecting one of the two options, we obtain particles with spin $-\frac{1}{2}\hbar$ along the x-axis. However, a second check of the s_z spin again gives two possible values. It looks as if, as a result of the second measurement, half of the electrons "flipped" their previously measured spins. Such behavior is difficult to explain from a classical point of view. We do not know any mechanism responsible for the reversal of magnetic moments. The explanation proposed by the new theory is that certain physical quantities cancel each other out, in the sense that the exact determination of the value of one of them distorts the value of the other. Such quantities are called "incompatible" – they include the components of spin in different directions, as well as momentum and position. In addition, note that the proposed analysis of the situation *implicitly* assumes that a given physical quantity (e.g. spin) may not take any well-defined value at a given moment. The particles leaving the second magnet are in no way characterized in terms of spin along the z-axis. Only the next measurement "creates" the appropriate values for individual electrons (at the same time, however, "destroying" the previously obtained spin value s_x).

We will conclude the review of experimental facts revealing the non-classical behavior of the components of the microworld with the two-slit experiment. In a sense, it is a thought experiment, the practical implementation of which is much more complex,[2] but for illustrative purposes it is used in this highly idealized form. Consider a barrier with two slits (holes) in it, and a screen behind it (Fig. 7.4). This is a typical setting for a wave interference experiment – for example, we can easily conduct such an experiment on water by setting up a properly prepared barrier with two holes and creating a wave on the surface of the water. Currently, however, the experiment will not concern waves, but particles, i.e. electrons. In the first stage,

[2] The closest to this idealization version is the Davisson-Germer experiment with electron scattering on crystal lattices.

we cover one of the two slits and send a beam of electrons towards a barrier with one open slit. In this way, we will obtain a smooth distribution of incident electrons on the screen with the maximum falling opposite the slit (Fig. 7.4 on the left). It is normal behavior of molecules that follow straight lines from the slit in a certain range of directions, due to possible slight reflections from the edge of the slit. A similar distribution will be obtained with the second slit open and the first one closed. But what happens when we open both slits?

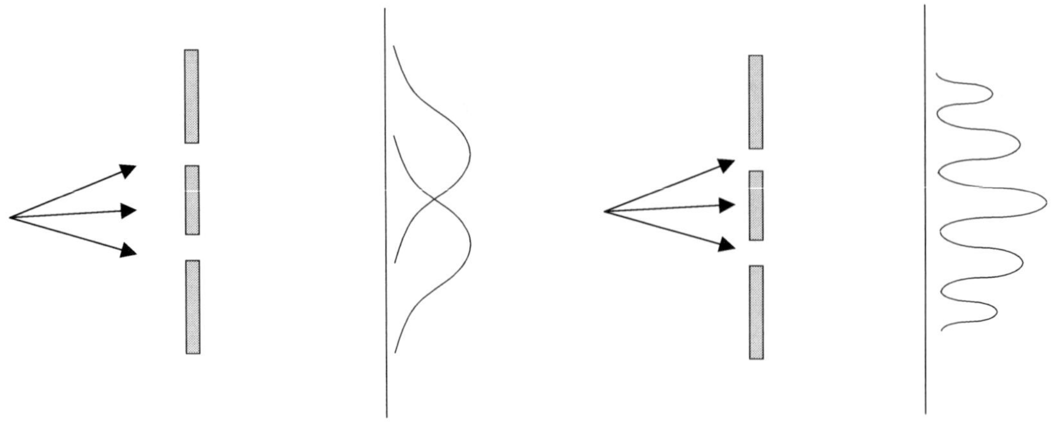

Fig. 7.4. A two-slit experiment. Distribution of electrons passing separately through single slits (left) and through both open slits (right)

It turns out that in such a situation, the distribution of electrons on the screen will be fundamentally different from the previous distributions (or their sum). Namely, the maxima and minima of intensities (the so-called interference fringes), well known from wave effects, will appear. This means that the electrons "prefer" certain directions, and disperse less willingly in other directions. This effect is interpreted as indicating that matter at the elementary level exhibits wave properties, analogous to those of light. Like light, which in some cases manifests wave-like properties, and in others corpuscular (e.g. in the photoelectric effect), electrons also reveal a peculiar "dual" nature. However, the issue of the so-called wave-particle duality is closely related to interpretations of quantum mechanics – it takes different forms in different interpretations. For example, in Bohm's mechanics, particles are completely classical objects with well-defined trajectories, while wave effects are explained by the existence of an additional "field", mathematically represented by the so-called wave function, which interferes with itself and at the same time affects the behavior of individual particles. On the other hand, according to the hypothesis of matter waves by the French physicist Ludwig de Broglie, electrons are "really" packets of waves, where the distribution of a given wave is the distribution of the density of matter. However, let's leave these interpretative issues aside for now and try to analyze the obtained measurement results in the language of particles.

The first possible explanation for the behavior of electrons is that perhaps the interference effects occur as a result of the interaction of particles in the stream passing through the slits. It is possible that electrons passing through one slit collide with the electrons from the other slit, which leads to a distribution different from the sum of the two distributions for each slit separately. To verify this hypothesis, a similar experiment was performed with very weak beams, where in practice each electron passed through the barrier individually. The effect remained unchanged – still, after collecting an appropriate number of recorded electrons, the familiar image of interference fringes appears. Therefore, it is not the interaction between electrons that is responsible for the wave effects. So, we can consider the behavior of a single electron. Classical physics tells us that an electron must pass through one of two slits – either the first or the second. However, each of the two paths leads to a smooth distribution, with no highs and lows. This leads to the surprising but difficult to avoid conclusion that the electron does not pass through any of the slits separately, but in a sense through two slits at once.

But can't this hypothesis be subjected to an empirical test? For example, we can illuminate one of the slits with a beam of concentrated light so as to locate an electron passing through it. In such a situation, we will be able to check each time whether a given electron registered on the screen had passed through the first or the second slit. It turns out that both the collected rich empirical material and theoretical considerations clearly indicate that in such a situation the distribution of electrons will be completely classical, i.e. no interference effects will occur. The occurrence of these effects is unequivocally correlated with our fundamental ignorance of which path the electron has chosen. The standard approach to this fact is presented by the concept of superposition. At the moment of passing through the barrier, the position of the electron is not well defined – it is neither near one slit nor the other. We can only talk about a certain probability of finding an electron in a given area. Such a state is referred to as a superposition of two states, each of which has a well-defined location. The superposition of states is "reduced" to one state during measurement, e.g. when one of the two slits is illuminated by radiation. Thus, the measurement inevitably leads to a change in the state of the observed object.

7.2 Quantum states, probabilities, superpositions

The path from empirical data to a full quantum theory was not easy. In order to explain the observed phenomena, various types of "semi-classical" models have been proposed, such as the well-known Bohr-Sommerfeld model of the atom. However, these solutions were half-hearted. Both physicists (Werner Heisenberg, Wolfgang Pauli, Niels Bohr, Erwin Schrödinger) and mathematicians (Johann von Neumann) joined forces to work on the foundations of a completely new theory. In this section, we will present in the most general outlines the basic concepts of this theory, emphasizing its differences from classical physics. In the section marked with an asterisk, we will take a closer look at the mathematical side of the new formalism. Let us start by recalling the concept of the state of a physical system from classical mechanics. As we remember, a detailed description of the system of N classical particles requires their instantaneous positions and velocities (momenta). With this data, we can calculate all other quantities characterizing the system at a given moment, for instance the total kinetic energy or the angular momentum. Therefore, in general, we can assume that the complete state of the system is given in the form of a function that assigns an exact value to each physical parameter. In contrast, the state of the system in quantum theory operates with the concept of probability. At a given moment, the probability distribution for all possible values of each physical quantity is determined.

Quantum states are represented mathematically by vectors in some abstract space called Hilbert space. The details are not important to us at the moment – it is only vital to remember that some basic mathematical operations are specified on vectors, such as vector addition, multiplication by a number and the scalar (inner) product of vectors, which gives a certain number. Thanks to the introduction of the product, we can talk about a very important concept of perpendicularity of vectors – in the language of linear algebra it is called orthogonality (two vectors are orthogonal when their product is equal to zero). A single vector encodes the information about all the probabilities involved, which can be recovered using a certain principle called the Born rule. We will not formulate this rule in general terms for now, but only for a special case. To begin with, we note that although in general the quantum state assigns "non-trivial" probabilities (i.e., probabilities other than 0 and 1) to the values of a given quantity, it is nevertheless possible under certain circumstances for one value to obtain the probability of 1. This means that a given quantity (energy, momentum or spin) has a well-defined value in this state. Such a state is called an eigenstate of this quantity. Of course, for each possible value of this quantity, the corresponding eigenstate will be different (vectors representing such states are orthogonal to each other).

Now consider two eigenstates for some quantity, corresponding to two different measured values. As an example, let us use spin in a certain direction discussed in the previous section. Let z_+ denote the state in which the spin of the particle along some z axis is directed upwards, and z_- the state with the z-spin downwards. The basic principle of quantum formalism, contained in the assumption of the vectorial nature of states, prescribes that any two states can be "combined" with each other (we represent this operation by addition). In this way, we will obtain a new state, the physical interpretation of which we will now deal with. The general formula for a linear combination of two vectors z_+ and z_- is as follows:

$$az_+ + bz_-$$

where a and b are some numbers.[3] The interpretation of this formula is that it describes a situation in which the probability of a spin-up result is $|a|^2$ and the probability of a spin-down result is equal to $|b|^2$. Of course, in order to maintain the principle that the probability of the certain event is equal to unity, we must assume that $|a|^2 + |b|^2 = 1$, which is referred to as the "normalization" of the vector (this is equivalent to saying that the length of the vector calculated by taking the inner product with itself is one). Thus, the vector written above defines a new state, unknown in classical physics, in which spin values are characterized by only certain probabilities. Such a state is often referred to as a superposition of states with well-defined spins.

A special case of superposition is when both superposed states enter it with equal probabilities. In this case, the above vector takes the form of an equally weighted combination:

$$\frac{1}{\sqrt{2}}(z_+ + z_-).$$

The triple Stern-Gerlach experiment considered in the previous section can be described by means of an analogous superposition. The electrons in the upper beam coming out of the first magnet are in a state of the superposition of the states "spin up in the x direction" and "spin down in the x direction", which shows in the result of the second measurement, where

[3] In general, these will be complex numbers, so when squaring them we have to take the so-called modulus.

exactly half of the electrons are deflected upwards and half downwards.[4] Incidentally, let us note that exactly the same state, which is a superposition of spins in the x direction, can be described as a state in which the spin in the z direction is well defined. This shows that there are strong relationships between spins in different directions. As it turns out, a specific spin state in any direction unambiguously determines the probability distributions for spins in any other direction.

We must distinguish the state of superposition from a statistical mixture composed of equal numbers of particles with spin up and spin down. Although in such a mixture it is true that a randomly selected particle has the 1/2 probability that its spin along e.g. the x-axis will be directed upwards (or downwards), the probabilities of the spin values along other axes will differ from those in the state of superposition: they will be equal to 1/2 for each axis. In contrast, in the state of superposition of spins, as we have already noted, there is always an axis along which the probability of obtaining a given spin result is equal to one. Mixed states are represented in the formalism of quantum mechanics not by vectors, but by some kind of operators, called density operators (or statistical operators).

Another example of superposition is provided by the previously described case of the two-slit experiment. During the passage through the barrier, the state of the electron is described by the superposition of two wave functions – one describing the passage through the first slit and the other describing the passage through the second slit. This mathematical fact also explains why interference effects appear on the screen, although it is a little too difficult to depict it in proper detail. Since the wave function is a "vector" describing the position of a given particle (more precisely, the so-called probability density of finding a particle), in order to calculate the probability of finding a particle in a given area, you need to take the square of this function and integrate (sum) it over this area. Since the square of the sum of numbers is not equal to the sum of their squares, the probability of finding an electron in a given place on the screen will not be a simple sum of the probability of finding it assuming that it passes through one slit and the probability that it has passed through the other.

We will deal more closely with the ontological aspects of superpositions when analyzing the paradox of Schrödinger's cat. For now, let's pay attention to an important fact about measurements on systems prepared in superposition states. Since, before the measurement of e.g. spin, the system was in a superposition of states with the values "up" and "down", and after the measurement each electron is in one of the two beams with a well-defined spin, it is natural to assume that as a result of the measurement, the electrons changed their initial state $az_+ + bz_-$ to one of the two states z_+ or z_-. Such a change is a probabilistic process – the probability that an electron will "jump" to the state z_+ equals $|a|^2$, and that it will jump to z_- – $|b|^2$. In the language of quantum mechanics, such a process is called a "collapse" or "reduction" of the initial state. As it will turn out, an interpretation of measurement collapse in quantum mechanics is one of the most controversial and hotly debated issues.

[4] It could be supposed that the electrons entering the first Stern-Gerlach magnet are also in a state of spin superposition along the "up" and "down" axis, since exactly half of them have deflected upwards and half downwards. Unfortunately, this is a wrong assumption. The initial state of electrons belongs to another category of so-called mixed states (see the entry in the box). To go from such a state to a "normal" state (called a pure state), a spin measurement must be taken in either direction. Such measurements are often called "preparatory".

7.3 Measurable quantities and uncertainty principle

Even a cursory contact with quantum physics must at some point come across Heisenberg's famous uncertainty principle. In the original version, this principle was derived from the wave description of quantum processes. From the classical analysis of wave phenomena, we know that each so-called wave packet (i.e. a wave that occupies a certain finite area in space) can be represented in the form of a sum (superposition) of sine waves of different frequencies and different wave numbers (the reciprocal of the wavelength). It can be proven that the width of a given packet is inversely proportional to the "spread" of wavenumbers in the distribution of this packet into sinusoidal components (so-called harmonics) – that is, the better located the packet in space, the wider the range of different wavenumbers for the harmonic components of this packet. However, since the width of the wave packet associated with a given particle (photon or electron) is a measure of the indeterminacy of its position in space, and the wave number determines the velocity (or momentum) of the particle-wave, we have the relationship that the spread, i.e. the indeterminacy of position, is inversely proportional to the indeterminacy of momentum. In other words, the better the momentum of a particle is determined, the more undetermined its position is (and *vice versa*). This principle is captured in the known inequality:

$$\Delta x \Delta p \geq \hbar,$$

where Δx and Δp denote the measurement uncertainty associated with determining the position and momentum, respectively, and $\hbar$ is the Planck constant divided by 2π.

It turns out that the above relationship between momentum and position is only one of many examples of the phenomenon of *incompatibility* between certain measurable quantities, which is common in quantum mechanics. As we mentioned in the previous section, it also affects the components of spin in different directions. The formalism of quantum mechanics provides an elegant way to express this relationship. To discuss this, however, we need to specify how measurable quantities (which in this theory are also called observables) should be mathematically represented in the conceptual apparatus of quantum theory. We already know that for a given observable, we can distinguish some special vectors, called eigenvectors. They define the states in which this observable has a certain well-defined value. We also mentioned that eigenvectors corresponding to different values of the same quantity are mutually orthogonal. This is due to the obvious fact that if a system is characterized by a given value of a specific observable, the probability of obtaining a different value in the measurement is zero.[5] For the sake of simplicity, let's assume that the selected observable has a finite number of different values (like half-spin, with two possible values). In this case, the observable remains unambiguously correlated with the set of mutually perpendicular eigenvectors in Hilbert space (such a set of vectors is also called the "spectrum" of a given observable). Note that any vector that does not belong to the spectrum can be represented as a combination (superposition) of eigenvectors. This means that in the state represented by this vector, a

---

[5] According to Born's rule, mentioned above, the probabilities of the results of measuring the quantity A in a given state are calculated by decomposing the vector representing this state into components corresponding to individual values of A, and then taking the squares of the coefficients of this decomposition. Since a vector perpendicular to a given vector does not enter the decomposition of that vector, the probability of obtaining the result related to it must be zero.

given observable does not have a specific value, but there are only some probabilities that the values will be appropriate.

If we now choose some other observable, it will also be represented by a set of perpendicular vectors, but generally different from the previous ones. What does it mean that two observables A and B have different eigenvectors? Let v_a be the eigenvector for observables A corresponding to some value of a. Suppose v_a is not an eigenvector for B. In this case, the system in the state v_a has a well-defined value for A, but not for B. So the quantity B is "fuzzy" in this state. This is exactly the situation described in Heisenberg's uncertainty principle – a well-defined position, a poorly defined momentum. Observables for which not all eigenvectors coincide are called incompatible. Conversely, if two observables have exactly the same eigenvectors, we define them as compatible.

There is an even more elegant and "synthetic" way to represent observables and their compatibility relationships. We will tell you more about it in the section marked with an asterisk. For now, we will only mention that the standard representation of measurable quantities in quantum mechanics are objects called linear operators on vectors (these are functions that transform one vector into another and satisfy the appropriate linearity conditions). It turns out that with every set of orthogonal eigenvectors and their corresponding values (these are called eigenvalues), a linear operator is correlated. Its action on any vector is as follows: first we decompose this vector into components parallel to the eigenvectors, then we multiply each such component by the appropriate eigenvalue and sum back the vectors obtained in this way. This seems quite complicated and hardly clear, but such a procedure leads to extremely simple and elegant representations of many physical facts and regularities. In particular, there is a very neat way to express compatibility relationships. Two observables are compatible if their corresponding operators commute, i.e. their operation is the same, regardless of the order in which they are executed: $AB = BA$. As it turns out, operators representing different spin components, as well as momentum and position operators, do not commute with each other.

7.1 The EPR argument and the incompleteness of quantum mechanics

From the above sketch emerges a picture of a new theory, which is based on the indelible use of probability, limiting the possibility of an accurate and complete description of the microworld. This form of scientific theory was unacceptable to Einstein. Einstein was fully aware of the important role that probability and statistical considerations play in science. We use statistical methods when we are dealing with a large number of objects that we are not able to track individually. We encountered such a situation during the statistical analysis of thermodynamic phenomena, where we had to use, for example, quantities averaged over a huge number of particles. However, we assumed that each individual particle possessed well-defined characteristics. In the case of quantum theory, on the other hand, measurement probabilities are applied directly to individual objects, assuming that these objects are not fully characterized by physical quantities before measurement. This approach aroused Einstein's deep suspicions.

In his polemics with the supporters of the new theory (mainly Bohr), Einstein tried to show that it is theoretically possible to overcome the limitations imposed on the possibility of knowing the exact properties of quantum systems (including those resulting from the uncertainty principle). These attempts were usually countered by extremely subtle arguments by Bohr. The most famous argument in this polemic was published in a pivotal 1935 paper by Einstein and his two collaborators, Boris Podolsky and Nathan Rosen. This argument is known to this day under the name EPR after the initials of the authors. Its main idea can be presented in the form of the following story. Let's imagine that a certain store starts selling toys advertised as "magic boxes". These are ordinary small boxes that, when opened, reveal the contents – a black or white marble. None of the sellers or manufacturers knows what color marble we will find in the box. Comparing the results of a large number of opened boxes, we come to the conclusion that statistically the chances of revealing a black or white marble are equal. What's more, the sellers assure us that there is no way to find out what the color of the marble is before opening. It is best to imagine that the marbles "decide" what color to take only when the box is opened.

One day, a store offering boxes introduces a novelty: instead of one, it sells two boxes. Although it is still unknown what color the marbles will be revealed when the boxes are opened, experience teaches us that in both boxes, the marbles reveal opposite colors. It has never happened that two black or two white marbles appeared in a given pair. Now let's imagine that the boxes from the same pair were offered to two people: Alice and Robert. Alice takes her box on a long journey, and then driven by curiosity, she opens it and sees a white marble. Robert, on the other hand, suppresses his urge to open his box and does not touch it at all. Of course, Alice already knows that Robert's box must contain a black marble. However, she found out about it without having to open the second box! Moreover, due to the distance between her and Robert, we can assume that opening her box could not have affected the condition of the other box in any way. So we conclude that Robert's marble must have been black even before opening the box. The whole story with the magic boxes turned out to be a scam (which we suspected from the very beginning). Each box simply contains a black or white marble, and the black-marble and white-marble boxes are randomly mixed in equal proportion to give the impression of mysterious and random color picking at the moment of opening.

We can guess that the marbles in our story are quantum objects (e.g. electrons), and the boxes and their opening represent measurements. According to the assumption promoted by the supporters of the new quantum theory, before measurement, a quantum particle does not have a feature that is revealed in it (e.g. electrons do not have a specific spin before entering the Stern-Gerlach device). However, the question remains as to what in quantum theory corresponds to the pairs of boxes containing marbles of the opposite color. Is it possible to select a pair of particles with perfectly correlated properties without finding out what those properties are? It turns out that quantum mechanics allows for the existence of such systems. We call them entangled.

Quantum entanglement is one of the most fascinating phenomena that the physics of the microworld has encountered. Let us first describe it from a theoretical point of view, and then present its specific implementations. Let's start with the question of how to express the state of a system consisting of two (or more) objects, e.g. two electrons, in the language of quantum mechanics. In the simplest case, when each of the particles taken separately has its own state, the state of the whole is a straightforward composition of the two states. Such a composition is basically nothing more than an ordered pair: if the state of the first particle is represented by vector a, and the state of the second particle by vector b, then their combined state can be written for instance as (a, b). In the quantum-mechanical formalism, such a pair is presented in the form of a certain product – not an ordinary dot product, but the so-called tensor product (this has something to do with the tensors we talked about in previous chapters). It is denoted by the symbol ⊗, although sometimes this symbol is omitted, as in the case of the ordinary product of two numbers, in writing of which we can drop the dot. Therefore, the state of our two particles will be written as $a \otimes b$, or ab for short.

It is important to note that the tensor product is not "reversible" (commutative). The state of the particles presented as $b \otimes a$ is different from $a \otimes b$ – it means that the first particle has the property b and the second has the property a.[6] For now, everything looks completely classical. However, we should remember that in quantum mechanics the principle of superposition applies – if two vectors represent certain states of a physical system, then by adding them we will get a new state, different from each separate state. We can therefore consider the following vector-state:

$$\frac{1}{\sqrt{2}}(a \otimes b \; + \; b \otimes a).$$

What does such a vector represent? It is a state in which with probability 50% the first particle is in state a while the second is in b, and also with probability 50% the first particle is in the b state while the second is in a. Therefore, the state of any of the particles taken separately is not precisely determined, but only given with a certain probability. However, there is an unambiguous correlation between the states of particles – if we reveal that the first particle is in state a, then the second particle must be in state b. And vice versa – if the first one is in state b, the second has no choice but to occupy state a. So we have found in quantum theory the magic boxes from the story!

The physical realization of this theoretical possibility, which Einstein, Podolsky, and Rosen used in their argument, involved two incompatible quantities: momentum and position. The entangled state of the two particles proposed by the authors was such that although neither the momentum nor the position of the particles was well determined, both the difference in their positions and the sum of the momenta were fixed. Thus, by measuring the momentum of one of the particles, it is possible to find out the momentum of the other particle. The situation is similar in the case of location. Niels Bohr even proposed a very ingenious quasi-technical possibility of realizing the EPR state by means of a diaphragm with two slits cut in it, placed on a spring, thanks to which the particles flying through it would be characterized by both a constant difference in position (equal to the distance between the slits) and the total momentum along the axis of the diaphragm (equal in value to the momentum transmitted to the entire aperture). However, the practical implementation of such an arrangement is fundamentally unfeasible.

Nowadays, the implementation of the entangled EPR state is achieved by using electrons with spins or photons and their so-called polarization (the latter method is the easiest to realize in the laboratory). So let us use our familiar z_+ spin states (spin z up) and z_- (spin z down) in the selected direction of the z axis. Consider the following two-electron state, called a singlet-spin state:

[6] In the case of particles of the same kind, such as two electrons, which are in some sense indistinguishable, this assumption can be questioned. We will return to this problem in later sections. For now, we tacitly assume that each of the two particles can be somehow "individualized" by putting the labels "number 1" and "number 2" on them.

$$\frac{1}{\sqrt{2}}(z_+ \otimes z_- - z_- \otimes z_+) \tag{7.1}$$

The minus sign in the above expression does not affect the probabilities of finding particles in the corresponding z-direction spin states, because we are squaring the coefficients anyway. However, it makes all the difference when it comes to calculating the probabilities of spins along other axes (e.g. x or y), but for our current purposes this is irrelevant. As in the general situation of states a and b described above, we are dealing here with a perfect correlation. The spins of the two particles are not specified, but it is certain that they will have opposite values. Whatever value we reveal on one of the electrons, the other electron will definitely have the opposite value.

We can now present the reasoning of Einstein and colleagues. Suppose that the electrons prepared in the above state have been separated at such a distance that no physical interaction propagating at a speed no greater than the speed of light can reach from one electron to another (remember that such a separation is called space-like in special relativity). Then we measure the spin of one particle. The result obtained allows us to precisely predict the measurement result for the second electron. At the same time, the state of the second electron was not physically disturbed in any way by the measurement. Thus, its state would be the same even if we did not subject our electron to measurement. This leads to an unequivocal conclusion: the distant electron has a fixed value of spin in a given direction, even if quantum theory is not able to predict it. Quantum mechanics therefore provides an incomplete description of reality.[7]

The logical structure of the EPR argument is extremely simple. Using the general assumption of locality (there is an upper limit to the velocity of physical interactions) and the predictions of quantum mechanics (the existence of entangled states), we conclude that a certain physical object has a property that our theory cannot predict. The only logical way to avoid the conclusion that quantum mechanics is incomplete is to reject the assumption of locality. This seems to have been the path taken by Niels Bohr in his response to the EPR argument. However, he recognized that the non-locality of the connection between distant particles is of a different nature than the ordinary physical interaction, such as the gravitational or electromagnetic interaction. Unfortunately, Bohr was very enigmatic about the details of this non-local quantum connection (he talked about the influence of local measurement only on the *conditions of the meaningfulness* of assigning certain values to the second particle). To this day, this issue remains highly debatable. However, in the sixties of the last century, things took a completely surprising turn thanks to a young and previously unknown physicist John Bell.

7.5 Bell's theorem and non-locality

As we saw in the previous section, an ideal physical theory should satisfy two conditions, according to Einstein. First, such a theory should in principle predict all possible measurement results. Secondly, it should not imply the existence of non-local interactions, i.e. those

[7] As it turns out, the above state of electrons (7.1) implies the existence of identical correlations for any spin directions. Thus, the analogous reasoning can be repeated for each direction of spin, which shows that all spins of the second particle are unambiguously fixed. This, of course, violates the uncertainty principle (the fact that the spins in different directions are mutually incompatible).

that propagate at arbitrarily high speeds or are even instantaneous. As we know, quantum mechanics does not meet the first of these conditions (we will talk about the second later). However, Einstein was convinced that in the course of the development of science, it could be replaced by a new theory that would adapt to both requirements. Einstein did not live to see the creation of such a theory, but the hope for its formulation remained even after his death. However, this hope was shattered as a result of a groundbreaking discovery, in the form of Bell's famous theorem In this section, we will present a version of this theorem known as the Clauser-Horne-Shimony-Holt theorem (CHSH)

Bell's reasoning was actually very simple and yet general. He hypothetically assumed that there is a new theory which amends the current quantum mechanics by using some unknown parameter λ (this parameter is called "hidden"). The use of a hidden parameter (or parameters) λ allows you to predict the measurement results for any quantity. Bell then considered an entangled system of two particles and a series of measurements made on these particles. In accordance with the principle of locality, Bell assumed that taking a measurement on one of the two particles could not change any properties (hidden or not) of the other. On the basis of these assumptions, Bell derived a certain inequality (named after him) that connects the statistical distributions of the results obtained in a series of measurements. It is important to note that although the new hypothetical theory uses data unknown to us, the statistical distributions calculated on its basis are experimentally verifiable. Bell then showed that standard quantum mechanics violates the inequality he derived for some selected measurable properties. So Einstein's program turned out to be unfeasible. Quantum mechanics cannot be consistently extended by certain hidden parameters while maintaining the principle of locality.

Of course, it is still theoretically possible that quantum mechanics is not an empirically correct theory and should be replaced by a better theory. If this were the case, the experiment should show that the Bell inequality is satisfied in reality, contrary to the predictions of quantum mechanics. Experimental research on this problem, for which the Nobel Prize was awarded in 2022, showed unequivocally that the Bell inequality does not hold in some quantum settings. Thus, Einstein's ideal theory is not only incompatible with currently accepted quantum theory, it is also incompatible with experience. The quantum world cannot be fully determinate and local at the same time.

Let us now present the reasoning leading to the Bell-CHSH theorem. Suppose we have two particles A and B prepared in a certain quantum state and spatially separated from each other. On each of them, we can make one of two measurements of certain quantities: A_1 and A_2 for the first of them, and B_1 and B_2 for the second (we assume that these quantities are pairwise incompatible, although strictly speaking this assumption is not necessary to derive the inequality itself). Each of the measurements can reveal one of two possible values: 1 or -1 (we can think of the measured quantities as spins in different directions). There are four "global" experimental settings: A_1B_1, A_1B_2, A_2B_1, and A_2B_2 (by a setting we understand a choice of quantities to measure on both particles, which can be implemented for instance by orienting appropriate Stern-Gerlach apparatuses). Let a_{11} and b_{11} denote the measurement results obtained respectively on particles A and B in the first setting, a_{12}, b_{12} in the second, and so on. We can now define the following parameter

$$\gamma = a_{11}b_{11} + a_{12}b_{12} + a_{21}b_{21} - a_{22}b_{22}\,.$$

Of course, the current quantum theory is not able to calculate the value of this parameter, because it operates with alternative measurement results that cannot be revealed together in experiment. For example, it is not possible to determine the result a_{11} and a_{22} at the same time, because this would require simultaneous measurements for two incompatible quantities A_1 and A_2.[8] Yet, the new theory should unequivocally determine these numbers. However, standard quantum mechanics can operate with averaged quantities. Let's calculate the average value of the parameter γ, denoted as: $\langle\gamma\rangle$

$$\langle\gamma\rangle = \langle a_{11}b_{11}\rangle + \langle a_{12}b_{12}\rangle + \langle a_{21}b_{21}\rangle - \langle a_{22}b_{22}\rangle.$$

The average of the sum of the given numbers is equal to the sum of the averages of these numbers. Note that each of the four values on the right side of the equation is already experimentally measurable. For example, the value of the average $\langle a_{11}b_{11}\rangle$ is obtained by repeating a large enough number of measurements in the A_1B_1 setting. Therefore, the number $\langle\gamma\rangle$ is experimentally determinable. It can also be calculated theoretically based on the rules of standard quantum mechanics, as long as the state in which the particles were prepared is known.

Now let us introduce the principle of locality. In the present case, it tells us that the value obtained in a measurement made on one of the particles should not depend on which measurement was made on a distant, spatially separated particle. This means that, for example, the numbers a_{11} and a_{12} should be equal, because the first of them determines the result of the measurement of the observable A_1 in a situation where B_1 was measured on the second particle, and the second determines the result of the same measurement with an alternative quantity B_2 measured for the second particle. However, the second measurement should not have any effect, so the two values should be equated. The same applies to the other parameters – e.g. b_{12} and b_{22} should be equal. As a result, we will get a simplified version of the formula for γ:

$$\gamma = a_1b_1 + a_1b_2 + a_2b_1 - a_2b_2.$$

We can now perform a simple algebraic transformation:

$$\gamma = a_1(b_1 + b_2) + a_2(b_1 - b_2).$$

The maximal possible value of the number $b_1 + b_2$ is 2, with $b_1 - b_2$ then equal to 0. On the other hand, the minimal value of the same sum is –2, in which case $b_1 - b_2 = 0$. It is easy to see that it follows that the value of the parameter γ must be in the range from –2 to 2. The same must of course apply to the average, and so we got the following inequality:

$$-2 \leq \langle\gamma\rangle \leq 2$$

This is Bell's inequality in the CHSH version. It turns out that if we substitute spins in appropriately selected directions for A_1 and A_2 and B_1 and B_2, and the state of the particles is the singlet state (7.1), then the mean value $\langle\gamma\rangle$ calculated on the basis of the principles of quantum mechanics will be –2.5, which is well outside the allowed range. Thus, quantum

[8] As we know, sequential measurements of first the quantity A_1 and then A_2 does not solve the problem, because performing the second measurement will cancel the value obtained in the first one (remember the sequential Stern-Gerlach experiments that we discussed in sec. 7.1).

mechanics is incompatible with one or more of the assumptions leading to the above inequality.

Usually, the blame for the above inconsistency is placed on the assumption of the complete determination of measurable quantities (also called realism of possessed properties or the hypothesis of hidden parameters). If we reject the assumption that quantum objects have well-defined values for all physical quantities, then we can at least avoid the second of Einstein's worrying features, namely non-locality. However, the matter is not so simple. Non-locality, understood as the dependence of the measurement result for one particle on the quantity measured on the other, is indeed avoidable. Yet standard quantum theory is characterized by a different type of non-locality, which, although in a sense weaker than the previous one, is still a departure from Einstein's ideal. As we have seen in the example of the singlet state, the revealed measurement result on one of the particles depends strongly on what outcome is obtained on the other particle. If one of the outcomes is "up", the other must be "down". If, on the other hand, the first one was "down", the second one would be "up". Therefore, the situation in a distant region somehow "influences" what happens to the other particle.

So we have two types of non-locality that can potentially appear in any quantum theory. Stronger non-locality, sometimes called "parameter dependence", manifests itself in the fact that the physical state of a given system depends on what quantity the distant experimenter decided to choose for measurement. For example, choosing the orientation of a Stern-Gerlach apparatus, that is, choosing spin along a certain axis, would change some property of the other object. On the other hand, weaker non-locality (called outcome dependence) is the effect of making the state of a given particle dependent on the measurement result obtained in a distant area. The main difference between these two types of non-locality is that in the first case we have control over the cause of the distant change, because the settings of the apparatus depend on the researcher's decision. So theoretically, we could use such a non-local relationship to transmit information at a distance, breaking the principle of the non-existence of faster-than-light signals. On the other hand, the second type of non-locality does not give us the ability to transmit signals. The experimenter has no control over what result his measurement will reveal. It can be presumed that information about the measurement result in one system must be sent to another system, but the observer is not able to use this information. Speaking tongue in cheek, this information is only for the internal use of quantum objects.

In quantum information theory, a simple theorem is proven, known as the no-signaling theorem. It states that it is not possible to send a message from one part of an entangled system to another. Such information could be given, for example, in the form of a measurable probability distribution on one of the two particles. If it were possible to perform a given measurement on one particle in such a way that the probability distribution of the results for the other particle changed, we would be able to communicate instantly at any distance. However, the probabilities of the measurement results are independent of which measurement was made on the other particle. In the case of the singlet state (EPR), the probability of any result for any selected spin of a single particle is always equal to 1/2, regardless of whether the other particle has been measured or not. Even if we measure spin in the same direction on a distant particle, then since the probabilities of two different outcomes are equal there too, our local particle will have equal probabilities as if nothing had happened (each single outcome will, of course, unambiguously determine the other, according to the strict correlation, but determination by an event that is itself indeterministic also results in an indeterministic event). The same applies to spin measurements along different axes on both particles.

7.6 The measurement problem

We will now proceed to discuss perhaps the most controversial problem underlying quantum mechanics. No popular approach to quanta can do without the famous creature that Schrödinger's cat undoubtedly has become. The popularity of this animal is so huge that the desperate Roger Penrose once said "When I hear about Schrödinger's cat, I cock my revolver". We will perhaps try to go a little beyond the discussion of the sad fate of Schrödinger's favorite pet, only briefly recalling its story, which is probably well known to the readers. This cat is trapped (fortunately only fictionally) in a devilish box, which contains a device that breaks a bottle of poison. This device is connected to some quantum system – for example, it can be an unstable atom undergoing radioactive decay. The connection is made in such a way that the state of the atom before decay is correlated with the intact bottle, while the state after decay is correlated with the broken bottle. The state of a decaying atom at a given moment is described as a superposition of two states, with the coefficients of this superposition changing with time – the coefficient of the state before decay decreases and the state after decay increases. This means that with the passage of time, the probability of finding an atom in its decayed state increases. However, until we open the box with the device and the cat, the state of the atom is indeterminate. This raises an intriguing question: what is the cat's condition before opening the box? The superposition of a living and dead cat is rather difficult to imagine, so there is wide room for various divagations.

The example with the cat illustrates the fundamental difficulty of reconciling the behavior of microobjects that can occupy superposed states with the behavior of familiar macroscopic objects, such as a cat, which are not known to be in indeterminate states. This problem becomes particularly acute in the context of measurements that rely on the correlation between certain quantum states and the states of macroscopic objects, such as the position of the pointer of a suitable measuring device. However, before we move on to a detailed analysis of the issue of measurement, we must fill in a fundamental gap in our presentation of quantum mechanics. So far, we have completely ignored the part of theory that is most important for every practicing physicist. The primary goal of any physical theory is to predict the future evolution of physical systems. What law governs the evolution of quantum objects? It will certainly not be Newton's second law of motion, which applies to objects with well-defined properties, such as position and momentum. We need to find a new rule determining how the quantum state of a system, described in the form of an abstract vector, will change over time. This rule is given in the Schrödinger equation.

Schrödinger's equation was proposed within the framework of the so-called wave mechanics, in which quantum phenomena were to be described in analogy to wave phenomena from classical mechanics. As we remember, many quantum effects, such as the interference of electrons in a two-slit experiment, can be explained by the assumption of the existence of a certain wave, which represents the probability of finding a particle in a given region. This wave is called the wave function, to emphasize its abstract nature (it is not any "tangible" wave propagating in a certain medium, such as air, or even – like electromagnetic waves – in the vacuum itself). In abstract terms, wave functions are simply certain vectors in infinitely multidimensional Hilbert space. In any case, relying on certain analogies with the classical situation and also postulating specific mathematical properties, Schrödinger came up with the following equation, which should be obeyed by wave functions, and more broadly by all representations of quantum states. Denoting by ψ the wave function of a given system, we have:

$$i\hbar\frac{\partial\psi}{\partial t} = H\psi.$$

The symbol H stands for a differential operator whose physical interpretation is that it represents the total energy of the system. It will therefore come as no surprise that this operator is called the Hamiltonian, as in classical mechanics. So, as you can see, the evolution of a system depends on its total energy, just like in Hamiltonian mechanics. The mathematical details of the Schrödinger equation are not important to us at the moment. However, it is extremely crucial that this equation can be rewritten in a much simpler form, which will be very useful for our current goals:

$$\psi(t) = U(t)\psi_0.$$

Symbol ψ_0 represents the state of the system at a particular fixed initial moment, while $\psi(t)$ is of course the state at a later moment t, and $U(t)$ is a certain linear operator that depends on the Hamiltonian H and also on the time t. This operator belongs to the class of so-called unitary operators, which do not change the length of a given vector. The meaning of the above equation is quite clear. It implies that if we have a given initial state and the total energy of the system (Hamiltonian), then the state of this system at any later moment t is uniquely determined by the action of the evolution operator $U(t)$.

Therefore, we are dealing with strict and non-exceptional determinism – the Schrödinger equation unambiguously determines the evolution of the state of any physical system. This may come as a surprise to many readers. Have we made a mistake here? How to reconcile the alleged deterministic nature of the evolution of quantum systems with the indeterminism present in the probabilistic rules of quantum mechanics? The solution to this apparent paradox lies in the special form of quantum states. As we remember, the state of a system at a given moment does not unequivocally determine all the parameters that can be measured, but only the probability of obtaining individual values. The determinism of Schrödinger's equation means that the probability distribution of measurement results at a given moment uniquely determines the probability distributions in later moments. However, this determinism does not allow us to predict exactly what measurement results we will obtain. As we remember, standard quantum mechanics does not have hidden parameters that would fix the values of all measurable quantities. Thus, the determinism of quantum mechanics only implies that "incomplete" information about a given system at some point determines incomplete information at later times. This is, in a sense, a "determinism of indeterminism".

However, a problem remains. Since the indeterminism of quantum mechanics manifests itself at the level of measurement processes, and the physical evolution of quantum systems is given by the deterministic Schrödinger equation, it can be concluded that measurements are not subject to this equation. It seems, therefore, that in quantum mechanics we are dealing with two types of processes: a calm, deterministic development of the wave function, according to the Schrödinger equation, and a violent, unpredictable and only probabilistic measurement process. This is also how this issue is presented in many textbooks and approaches to quantum mechanics. However, this raises justified doubts. Doesn't limiting the applicability of the Schrödinger equation deprive it of the status of a universal law of nature? In addition, the question must be asked where exactly the boundary between Schrödinger processes (also called "unitary") and indeterministic measurement processes lies. What physically distinguishes measurements from other processes? What is so special about measurements? Different answers to this question have been offered, especially in different interpretations of quantum mechanics. However, we will try to treat measurements like any other physical processes. In principle, they should be describable in terms of unitary evolution. We'll see where such an idea will lead us.

Let us use the example of measuring the spin of an electron again. Let's treat the measuring device as a "black box", the inner workings of which are unknown to us. All we know is that when an electron is fed into our box, we get one of two results at the output: either an upward spin or a downward spin. Thus, the spin measurement box must have two possible states, corresponding to these results. Let us denote them as S_+ and S_-. We can add to this the third state S_0 before the measurement, which signals the readiness of the device. Now suppose that the absorbed electron had an upward spin. Thus, the initial state of the entire system electron plus the measuring device will be given in the form of the product $z_+ \otimes S_0$. If the device works properly, the output should show the result S_+. That is, the evolution operator U describing the measurement process must act on the input state in the following way:

$$U(z_+ \otimes S_0) = z_+ \otimes S_+ \,.$$

The situation should be similar with the downward spin of the electron. But how will the measuring device behave when the input electron is in a state of superposition of spins up and down? The answer to this question is contained in the way linear operators act on vectors. As the name suggests, they satisfy the condition of linearity, i.e. the action of an operator on a vector sum is equal to the sum of the actions of this operator on each vector separately. This can be written as follows:

$$U(z_+ \otimes S_0 + z_- \otimes S_0) = U(z_+ \otimes S_0) + U(z_- \otimes S_0) = z_+ \otimes S_+ + z_- \otimes S_- \,.$$

Thus, the final state of a system consisting of an electron and the measuring device will be a superposition of states with a "spin up" result and a "spin down" result. This is nothing more than a repetition of the problem of Schrödinger's cat. According to the current rule of evolution of quantum systems, the macroscopic measuring device should be in a superposition. However, in reality, we are seeing exactly one of the two outcomes, not a mysterious "combination" of two different outcomes.

The above reasoning, which leads to a conflict with experience, is based on three fundamental assumptions. Let us list them below in bullet points, as they will be the starting points for various alternative interpretations of quantum mechanics.

1. The Schrödinger equation is a non-exceptional law valid for all physical systems.
2. A superposition of states with well-defined but different values of a given quantity is a state in which this quantity is not well defined.
3. Each physical measurement ends with exactly one result.

As we have seen above, these three assumptions lead to contradiction. Therefore, one of them must be rejected. In the next section, we will discuss the most important attempts to avoid the resulting paradox.

7.7 Interpretations of quantum mechanics

As we mentioned earlier, the most common method of dealing with the measurement problem is to reject premise number 1. Measurements introduce a new quality, because they do not follow the unitary Schrödinger evolution. They are an example of essentially non-deterministic processes, in contrast to the deterministic evolution described by the Schrödinger equation. Since the rapid change of state caused by the measurement is often referred to as the collapse of the wave packet, the proposed solution bears the name of the collapse theory. The fundamental challenge for the collapse theory is how to draw a clear boundary between quantum deterministic processes and indeterministic measurements. The founders of quantum mechanics, primarily Bohr, preferred a solution to this problem based on a division between micro- and macroobjects. A characteristic feature of the measurement processes, distinguishing them from the "ordinary" phenomena of the microworld, would be the interaction of macroscopic measuring devices, consisting of a huge number of particles, with small, microscopic systems subjected to measurements. It is this interaction that would be responsible for breaking the rule governing the very phenomena of the microworld. Bohr claimed that all interpretative problems of quantum mechanics appear when we want to apply classical concepts developed by contact with macroscopic objects to the phenomena of the microworld. This thesis became the basis for the so-called orthodox interpretation of quantum mechanics, also known as the Copenhagen interpretation.

However, making a sharp division between the macro and micro worlds encounters fundamental difficulties. First of all, it is not clear where exactly the boundary of this division should be drawn. Is a molecule consisting of two hundred atoms a micro or macroobject? And a thousand such molecules? Another problem is even more serious. Explaining the special character of the measurement collapse by reference to the macroscopic character of the measurement device does not explain why macroscopic objects are not fully subject to the Schrödingerian dynamics. What is so special about systems consisting of a large number of elementary components that they cannot be in superposed states? Experience tells us that this is indeed the case, but our theory should be able to justify it, not simply take it for granted. We will return to this problem in the box below.

There are alternative approaches to the measurement collapse. For some time, for example, the concept that collapse is the result of the interaction of the observer's consciousness with physical reality was popular. Until the observer sees the cat in the box, the animal's condition is indeterminate. This solution is, of course, exposed to various criticisms. How about the consciousness of a cat or other animals – shouldn't they also have a "destructive" effect on quantum superpositions? In addition, of course, it is necessary to adopt a dualistic division into physical and mental reality, as well as to assume the mysterious influence of the mind on the physical world. Currently, few physicists follow such a radical position (the American physicist Henry Stapp is one exception).

The most interesting interpretation of quantum mechanics, rejecting the first premise of the measurement paradox, was proposed at the end of the last century by a group of Italian physicists led by the late GianCarlo Ghirardi (the other two being Alberto Rimini and Tulio Weber). This interpretation is called the theory of spontaneous localization or the GRW theory. Its basic idea is very simple. It is postulated that for any quantum object whose location is characterized by a "spread" wave function, there is some very small probability that this function will spontaneously locate itself around a single point. Such a phenomenon is completely indeterministic and is not subject to the Schrödinger equation. It is also extremely rare, so that the average waiting time for this spontaneous localization for a single particle to occur is comparable to or even greater than the lifetime of the universe. However, if we consider a macroscopic body that contains a huge number of particles (on the order of 10^{20}), it is of course almost certain that one of the particles will undergo such a localization. Due to the fact that quantum particles belonging to a given system remain in complex entanglement relationships with each other, the localization of even one of them causes the other particles to immediately follow it and also undergo localization (this is an effect analogous to "agreeing" of spin outcomes in an experiment with electrons in an entangled singlet state). Thus, macroscopic bodies, even if they are in a superposition of states with different positions for a tiny fraction of a second, will almost immediately localize themselves in one well-defined place. The same, of course, applies to measuring devices, i.e. the problem of the determinate character of measurement outcomes has been solved.

The GRW theory explains the difference in the behavior of individual quantum objects and their huge collections in a very elegant way. Unfortunately, its main difficulty is that there is no experimental data to support the existence of such a completely stochastic localization effect. Moreover, some experiments suggest that if such an effect exists, the average waiting time for localization will be too long to explain the observed behavior of macroscopic bodies. Therefore, we must look for further solutions to the paradox of measurement. The second group of interpretations questions the second assumption, i.e. the thesis that quantum systems in superposition do not have well-defined features. This is, of course, the good old concept of hidden parameters, which, although under heavy stress from Bell's theorem, has not been completely refuted. The best-developed version of this concept, with a large group of ardent supporters among both physicists and philosophers, is Bohmian mechanics, whose creator was David Bohm.[9] Bohm's interpretation is based on the assumption that all quantum objects have well-defined locations and trajectories – that is, that they essentially look like classical objects from Newtonian theory. However, their behavior is not subject to Newton's laws of mechanics, but to new quantum laws. Bohm's theory introduces an additional equation that makes the velocity of a given body (and not the acceleration, as in Newton's) dependent on the quantum wave function. Thus, the wave function plays the role of a force field here, acting on the particles and "pushing" them through space. At the same time, the wave function, as in standard quantum mechanics, contains probabilistic information about the chances of finding a particle in the right area. Of course, now this probability only characterizes our ignorance, because the particles really have a well-defined location, which, however, for fundamental reasons we do not know.

In the Bohmian interpretation, the results of measurements are unambiguously determined by the position of the particle within its wave function. Basically, all measurements for Bohm come down to measuring position in space (e.g. we measure spin in the Stern-Gerlach experiment by locating the electron either in the upper or lower beam). The superposition of the input state means that the wave function consists of a part representing the spin up and a part of the spin down. If an electron at the beginning of the interaction is in the area where the "spin up" part of the function dominates, then as a result of the interaction

[9] Similar ideas were also formulated by the aforementioned de Broglie, which is why this theory is sometimes called the de Broglie-Bohm theory.

with the magnet, this electron will fly up and we will register it in the right place. This result is unambiguously determined by the previous position of the electron, which, however, we are not able to know.

As we remember, Bell's theorem teaches us that any theory that assumes the determination of all measurement results must be strongly non-local. This is indeed the case with Bohm's theory. The non-locality of this theory is due to the fact that the evolution of a given particle is influenced by the entire wave function extended in space, which in turn depends on the distribution of other, distant objects. For example, in the case of the two-slit experiment, whether the second slit is closed or open has an impact on the shape of the wave function and therefore also on the behavior of the electron passing through the first slit. So, in a sense, the electron "knows" without any delay whether the other slit is open and adjusts its trajectory accordingly.

Finally, the third condition may be rejected. On the surface, this may seem ridiculous. How can one question the obvious fact that, upon entering the laboratory, the physicist can see whether the electrons, after passing through the Stern-Gerlach device, have been registered in the upper or lower part of the photographic plate? And what would it mean that the measurements do not have unique results? However, there is a very popular conception that explains how this is possible. This interpretation is called the many-worlds theory or Everett's interpretation. Hugh Everett III, in his doctoral dissertation, suggested that the Schrödinger equation should be satisfied by all physical processes, regardless of whether we call them measurements or not. In addition, Everett was adamant not to introduce any additional equations or relationships into quantum theory, as in the case of Bohm's theory, for example. His interpretation is, in a sense, the "purest" interpretation of the quantum-mechanical formalism without any additions. On the other hand, Everett interpreted the measurement results correlated with the individual components of the superposition as existing "relatively" to the selection of this and not another component. But what does it mean to "exist relatively"? Brian de Witt has suggested a solution that is still considered standard today: each measurement result exists in a different reality, called a world. Thus, as a result of measurement, our universe breaks up into as many worlds as there are possible results. In each of these worlds there is an observer who sees one outcome, but really all the outcomes exist "at once." For example, if we measure the spin of an electron, this measurement results in a splitting into two worlds, one of which contains the result of "spin up" and the other "spin down".

It must be admitted that this is a bold hypothesis, although it does not impress many philosophers (for instance the philosopher David Lewis did not hesitate to postulate the real existence of a huge number of so-called possible worlds to explain the meaning of modal concepts of possibility and necessity). However, there are some difficulties that must be solved immediately. First, there is the formal problem of arbitrariness in the selection of appropriate components of the superposition of a given state, which are to correspond to separate worlds. As we remember, as a result of the measurement process, the measured system and the measuring apparatus will be in a superposition of the orthogonal states corresponding to different results. It turns out, however, that mathematically speaking there are many (even infinite) alternative distributions of the same state into other orthogonal components, which do not correspond at all to the obtained measurement results, but only to their superpositions.[10] Why, then, do possible worlds correspond only to the distinguished decomposition of a given quantum state? This problem is known as the preferred basis problem. Its typical solution refers to a certain process called decoherence. More information on this topic can be found in the box.

> In the analysis of the interaction of a single physical system (e.g. a particle) with a macroscopic object, the process of decoherence plays an important role. Decoherence is roughly based on the fact that the states of a microscopic system corresponding to well-defined values of a certain measurable quantity (mainly position) are correlated with certain states of the macroscopic system. In other words, microscopic eigenstates are "recorded" in the global state of a system composed of a very large number of microobjects. As a result of this process, the state of the system subjected to such interaction changes from a superposition to a mixed state (cf. the entry in the box on p. 247). This means the disappearance of interference effects typical of the superposition of two or more coherent wave trains (hence the name "decoherence"). There are theoretical models showing that the most likely type of decoherence for typical measurement systems is decoherence due to position eigenstates. Therefore, it can be assumed that as a result of measurement, the state of the particle "decomposes" into states with a well-defined position (alternative "worlds") that do not interfere with each other.

Another problem affecting the many-worlds interpretation is more philosophical. As we know, the basic concept in quantum mechanics is the concept of probability. Before measuring e.g. spin, we can only say that the probability of the "up" result is such and such, because we do not know what the result will be. In the many-worlds theory, however, this fundamental ignorance disappears. Since for every possible outcome there is a world in which that result will occur, there is no need to introduce an objective notion of probability. The evolution of the system is unambiguously determined – this is due to the fact that in Everett's interpretation the Schrödinger equation is universally valid, and yet, as we have already seen, it is completely deterministic. Some try to solve this problem by introducing a "measure of reality": for each world, there is a measure, given by the square of the amplitude of the superposition component corresponding to that world, which determines how real the world is. However, the concept of a measure of reality is in itself very mysterious (can there be more and less real objects?).

There is another approach, more interesting from the point of view of a philosopher. It refers to the interpretation of probability as a measure of ignorance of individual personal experience. The division of the universe into a number of alternative possible worlds is related to the fact that in the future the observer will have a number of copies or continuations. Despite this, it remains unknown which of the continuations a given observer will see. This is analogous to the following problem: we give the observer a sedative and take him to one of two places: Paris or London. When we wake up, we ask him: "Where are you?" The observer can only say that with some probability it will be Paris or London. This is the so-called self-location problem. In the quantum-mechanical case, each observer, after the measurement, but before seeing its result, should conclude that the probability of seeing a given result

[10] This is easy to explain geometrically: as we know, for each vector there are infinitely many ways of decomposing it into component vectors perpendicular to each other.

is as predicted by formalism. However, it remains questionable whether these probabilities also make sense before the measurement, when there is only one observer.

7.8 Quantum statistics and indistinguishability

The last issue that we will discuss in this chapter will be the question of the behavior of large groups of quantum particles. As we know, when we are dealing with a large number of objects, their description must be given in the language of statistics or probability. This applies equally to bodies subject to classical physics and to quantum objects. However, there is a fundamental difference between classical and quantum objects in the details of such a description. This is related to the fundamental problem of indistinguishability. Let us start by considering the case of classical particles, such as atoms or molecules of statistical physics. In classical physics, each individual particle is endowed with a well-defined position and trajectory, which in principle (although not in practice) makes it possible to distinguish and "label" each particle separately. If we now want to describe the state of a very large number of such particles, we can do this by assigning to each label of a particle its individual state. We remember from the chapter on statistical mechanics that such an assignment is called an arrangement. Individual arrangements lead to the so-called distribution, i.e. to determining how many particles occupy each state, but without specifying which particle occupies which state. The number of arrangements implementing a given distribution gives us a measure of the probability with which we can expect a given distribution to appear.

Let us illustrate these considerations with a very simple example. Suppose we have two particles 1 and 2 and two states A and B. There are four possible distributions of these particles: both can go to state A, both to B, the first to A and the second to B, and the first to B and the second to A. The last two arrangements carry out the same distribution: one particle in A, one in B. If we assume that every arrangement is equally probable, then we get the following distribution probabilities: the distribution "two particles in A" – 1/4, the distribution "two particles in B" – 1/4, the distribution "one particle in A, one in B" – 1/2. This is an example of a classical probability distribution, called Maxwell-Boltzmann statistics (Table 7.1).

<table>
<tr><th>Arrangement</th><th>Maxwell-Boltzmann statistics</th><th>Bose-Einstein statistics</th><th>Fermi-Dirac statistics</th></tr>
<tr><td>A(1, 2)</td><td>¼</td><td>⅓</td><td>0</td></tr>
<tr><td>B(1, 2)</td><td>¼</td><td>⅓</td><td>0</td></tr>
<tr><td>A(1), B(2)</td><td>¼</td><td rowspan="2">⅓</td><td rowspan="2">1</td></tr>
<tr><td>A(2), B(1)</td><td>¼</td></tr>
</table>

Tab. 7.1. The probabilities of arrangements according to the three statistics.

However, as it turns out, quantum particles are not subject to this statistics. Their observed statistical behavior is different. Particles such as photons, for example, occupy states A and B in such a way that the probability of the distribution in which particles will end up in different states is the same as the probability that they will be in the same state. All three probabilities are 1/3. This can be explained by the fact that the arrangements "particle 1 in A, particle 2 in B" and the swapped arrangement "particle 1 in B, particle 2 in A" are in fact one and the same arrangement. Situations that differ only with respect to an exchange, i.e. permutation of particles, are identical. There is no way to know which particle went to which state, because they are indistinguishable (e.g. they do not have trajectories to identify them).

There is a group of quantum objects whose behavior is still different. These include all the elementary particles that make up matter – electrons, protons, neutrons, as well as smaller objects such as quarks. For these objects, there is only one possible distribution: the one in which one particle is in A and the other in B. For mysterious reasons, these particles "refuse" to occupy the same states. If one state is already occupied, the other particle must look for another state. Such behavior is described by a rule called the Pauli exclusion principle, which precludes more than one particle from occupying the exact same state. Such objects are called fermions, and the statistics to which they are subject is called Fermi-Dirac. Photons and other particles that behave similarly to them are called bosons. Their behavior is described by Bose-Einstein statistics, described earlier.

The aforementioned statistics are explained by the particular form of joint states that bosons or fermions can occupy. As we remember, the state of the entire system of two or more quantum particles is represented by the tensor product of the states of individual particles, or by any linear combination (superposition) of such products. Consider as an example a state in which one particle has the characteristic A and the other has the characteristic B. In principle, this condition could be described as the product $A \otimes B$. However, such a product is mathematically different from the "swapped" product $B \otimes A$, whereas quantum particles do not distinguish the states related by permutation. To solve this problem, some constraints are placed on the possible states of multi-particle systems. The basic requirement is that the physical state of the system of many particles remain the same during an exchange, i.e. permutation of the particles. This postulate can be implemented in two ways. First, the permutation-induced representation of the particle state can be exactly identical to the initial state representation. It is also permissible that the permutation changes the sign of the state vector. This has no effect on the physical state, because vectors can be multiplied by any number and they still represent the same state.[11]

State vectors that do not change with particle permutation are called symmetric. They characterize bosons. Fermionic states, on the other hand, are called antisymmetric. This is a bit of a misleading terminology, because it can suggest that antisymmetry is the lack (opposite) of symmetry. In fact, it is symmetry, but of a different kind. It consists in the fact that a permutation of particles can "at most" change the sign of a given state vector. Simple permutations of two particles (called transpositions in mathematics) always change the sign of the

[11] One may ask why not introduce the possibility of a permutation that multiplies the initial state by any complex number, rather than just +1 or –1. The usual answer is that applying the same permutation of two particles twice should give us the starting state, and only the numbers 1 and –1 squared give one. However, it is theoretically possible that a permutation of the state of three or more particles multiplies the state vector by a complex number of modulus 1 (the nth root of 1). Composing together several such permutations should get us back to the initial state as a result. The particles described by these types of symmetries are called "paraparticles". So far, no empirical evidence has been found for the existence of paraparticles – all known particles are divided into bosons and fermions without exception.

antisymmetric vector, while some more complex permutations (e.g. the so-called cyclic permutations, changing the arrangements of three particles from 123 to 231) leave the sign unchanged.

Let us apply the concepts introduced above to the analysis of a situation with two particles distributed between two states A and B. In the case of bosons, three combined symmetric states constructed with the help of such single states are possible. These are:

$$A \otimes A$$

$$B \otimes B$$

$$A \otimes B + B \otimes A.$$

As you can see, we have three, not four, possibilities for occupancy in the case of two particles and two states. This reproduces Bose-Einstein statistics, discussed above.

In the case of fermions, on the other hand, there is only one state built out of two states A and B, which satisfies the condition of antisymmetry:

$$A \otimes B - B \otimes A.$$

It is easy to see that replacing A with B will change the sign of the whole to the opposite. By the way, we can see that there can be no antisymmetric state of the whole composed of the same single-particle states. This immediately implies Pauli's exclusion principle, mentioned earlier. One state can only be occupied by one fermion.

Limiting the states of many particles of the same kind to symmetric or antisymmetric functions of single states has consequences beyond the statistical behavior of large groups. There are arguments showing that the symmetrization postulate implies that particles of the same type (e.g., electrons or photons) become completely indistinguishable with respect to all their measurable properties. It is quite obvious that, for example, two electrons are indistinguishable from each other if we limit ourselves to their identifying, characteristic properties, such as mass or electric charge (as well as the total spin, which is $\frac{\sqrt{3}}{2}\hbar$). The category of electrons does not break down into smaller subcategories with different characteristics. It seems, however, that individual electrons can differ from each other in "acquired" (contingent) features, which do not constitute their essence, such as position, momentum or energy. After all, we often deal with electrons located in a suitable measuring device in the laboratory, which certainly differ in location from electrons on Mars or in the Andromeda galaxy.

It turns out that the formalism of antisymmetric states (electrons are fermions) and symmetric states leads to a different and surprising conclusion. Consider two particles of the same kind, which we will number 1 and 2 respectively. Their state, as we already know, is described by a certain vector $\psi(1,2)$, which behaves appropriately when using the operations of exchanging labels 1 and 2:

$$\psi(2,1) = \pm\psi(1,2)$$

where the sign depends on whether we are dealing with fermions or bosons. If we assume that all the information about the physical properties of particles 1 and 2 is exhausted in the vector ψ, then such symmetry clearly indicates that everything that is physically true about particle 1 must also be true about particle 2. The labels of particles 1 and 2 occupy exactly the same places in the state vector, and thus any true sentence about particle 1 can be "transposed" into an identical sentence about particle 2. In other words, sentences attributing properties to individual particles should be invariant with respect to permutations. But this means that if the proposition "Particle 1 has the property *P*" is true, then the proposition "Particle 2 has the property P" must also be true.

So we have the surprising consequence that particles of the same kind cannot differ in any properties. This is a clear violation of the principle of the identity of indiscernible objects, which we encountered in Leibniz's argument against the substantial treatment of space. According to this rule, each two entities should differ in at least one property. It can be an "intrinsic" property, characterizing only the object itself, or an "extrinsic" property, relating a given object to other objects. In the case of symmetric or antisymmetric quantum states, neither the intrinsic nor the extrinsic properties can distinguish between the particles. So what causes us to deal with two particles, not one?

Many philosophers and physicists believe that in such a situation we should reject the assumption that quantum particles are individual objects, numerically and qualitatively distinct from the rest of the world. As an illustration of this approach, the example of money in a bank account is often cited. One day John deposits one dollar to his account, the next another dollar, and after a few days he withdraws the amount of one dollar. Which dollar did John pay out? The question is, of course, absurd. The dollars in the account are no different from each other – they are virtual items with the same characteristics. Electrons behave in a similar way – their identity is not clearly defined. Some even argue that quantum objects are not subject to the usual law of logic, known as the principle of identity, which states that each object is identical with itself. Proponents of this approach develop a non-classical theory of non-individuals, according to which collections of such non-individual objects are characterized by "cardinality" (the number of elements) but cannot be ordered. It makes sense to ask how many dollars there are in the account, but you cannot send an order to the bank for the first and third dollar in the account to be swapped with each other.

However, not everyone thinks that the basic laws of logic in relation to quantum objects should be rejected. There is an approach according to which the concept of the so-called weak discernibility, which is based on relations, not properties, can be applied to quanta. An example is the relation of having oppositely directed spins in the EPR singlet-spin state, which can connect numerically distinct objects, but never applies when we compare an object with itself (no object can have a spin opposite to its spin – this is a logical impossibility).[12] An even more "conservative" approach emphasizes that even in the symmetric language of quantum mechanics, it is possible to express absolute distinguishability using ordinary properties. Let us illustrate this observation with a simple example. The sentence "Particle number 1 has spin up, and particle 2 has spin down" is not invariant with respect to the permutation, because replacing the labels 1 and 2 changes this sentence to another one, which will not be true if the

[12] An attentive reader may protest at this point (if you are that reader – congratulations!). How is it possible for particles to have opposite spins, when we have just stated that particles of the same kind cannot differ from each other with respect to properties? The matter is a bit complicated. In fact, in the singlet state (EPR), the particles do not have any well-defined spins—the spin in each direction is determined only with the probability of 1/2. When we talk about the anticorrelation of spins, we do not mean that one electron really has an upward spin and the other a downward spin, but only that their sum is zero. It is of course bizarre that we can add spins that are not well defined, and the result will still be well defined. But this is precisely the quantum entanglement that we have already encountered in the EPR argument.

former is true. But consider the following, slightly more complicated sentence: "Either particle number 1 has spin up and particle number 2 has spin down, or particle number 1 has spin down and particle number 2 has spin up." This sentence is fully symmetrical with respect to the permutation – the swapping of labels 1 and 2 retains its meaning, but at the same time it expresses quite clearly the idea that particles are distinguishable by spins. It turns out that in the formalism of quantum mechanics, it is possible to formulate analogous statements that turn out to be true, so it can be maintained – contrary to previous suggestions – that particles of the same kind are nevertheless distinguishable.

The above considerations concerned the issue of distinguishability and identity of quantum objects at a certain point in time – in this case, we speak of the so-called synchronic identity and discernibility. Another but related concept is that of diachronic identity, i.e. identity in time. It is a well-known philosophical question how to determine whether a given object at a moment t_1 is the same as an object at another moment t_2. In the case of quantum objects, it may be impossible to answer such a question for fundamental reasons. This is the case, for example, when two particles of the same kind interact with each other in such a way that their wave functions first overlap and then diverge (Fig. 7.5). In such a situation, quantum formalism requires us to take into account two possible "ways" of reaching the final state, and to calculate the probability of finding particles in specific areas, these paths, not the probabilities, are added together. This is analogous to the two-slit experiment. An interference effect is also created, which is an indicator of the loss of diachronic identity by the particles. Particles in the final state cannot be uniquely linked with the particles before the interaction.

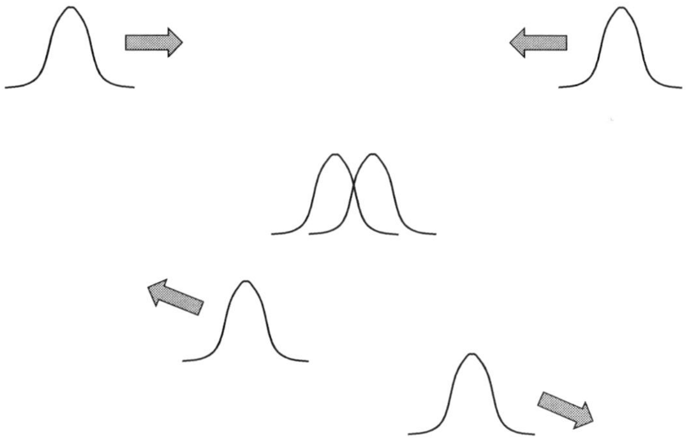

Fig. 7.5. Interaction of two indistinguishable particles

7.9* Elements of the formalism of quantum mechanics

In its mathematical and formal aspect, quantum mechanics is based on very elegant foundations, which are worthy of the attention of even a wider group of philosophers. The basic mathematical theory used in quantum description is linear algebra, supported by mathematical analysis, especially when applied to specific problems. We will focus mainly on the first theory and its use in quantum theory. In addition, we will introduce Dirac's notation, widespread in quantum mechanics, which is an extremely convenient and intuitive, although not necessary way of writing formulas (there are textbooks that do not use it). The basic concept of quantum theory is the concept of a vector space. It consists of vectors, of course, written in Dirac's notation as $|u\rangle$, $|v\rangle$ etc. These vectors are also called "kets" in quantum jargon (we will explain the meaning of this terminology later). Two operations are defined on vectors: addition and multiplication by a number. In the case of spaces used in quantum mechanics, unfortunately, these are complex numbers, i.e. numbers of the form $a + ib$, where a, b – real numbers, and $i = \sqrt{-1}$. Both operations on vectors satisfy a number of intuitive conditions, such as associativity or distributivity of multiplication over addition. (The existence of a zero vector is also assumed.) We will not list here all the axioms that can be found in every textbook. In any case, it is crucial to note that the operations introduced in these axioms allow us to create linear combinations of vectors, such as $a|u\rangle + b|v\rangle$.

A very important operation defined in vector spaces is the scalar (dot) product of vectors, which we have already learned about (in quantum mechanics it is usually referred to as "inner product"). The inner product of vectors $|u\rangle$ and $|v\rangle$ is denoted in Dirac's notation as $\langle u|v\rangle$. It is generally a complex number. The inner product satisfies the condition of linearity in the second argument, which means the following rule:

$$\langle u|av + bw\rangle = a\langle u|v\rangle + b\langle u|w\rangle. \tag{7.2}$$

It is usually assumed that the product is commutative, i.e., the order of the factors does not play a role. However, in the case of vector spaces over complex numbers, changing the order in the product alters the result of the multiplication to a small extent. Instead of the original number, we get its complex conjugate. The conjugate of the complex number $a + ib$ is $a - ib$. This operation is represented by an asterisk *. So we have the following rule of "almost commutativity":

$$\langle u|v\rangle = \langle v|u\rangle^*.$$

It follows that the inner product will be *antilinear* in the first argument, which means a similar formula to (7.2), except that the coefficients a and b will be conjugate (write it down as an exercise). Let us also note an important property of the complex conjugate, which will be useful in a moment: the complex conjugate of a real number is identical to this number, which gives the definition of a real number in the form of the condition $c^* = c$. In addition, the product of the complex number $a + ib$ with its conjugate also forms a real number of the form $a^2 + b^2$. We call this number the square of the modulus and denote it as follows: $|c|^2 = cc^*$.

The dot product allows us to define the concept of perpendicularity, called orthogonality, as well as the length of the vector. Two vectors are orthogonal when their product is zero. In turn, the length of the vector is the (positive) root of its product with itself: $\sqrt{\langle u|u\rangle}$. Of course, it will be a real number, because the product under the root satisfies the condition $\langle u|u\rangle^* = \langle u|u\rangle$. Therefore, for vectors of unit length, we have $\langle u|u\rangle = 1$.

Each vector space has a specific dimension. The dimension of a given space is the maximal number of linearly independent vectors, i.e. those that cannot be expressed by their mutual linear combinations. In a finitely dimensional space (and these are the only ones we will deal with here) we can distinguish N independent vectors such that any other vector can be represented by their combination. Such vectors are called basis vectors. If, in addition, the basis vectors are orthogonal to each other and have a unit length, the resulting basis is called orthonormal. Basis vectors are often symbolized as $|e_1\rangle$, $|e_2\rangle$, ..., $|e_N\rangle$. Of course, the number of basis vectors is equal to the dimension of space.

Let us now introduce the key concept of a linear operator on a vector space. A linear operator is a function that transforms vectors into vectors, satisfying the condition of linearity:

$$A(a|u\rangle + b|v\rangle) = aA|u\rangle + bA|v\rangle.$$

An extremely important category of linear operators are the so-called Hermitian operators, also known as self-adjoint. Their loose definition is as follows: they are operators that act in the same way on the first and second elements in the inner product. More precisely, for a Hermitian operator A the following equality is satisfied by any two vectors $|u\rangle$ and $|v\rangle$:

$$\langle u|Av\rangle = \langle Au|v\rangle.$$

This definition can be reformulated by swapping the order of the multiplication factors on the right:

$$\langle u|Av\rangle = \langle v|Au\rangle^*.$$

Now let us introduce the concept of taking an adjoint for operators. An operator adjoint to a given operator A is the operator $A^\dagger$ that satisfies the following equality:

$$\langle u|A^\dagger v\rangle = \langle v|Au\rangle^*.$$

Thus, we can see that the condition of being a Hermitian operator can be intuitively rewritten in the form of the equation $A = A^\dagger$, which justifies the name "self-adjoint operator", i.e. one whose adjoint is identical with it.

The next concepts are the concepts of eigenvectors and eigenvalues for a given operator. A given vector is called an eigenvector for operator A when the action of A on this vector is equivalent to multiplying it by a number. This number will be called the eigenvalue corresponding to this vector:

$$A|u_a\rangle = a|u_a\rangle.$$

Not all operators have eigenvectors (e.g., rotation operators do not), but if an operator does, there is more than one. A very important fact characterizing Hermitian operators is that eigenvectors corresponding to different eigenvalues must be orthogonal to each other. Let there be vectors $|u_a\rangle$ and $|u_b\rangle$ such that the following equalities for $a \neq b$ hold:

$$\begin{aligned} A|u_a\rangle &= a|u_a\rangle \\ A|u_b\rangle &= b|u_b\rangle \end{aligned} \tag{7.3}$$

Let us multiply both sides of the first equation by the vector $|u_b\rangle$, and the second by $|u_a\rangle$:

$$\langle u_b|A|u_a\rangle = a\langle u_b|u_a\rangle$$

$$\langle u_a|A|u_b\rangle = b\langle u_a|u_b\rangle.$$

Taking the complex conjugate of both sides of the first equation and using the definition of a Hermitian operator, we transform this equation to the form:

$$\langle u_a|A|u_b\rangle = a^*\langle u_a|u_b\rangle.$$

Comparing it with the second equation, we get:

$$a^*\langle u_a|u_b\rangle = b\langle u_a|u_b\rangle.$$

This equation can only be satisfied if $\langle u_a|u_b\rangle = 0$, which means the orthogonality of the vectors $|u_a\rangle$ and $|u_b\rangle$.

In a similar way, it can be proved that eigenvalues for Hermitian operators must be real numbers. I'll leave it as a simple exercise to do (hint: multiply the first of equations (7.3) by the vector $|u_a\rangle$ and take the complex conjugate of both sides). This feature of Hermitian operators will be very useful in their physical interpretation. We shall conclude this review of the basic facts about operators by quoting without proof an extremely important spectral theorem. It states, roughly speaking, that the eigenvectors of the Hermitian operator form an orthogonal basis of a given space. This means that the action of such an operator on any vector can be done by decomposing this vector in this eigenvector basis, multiplying each component by its corresponding eigenvalue, and summing the results.

The physical interpretation of the introduced elements of mathematical formalism is as follows. Vectors in a given space represent the physical states of a system, and their sum represents the superposition of states. Hermitian operators, on the other hand, are formal counterparts of measurable quantities (called observables). An eigenvector of the operator representing a particular observable denotes a state in which the system has a well-defined value of this observable – this value is precisely equal to the eigenvalue of this eigenvector. Now we can see how important it is to see that Hermitian operators have only real eigenvalues, because the measurement results are given in the form of such numbers, and not in general complex numbers. But what about states that are non-trivial combinations of eigenvectors? In such a situation, as we know, there are only probabilities that the measurement will reveal one of the possible eigenvalues. The rule for calculating these probabilities is simple: they are given by the squares of the coefficients of the vector's decomposition into individual components along the eigenvectors. More precisely, it can be presented like this. Let $|u\rangle$ be the state of a given physical system, and $|v_a\rangle$ be the eigenvector of the operator-observables A corresponding to the eigenvalue a. We also assume that vectors $|u\rangle$ and $|v_a\rangle$ are normalized, i.e. their length is 1. In such a situation, the probability that the measurement of the system in the state $|u\rangle$ will reveal the value of a is (this is the so-called Born rule):[13]

$$P_u(A = a) = |\langle u|v_a\rangle|^2$$

We are often interested not in the probabilities of individual results, but in the "global" information about the average of the results of measurements in a large series of them. Such an average value (also called expected value, which is quite misleading, however[14]) can be easily calculated by taking the probability of each outcome individually, multiplying it by

[13] In addition we have to assume that the operator A is non-degenerate, which means that no two orthogonal eigenvectors can have the same eigenvalue.

[14] Why is this misleading? The answer is simple – very often the "expected value" of a given quantity is a value that we should not expect in the experiment at all. For example, the expected value of spins of individuals particles in the singlet state is 0, although it is known that the spin of a fermion can never be zero. This brings to mind the well-known joke about the number of children in an average family being equal to 1.5 – you cannot expect to see such a value in real life! However, the use of the term "expected value" has become commonly accepted to describe the average of many repeats of an experiment.

that result, and adding it up. It is not difficult to show that the result is the following formula representing the averaged value of the quantity A in the state $|u\rangle$:

$$\langle u|A|u\rangle \tag{7.4}$$

Let us dwell for a moment on the issue of Dirac's notation and its usefulness, using the example of an additional formal concept that can render certain services, although perhaps it is not necessary to describe the basic features of the quantum world. In Dirac's notation, we can treat the expression for the inner product of vectors $\langle u|v\rangle$ as consisting of two independent components: $\langle u||v\rangle$. We recognize the second element $|v\rangle$ as a ket. What is the first component $\langle u|$, which is a kind of mirror reflection of the ket? It turns out that it can be interpreted as a new object, called a "bra". This name does not come from a piece of women's clothing, but is derived from the word "bracket". As you can see, bra vectors are the first components of the parenthesis denoting the inner product, and kets are the second.

What is the role of the new bra objects? The formal definition is as follows: for each ket $|u\rangle$, the corresponding bra vector $\langle u|$ is a linear functional that assigns to each vector from the vector space (i.e., to each ket) a number, which is the result of the product of this ket and $|u\rangle$. In mathematical notation, we simply have:

$$\langle u|(|v\rangle) = \langle u|v\rangle.$$

To the left of this equality we have an object $\langle u|$ acting on $|v\rangle$, and to the right a number. Those readers who have bravely fought their way through the sections with an asterisk from the previous chapter should recognize that bra objects defined in this way form the so-called dual space to the initial space, which we talked about on the occasion of the definition of tangent space and the concept of covariant and contravariant vectors. The dual space to a given is, in a sense, its mirror reflection. To "transition" from one space to another, one must first replace each ket $|u\rangle$ with the corresponding bra vector $\langle u|$. Second, we also need to convert all the numbers in the respective combinations of kets into their complex conjugates. For example, the equivalent of the ket $a|u\rangle + b|v\rangle$ would be the following bra vector: $a^*\langle u| + b^*\langle v|$. Finally, an operator A in kets' space corresponds to the conjugate operator $A^\dagger$ in the dual space of bra vectors.

This is not the end of the extraordinary flexibility of Dirac's notation. Consider, for example, the following strange creature:

$$|u\rangle\langle v|.$$

Let's try to guess what such an expression may mean. On the right side, we have a bra vector which is a linear functional and hence when applied to any vector gives us a number. The left part of the above formula is a ket $|u\rangle$. Multiplying a ket by a number (the order doesn't matter) gives us a new ket. Therefore, the entire formula written above denotes an operator that converts kets into kets. You can easily check that it will be a linear operator. This is a so-called projection operator (projector in short) which projects any vector onto the line along which the vector $|u\rangle$ lies, in the direction perpendicular to the direction of the vector $|v\rangle$. It is easy to check that every vector is transformed into a vector that is a multiple of $|u\rangle$, and that vectors perpendicular $|v\rangle$ to are transformed to a zero vector.

A special case of a projection operator is an orthogonal one-dimensional projector, which has the form:

$$|u\rangle\langle u|$$

It can be shown that it is a Hermitian operator, and thus represents a measurable quantity.[15] It is a quantity that takes only two values: 0 and 1. The value of 1 is connected to the vector $|u\rangle$ and 0 corresponds to all vectors orthogonal to it. This means that this operator represents an observable "measuring" whether the system is in state $|u\rangle$ or in an orthogonal state. In particular, for any observable A, the projective operator on one of its eigenvectors measures whether the system has a value correlated with that vector or not.[16]

We should remember that the values of some measurable quantities in quantum mechanics cannot be precisely determined at the same time. The operators representing these quantities are called incompatible. Two operators A and B are incompatible when there is some eigenvector of one of them that is not an eigenvector of the other. Conversely, A and B are compatible when they have exactly identical eigenvectors. This condition can be represented equivalently as follows: compatible operators commute with each other (i.e. $AB = BA$). Let us take any eigenvector $|u\rangle$ of operator A, which is also assumed to be an eigenvector for B. Therefore, $AB|u\rangle = ab|u\rangle$ and $BA|u\rangle = ba|u\rangle = ab|u\rangle$, where a and b are the corresponding eigenvalues. Since any vector can be represented as a linear combination of eigenvectors, the operation AB on any vector is the same as BA, i.e. they commute.

Let us illustrate the above abstract considerations with a concrete example. The best for this purpose will be spin one-half, which we have already used on several occasions to explain the complexity of the behavior of the quantum world. Vectors in a finitely-dimensional space can be represented by columns of numbers, which, as we remember from the previous chapters, represent their components in a certain coordinate system – that is, in a certain basis. For example, if a vector in a two-dimensional space is equal to the sum $2|e_1\rangle + 3|e_2\rangle$, then its representation as a column in the basis $\{|e_1\rangle, |e_2\rangle\}$ will be as follows: $\binom{2}{3}$. A two-dimensional Hilbert space is sufficient to describe spins-1/2 that have two values "up" and "down". Our task will be to determine the form of the operators for spins in three directions in physical space: s_x, s_y and s_z. As a basis of the state space (which should not be confused with the three-dimensional physical space), let us choose two eigenvectors for the s_z operator. Of course, the column representations of basis vectors are very simple: they are $\binom{1}{0}$ and $\binom{0}{1}$. Linear operators, on the other hand, will take the form of a matrix with two rows and two columns, whose action on the vector is given in the form of a known rule of multiplication of a column by a matrix. So we are looking for a matrix that reproduces the same vector multiplied by its eigenvalue in the action on the vector $\binom{1}{0}$ – and similarly for the other vector $\binom{0}{1}$. We take the numbers 1/2 and –1/2 as eigenvalues (for simplicity, we omit Planck's constant $\hbar$). Thus, the equations defining the s_z operator are as follows:

---

[15] Proof of this fact shows how immensely useful and intuitive Dirac's notation is. I will give it here for the purpose of illustration. Let's take any two vectors $|a\rangle$ and $|b\rangle$. If P is our projective operator, then the expression $\langle a|Pb\rangle$ can be written as $\langle a|u\rangle\langle u|b\rangle$. Notice that both elements of the symbol $|u\rangle\langle u|$ have been "disconnected" and "attached" each to one of the two vectors $|a\rangle$ and $|b\rangle$ to form two inner products! This is entirely mathematically correct. In order to calculate $\langle Pa|b\rangle$, let's replace this expression with $\langle b|Pa\rangle^*$, and then put $|u\rangle\langle u|$ in place of P again. That way we get $\langle b|u\rangle^*\langle u|a\rangle^*$, i.e. $\langle u|b\rangle\langle a|u\rangle$, which is identical to the previous result. Therefore P is Hermitian.

[16] Projection operators can also project onto many-dimensional spaces. They can be represented in the form of a sum of operators $|u_i\rangle\langle u_i|$ where $|u_i\rangle$ are orthogonal to each other.

$$s_z \begin{pmatrix} 1 \\ 0 \end{pmatrix} = \frac{1}{2} \begin{pmatrix} 1 \\ 0 \end{pmatrix}$$

$$s_z \begin{pmatrix} 0 \\ 1 \end{pmatrix} = -\frac{1}{2} \begin{pmatrix} 0 \\ 1 \end{pmatrix}$$

It is not difficult to check that the matrix that satisfies the above two conditions has the following form:

$$s_z = \frac{1}{2} \begin{pmatrix} 1 & 0 \\ 0 & -1 \end{pmatrix}$$

This can be seen by multiplying the numbers in the first row by the first and second elements of a given column, respectively, to obtain the first element of the new vector, and then repeating this for the second row. The above matrix is one of the three so-called Pauli matrices, representing spin. Note that it has the so-called diagonalized form, in which the only non-zero elements are on the diagonal. This is always true if the eigenvectors of a given matrix are selected as basis vectors. The numbers on the diagonal are then eigenvalues.

In order to determine the other two matrices, we will refer to the empirical fact that the probabilities of both values for spins in the x and y directions are equal to 1/2 in states with a well-defined z-spin. This means that the eigenvectors for the s_x and s_y operators should be linear combinations of eigenvectors for s_z with coefficients whose squares of the moduli are equal to each other. Suppose that the eigenvector for operator s_x corresponding to the result 1/2 ("up") is of the form

$$\frac{1}{\sqrt{2}} \begin{pmatrix} 1 \\ 0 \end{pmatrix} + \frac{1}{\sqrt{2}} \begin{pmatrix} 0 \\ 1 \end{pmatrix} = \begin{pmatrix} \frac{1}{\sqrt{2}} \\ \frac{1}{\sqrt{2}} \end{pmatrix}$$

The second eigenvector corresponding to the result "down" will be written as follows:

$$\frac{1}{\sqrt{2}} \begin{pmatrix} 1 \\ 0 \end{pmatrix} - \frac{1}{\sqrt{2}} \begin{pmatrix} 0 \\ 1 \end{pmatrix} = \begin{pmatrix} \frac{1}{\sqrt{2}} \\ -\frac{1}{\sqrt{2}} \end{pmatrix}$$

It is easy to check that the above vectors are orthogonal to each other. Their product, which is calculated as the sum of the products of the first components and the second components, is equal to zero.[17] We can now write down equations analogous to those for spin s_z, characterizing the action of the searched matrix s_x on the above vectors:

$$s_x \begin{pmatrix} \frac{1}{\sqrt{2}} \\ \frac{1}{\sqrt{2}} \end{pmatrix} = \frac{1}{2} \begin{pmatrix} \frac{1}{\sqrt{2}} \\ \frac{1}{\sqrt{2}} \end{pmatrix}$$

$$s_x \begin{pmatrix} \frac{1}{\sqrt{2}} \\ -\frac{1}{\sqrt{2}} \end{pmatrix} = -\frac{1}{2} \begin{pmatrix} \frac{1}{\sqrt{2}} \\ -\frac{1}{\sqrt{2}} \end{pmatrix}$$

The solution of simple linear equations provides us with information about the form of the matrix we are looking for:

[17] Remember, however, that if the components of the first vector contain complex numbers, then their complex conjugates should be taken into account when calculating the inner product.

$$s_x = \frac{1}{2}\begin{pmatrix} 0 & 1 \\ 1 & 0 \end{pmatrix}$$

In the case of the third matrix, representing spin in the y direction, we should impose analogous conditions on its eigenvectors – i.e. they should be linear combinations of the eigenvectors for spins s_z and s_x with coefficients having the same modulus. It turns out that for real numbers there are no solutions to the corresponding equations representing these imposed conditions. Thus, the two-dimensional Hilbert space over real numbers is not sufficient to describe all the components of half-spin. Complex numbers are needed for this purpose. Here is the column form of the eigenvectors of the sought-after operator s_y, which remain in the required relations to the eigenvectors of s_z and s_x:[18]

$$\frac{1}{2}\begin{pmatrix} 1-i \\ 1+i \end{pmatrix}, \frac{1}{2}\begin{pmatrix} 1-i \\ -1+i \end{pmatrix}.$$

These are the eigenvectors of the following matrix (this is worth checking):

$$s_y = \frac{1}{2}\begin{pmatrix} 0 & -i \\ i & 0 \end{pmatrix}.$$

So we have already found a matrix form of three operators s_x, s_y and s_z representing spins in three perpendicular directions.[19] We can see that these operators have different eigenvectors, so a particle cannot be in a state where the spins in different directions are well defined. This can be expressed in another way, showing that the matrices s_x, s_y and s_z do not commute with each other. For example, let's calculate the product of $s_x s_y$ followed by $s_y s_x$:

$$s_x s_y = \frac{1}{4}\begin{pmatrix} i & 0 \\ 0 & -i \end{pmatrix}$$

$$s_y s_x = \frac{1}{4}\begin{pmatrix} -i & 0 \\ 0 & i \end{pmatrix}.$$

As you can see, the results are different. Yet another way to express this fact is with the help of the concept of a commutator, i.e. the combination $s_x s_y - s_y s_x$. The commutator for non-commutative operations is always non-zero. In our case, we have an interesting regularity that connects the commutator of two Pauli matrices with the third matrix:

$$s_x s_y - s_y s_x = i s_z.$$

It turns out that this relation holds "cyclically" for each commutator of two Pauli matrices – it is equal to the third matrix multiplied by i.

Let us also show what will be the result of calculating the following combination of the Pauli matrix:

$$s_x^2 + s_y^2 + s_z^2.$$

[18] Check that these vectors decompose as required into equally weighted sums of the eigenvectors of operators s_z and s_x. The best way to do this is to calculate appropriate inner products, which will give us the coefficients of respective decompositions.

[19] It can also be shown that the operators represented by Pauli matrices are Hermitian, i.e. self-adjoint. There is a simple criterion for the Hermiticity of a matrix: this is the case when the numbers that are "mirror images" of each other with respect to the diagonal are each other's conjugates. It can be seen that the Pauli matrices meet this condition.

It is easy to see that squaring each of the three Pauli matrices gives the following simple result:

$$\frac{1}{4}\begin{pmatrix}1 & 0\\ 0 & 1\end{pmatrix}.$$

So the sum of these three elements is:

$$\frac{3}{4}\begin{pmatrix}1 & 0\\ 0 & 1\end{pmatrix}.$$

It is the identity matrix multiplied by $\frac{3}{4}$. Thus, as you can see, every vector is its eigenvector with an eigenvalue of $\frac{3}{4}$. This means that the square of the total spin vector always has the same well-defined value, independent of the state of the particle.

Now let us move on to the basic equation of non-relativistic quantum mechanics: Schrödinger's equation. As we have written before, this equation has the following general form (in Dirac's notation):

$$i\hbar\frac{\partial}{\partial t}|\psi\rangle = H|\psi\rangle.$$

Let's take a closer look at the form of the H operator, or Hamiltonian. In classical physics, H can be written as the sum of kinetic and potential energy:

$$H = \frac{p^2}{2m} + V.$$

In quantum mechanics, quantities such as momentum are represented not by functions but by operators. There are arguments based on wave phenomenon analysis which show that the operator representing momentum in the direction of the x-axis should be of the following form:

$$-i\hbar\frac{\partial}{\partial x}.$$

Inserting this formula in place of momentum, we get the following expanded form of the Schrödinger equation:

$$i\hbar\frac{\partial}{\partial t}|\psi\rangle = \left[-\frac{\hbar^2}{2m}\left(\frac{\partial^2}{\partial x^2} + \frac{\partial^2}{\partial y^2} + \frac{\partial^2}{\partial z^2}\right) + V(x, y, z)\right]|\psi\rangle.$$

Suppose the wave function $|\psi\rangle$ is separable, i.e., it can be represented as the product of the time-dependent part and the space-dependent part: $|\psi\rangle = U(t)|\psi_0(x, y, z)\rangle$. Inserting this into the above, we will get:

$$i\hbar|\psi_0\rangle\frac{d}{dt}U(t) = U(t)H|\psi_0\rangle. \tag{7.5}$$

If we additionally assume that the Hamiltonian of a given system does not depend on time, then we can determine the form of the function (operator) $U(t)$ from the above equation. Namely, it will be an exponential function:

$$U(t) = e^{-\frac{iH}{\hbar}t}.$$

Those readers who know a bit about the rules of differentiation of typical functions may know that the derivative of the exponential function e^x is exactly the same function (a very neat proof of this fact refers to the expansion of this function into an infinite power series). Therefore, differentiating the above function with respect to time and inserting the result into the left side of the Schrödinger equation gives us exactly the right side.

Remember that *U*(*t*) is an operator, as it is defined as a function of another operator, the Hamiltonian *H*. The operator *U*(*t*) is called the evolution operator, because its application to the initial state vector gives the state at the later moment *t*. In addition, *U*(*t*) has an interesting property called unitarity. Let us take the inner product of two vectors $\langle u|v\rangle$ and act on each of them with the operator *U*(*t*):

$$\langle U(t)u|U(t)v\rangle = \left\langle e^{-\frac{iH}{\hbar}t}u\middle|e^{-\frac{iH}{\hbar}t}v\right\rangle.$$

From the properties of the product, as well as from the definition of the adjoint for operators, we know that an operator acting on the first term can be "shifted" to the second term, provided that we take its adjoint. But the adjoint of the operator $e^{-\frac{iH}{\hbar}t}$ simply yields $e^{\frac{iH}{\hbar}t}$. So we get from the above:

$$\left\langle u\middle|e^{\frac{iH}{\hbar}t}e^{-\frac{iH}{\hbar}t}v\right\rangle = \langle u|v\rangle.$$

As you can see, the operator *U*(*t*) preserves the inner product, and thus also the length of the vectors. This property is called unitarity, and the operator itself is called unitary. Therefore, it can be said that the Schrödinger evolution of a given state vector is a kind of rotation – the vector does not change its length, but only its direction. It also follows that if the initial state is an eigenvector of a given operator, then the Schrödinger evolution transforming it into another eigenvector must pass through intermediate states, which are superpositions of these two. That is, quantum systems have a "tendency" to lose well-defined values of given quantities.

Let us formulate one more interesting observation about the Schrödinger equation with the time-independent Hamiltonian. A simple transformation of equation (7.5) gives us:

$$H|\psi_0\rangle = i\hbar\frac{1}{U(t)}\frac{d}{dt}U(t)|\psi_0\rangle.$$

The operator on the left side of the equation depends only on its position in space, and the operator on the right side on time. If this equation is to be satisfied for every value of *x*, *y*, *z*, and *t*, both sides must be equal to some constant number. Let's write it down concisely:

$$H|\psi_0\rangle = E|\psi_0\rangle.$$

This is nothing but an eigenequation for the operator *H*, which is an energy operator. By solving this equation, we will obtain the eigenvectors of the energy operator and its eigenvalues, i.e. the allowed energy values that a given system can take. In this way, we can learn, for example, that a particle in a so-called potential well, i.e. placed in a potential that takes a finite value in a certain area, and is infinite outside this area, will have quantized energy levels. The analysis of the energy levels of the hydrogen atom with the Coulomb potential $k\frac{q}{r}$ is similar. In addition, we will obtain the mathematical form of the wave functions of the electron in the hydrogen atom for subsequent energy levels, i.e. the shape of the so-called orbitals.

Finally, let us take a look at the mathematical side of the description of systems composed of many particles, including extremely important entangled states. In order to describe the states of two physical systems, the tensor product of the state spaces for each system separately is introduced: $\mathcal{H}_1 \otimes \mathcal{H}_2$. It is a space composed of objects of the form $|u\rangle_1 \otimes |v\rangle_2$, where $|u\rangle_1 \in \mathcal{H}_1$ and $|v\rangle_2 \in \mathcal{H}_2$, plus all linear combinations of such vectors. Particularly noteworthy are combinations that cannot be represented in the form of a simple product, such as:

$$|u\rangle_1 \otimes |v\rangle_2 + |v\rangle_1 \otimes |u\rangle_2.$$

Such states are called entangled. An example considered earlier in the chapter is the singlet state of the spin of two fermions, e.g. electrons (also called the EPR state):

$$\frac{1}{\sqrt{2}}(|z_+\rangle \otimes |z_-\rangle - |z_-\rangle \otimes |z_+\rangle). \tag{7.6}$$

Let us try to calculate the probabilities of obtaining spin measurement results for both particles in the above state. To do this, we need to formally represent in $\mathcal{H}_1 \otimes \mathcal{H}_2$ operators corresponding to the properties of each particle separately. In general, the rule is as follows: if A is an operator acting in $\mathcal{H}_1$ and representing some quantity of the first particle, then the product $A \otimes I$, where I is the identity operator in space $\mathcal{H}_2$, represents exactly the same quantity of the first particle in $\mathcal{H}_1 \otimes \mathcal{H}_2$. The principle of action for the products of operators $A \otimes B$ in the joint space $\mathcal{H}_1 \otimes \mathcal{H}_2$ is simple: each of the operators A and B acts in its own space, and the results of their action are then tensor-multiplied. Thus, the result of the action of the operator $A \otimes B$ on a vector $|u\rangle \otimes |v\rangle$ is simply $A|u\rangle \otimes B|v\rangle$.

We will want to calculate the probabilities of getting a given spin value for both particles separately, as well as the probability of getting two values at once. The most convenient way to do this is to use appropriate projection operators. The operator $|z_+\rangle\langle z_+|$ represents the "spin up" property, and thus the product $|z_+\rangle\langle z_+| \otimes I$ will encode this property with respect to the first particle, while the product $I \otimes |z_+\rangle\langle z_+|$ the same property with respect to the second. In turn, the product of two spin projection operators $|z_+\rangle\langle z_+| \otimes |z_+\rangle\langle z_+|$ expresses the same "spin up" property assigned to both particles. We can now compute the expectation values in state (7.6) for all three projection operators, according to the formula (7.4) given earlier. Since a projector can only take values of 1 (when the system has the appropriate property) or 0 (when it does not), its expected value is automatically equal to the probability that the system actually has the corresponding property.

Acting with the operator $|z_+\rangle\langle z_+| \otimes I$ on the vector (7.6) we get:

$$\frac{1}{\sqrt{2}}(|z_+\rangle\langle z_+|z_+\rangle \otimes |z_-\rangle - |z_+\rangle\langle z_+|z_-\rangle \otimes |z_+\rangle) = \frac{1}{\sqrt{2}}|z_+\rangle \otimes |z_-\rangle$$

(we take into account the fact that $\langle z_+|z_-\rangle = 0$ due to orthogonality and $\langle z_+|z_+\rangle = 1$ due to normalization). Calculating the inner product of the vector (7.6) with the above, we obtain:[20]

[20] The inner product of two vectors $|u\rangle \otimes |v\rangle$ and $|z\rangle \otimes |w\rangle$ is calculated by multiplying the products of the first term with the first and the second with the second: $\langle u|z\rangle\langle v|w\rangle$.

$$\frac{1}{2}(\langle z_+|z_+\rangle\langle z_-|z_-\rangle - \langle z_+|z_-\rangle\langle z_-|z_+\rangle) = \frac{1}{2}.$$

We will get the same result for the second projector $I \otimes |z_+\rangle\langle z_+|$. The conclusion is that for both particles, the probability of obtaining the spin-up result is the same as for the spin-down outcome. Now let's calculate the same for the third operator $|z_+\rangle\langle z_+| \otimes |z_+\rangle\langle z_+|$, acting on (7.6):

$$\frac{1}{\sqrt{2}}(|z_+\rangle\langle z_+|z_+\rangle \otimes |z_+\rangle\langle z_+|z_-\rangle - |z_+\rangle\langle z_+|z_-\rangle \otimes |z_+\rangle\langle z_+|z_+\rangle\,) = \mathbf{0}.$$

Since the result turned out to be the zero vector, its product with any other vector is also zero. So the probability that both particles will have spin up is zero. This is an example of quantum nonlocality (of the "outcome-dependence" type). Although each particle taken separately can reveal both spin values (up or down), they must always be opposite to both particles. If one of them shows the result "up", the other must show "down".

Questions and problems

1. What are the experimental facts supporting the quantization hypothesis of light (electromagnetic waves)?

2. Discuss the Stern-Gerlach experiment (singular and sequential) and the electron interference experiment (two-slit experiment). What conclusions can be drawn from them regarding the possession of properties by quantum objects?

3. Discuss the concept of superposition of states and its probabilistic interpretation.

4. Present the quantum interpretation of measurable properties (observables), the concept of eigenstates and the concept of incompatibility for observables. What are the empirical consequences of the incompatibility of observables, e.g. momentum and position?

5. How do we define entangled states in quantum mechanics?

6. Present the EPR argument for the incompleteness of the probabilistic description of quantum mechanics. What role does the principle of locality play in this argument?

7. Formulate two versions of the principle of locality: the principle of parameter independence and the principle of outcome independence. Which of these versions is assumed in the EPR argument?

8. Analyze the argument leading to the Bell inequality (in the CHSH version). Is it possible to replace quantum mechanics with a non-probabilistic and local theory that gives the same empirical predictions as the standard theory?

9. How to reconcile the deterministic nature of the Schrödinger equation with the occurrence of the irreducible concept of probability in quantum mechanics?

10. Discuss the measurement problem using an abstract approach to the measurement processes with the help of the unitary evolution operator. Present three theses about measurement, which together lead to contradiction.

11. Compare the standard interpretation of quantum mechanics based on the concept of collapse with the GRW interpretation. How does the latter explain the fact that the wave function of the measured object is reduced when it comes into contact with a macroscopic measuring device?

12. What are the main assumptions and difficulties of the many-worlds interpretation of quantum mechanics? How can we explicate quantum probabilities under this interpretation?

13. Discuss the statistical behavior of two types of quantum particles (fermions and bosons) and compare them with the behavior of classical particles. What are the implications of the existence of quantum statistics for the question of particle distinguishability?

Further reading

There is a myriad of popular and semi-popular introductions to quantum mechanics on the market. Personally, I would recommend the following book for inquisitive beginners: D. Albert, *Quantum Mechanics and Experience*, Harvard University Press, Cambridge, Mass 1992.

The following textbook contains a very clear and accessible presentation of the quantum mechanical formalism in Dirac's notation: R.B. Griffiths, *Consistent Quantum Theory*, Cambridge University Press, Cambridge 2002.

This joint work covers almost every aspect of the philosophical problems present in quantum mechanics: C. Friebe , M. Kuhlmann, H. Lyre, P.M. Näger, O. Passon, M. Stöckler, *The Philosophy of Quantum Physics*, Springer, Cham 2018.

The problem of the indistinguishability of quantum objects is analyzed in detail in the monograph: T. Bigaj, *Identity and Indiscernibility in Quantum Mechanics*, Palgrave-Macmillan (Springer Nature), Cham 2022.

THE ENDING IS JUST THE BEGINNING

Our journey through physics and its philosophical aspects has come to an end. We have come a long way, measured both by centuries and by the degree of complexity of physical theories and their formal and conceptual apparatuses. However, instead of attempting a synthesis or formulating general conclusions, let us note that we are in fact only at the beginning of the road. This introduction has not covered many extremely important areas of modern physics, which are of paramount importance for philosophical discussions conducted in recent years. Let us conclude by mentioning these "great absentees", to which I hope there will be an opportunity to return in the future. One field that has always been at the intersection of science and philosophy (as well as religion) is cosmology, i.e. the study of the universe as a whole. The first cosmological questions concerned questions such as whether the universe is temporally and spatially bound, or whether it extends indefinitely. These were mainly speculative issues, as there was no technical possibility to study them empirically. It was not until the development of astronomical observational techniques at the end of the nineteenth century that it became feasible to take a closer look at the megascopic structure of the universe.

The most famous discovery of twentieth-century cosmology was the observation of the phenomenon of the expansion of the universe, and then the acquisition of empirical data in the form of the so-called relic (background) radiation suggesting the existence of an initial singularity, called the *Big Bang*. According to the current state of knowledge, our universe is just under 14 billion years old, during which it has gone through many phases of expansion that continues to the present. It is important to remember that the expansion of the universe is not about the escape of stars and galaxies, but about the "bloating" of space itself in such a way that no point in the universe is singled out as the center from which all objects run, as is the case with ordinary explosions. The theoretical basis of modern cosmology is the general theory of relativity, which allows the introduction of the concept of global curvature. In standard models of the universe (the so-called FLRW, Friedman-Lemaitre-Robertson-Walker models), it is possible to describe the universe as spatially bounded with a positive global curvature (as in the case of a sphere), or unbounded with zero or negative curvature (as on the surface of a saddle). In the positive curvature model, the universe does not expand indefinitely: the expansion stage is followed by the "contraction" stage, leading to the "Big Crunch". Other models allow unlimited expansion (the matter becomes a bit more complicated if we add the so-called cosmological constant to the equations). The basic theory of the

Big Bang has encountered many theoretical and empirical problems, which has led, among other things, to the formulation of a new hypothesis of the so-called inflationary universe, or to the introduction of the puzzling and controversial concepts of dark matter and dark energy.

Moving from the level of megaphenomena to the microscopic level, we must first note that standard quantum mechanics (also called non-relativistic) is not the last word in physics. Already at the time of its creation, serious limitations were noticed related to the fact that the Schrödinger equation fundamental to this theory is not relativistically invariant. Attempts to "relativize" quantum mechanics have led to the emergence of so-called quantum field theory, which is even more astonishing than its predecessor. For example, it can describe situations in which the number of particles of a certain type is not well defined (a state is a superposition of states with different numbers of particles). The puzzling concept of virtual particles is also introduced here, the creation of which is fundamentally excluded by the laws of conservation of mass and energy, and yet they can appear for a "blink of an eye", to be gone a moment later. The basic feature of quantum field theory is that it describes fields known from classical physics, such as electromagnetic fields, using discrete quanta – particles carrying a given interaction. In the case of electromagnetic interactions, such particles are photons, while strong and weak interactions typical of the microworld are transmitted by appropriate bosons (such as gluons, mesons or intermediate bosons).

The fundamental philosophical problem with quantum field theory is the question of what is more fundamental in this theory: an extended field or discrete particles. This is particularly important in the context of the issue of individuality and identity of quantum objects, as the respective identity criteria for fields and particles are different. Another problem is related to the notorious mathematical difficulty that arises when trying to accurately describe the relevant interactions. It turns out that certain quantities characterizing these interactions "escape" to infinity, as a result of which the corresponding mathematical equations become meaningless. Physicists have devised clever ways to deal with this problem, known as "renormalization procedures", which nevertheless are often scorned by mathematicians as lacking firm foundations. Moreover, such procedures are not always available (the basic theory for which renormalization has been possible is quantum electrodynamics QED). Of course, the question remains whether renormalization is not a purely formal ad *hoc* procedure, without deeper theoretical justification. There are many arguments for and against this.

A detailed study of the world of elementary particles has revealed a huge variety of objects occurring in it, far beyond the familiar electrons, protons or neutrons. The so-called standard model classifies particles into appropriate groups. For example, particles called leptons include electrons, muons, and tau, and their corresponding neutrinos. Heavier particles, on the other hand, are classified as hadrons, among which there are baryons (including nucleons, i.e. protons and neutrons) and mesons. Heavy particles are not actually elementary, but are made up of smaller objects called quarks. Although the properties of quarks can be studied by appropriate interactions with the particles containing them, it is impossible to isolate individual quarks for fundamental reasons. In addition, it is worth noting that every existing particle corresponds to its antiparticle, which in the case of electrically charged particles will always have the opposite sign (for some particles, such as a photon, the particle itself is its own antiparticle). In order to introduce some order to the chaos of various types of particles, mathematical group theory is used, thanks to which bizarre and not always easily explainable symmetries between particles have been discovered (such as the formation of so-called octets containing eight particles by certain groups of baryons and mesons). One may ask whether the existence of such symmetries (or even "supersymmetries", as in so-called string theory) has any deeper philosophical or ontological justification.

The most serious challenge of modern physics is to reconcile two fundamental but mutually incompatible theories: general relativity and quantum mechanics. The result of such a synthesis should be a new theory called quantum gravity. Unfortunately, so far, attempts to quantize gravitational interactions have not been successful, due to the mounting mathematical and conceptual difficulties. Some physicists believe that philosophers should also join in the work on creating such a theory. This is, of course, a task that could only be accomplished by a small group of philosophers who also have an excellent physical and mathematical background. In any case, it should be noted that many attempts to create quantum gravity touch on fundamental problems of a philosophical nature, such as the nature of time and space. It is said that according to quantum gravity, time and space would be emergent properties, "appearing" only at a certain stage of the development of matter. Of course, it remains an open question whether it is possible to give expressions such as "appear" or "emerge" a meaning other than temporal or spatiotemporal in order to avoid an obvious vicious circle.

However, let us interrupt these musings, which could go on for a long time. Perhaps in my next book I will attempt to bring some of the issues outlined here closer to philosophically oriented readers. For now, I can recommend the monumental work of the famous British physicist, which, although published twenty years ago, has not lost much of its relevance: R. Penrose, *The Road to Reality. A Complete Guide to the Laws of the Universe*, Vintage Books, London 2005. It is an extremely demanding read, but its careful studying, even in fragments, will be well worth the effort.

Index

D

E

F

G

H

I

J

K

L

M

Q

R

S

T

U

V

W

Z

Made in the USA
Las Vegas, NV
11 January 2026

7579b563-55d9-4a07-be82-7cb5a4b9cd21R01